Patch-Clamp Methods and Protocols

METHODS IN MOLECULAR BIOLOGY™

John M. Walker, SERIES EDITOR

288. Patch-Clamp Methods and Protocols, edited by *Peter Molnar and James J. Hickman, 2007*

399. Neuroprotection Methods and Protocols, edited by *Tiziana Borsello, 2007*

286. Transgenic Plants: *Methods and Protocols*, edited by *Leandro Peña, 2004*

285. Cell Cycle Control and Dysregulation Protocols: *Cyclins, Cyclin-Dependent Kinases, and Other Factors*, edited by *Antonio Giordano and Gaetano Romano, 2004*

284. Signal Transduction Protocols, *Second Edition*, edited by *Robert C. Dickson and Michael D. Mendenhall, 2004*

283. Biconjugation Protocols, edited by *Christof M. Niemeyer, 2004*

282. Apoptosis Methods and Protocols, edited by *Hugh J. M. Brady, 2004*

281. Checkpoint Controls and Cancer, Volume 2: *Activation and Regulation Protocols*, edited by *Axel H. Schönthal, 2004*

280. Checkpoint Controls and Cancer, Volume 1: *Reviews and Model Systems*, edited by *Axel H. Schönthal, 2004*

279. Nitric Oxide Protocols, *Second Edition*, edited by *Aviv Hassid, 2004*

278. Protein NMR Techniques, *Second Edition*, edited by *A. Kristina Downing, 2004*

277. Trinucleotide Repeat Protocols, edited by *Yoshinori Kohwi, 2004*

276. Capillary Electrophoresis of Proteins and Peptides, edited by *Mark A. Strege and Avinash L. Lagu, 2004*

275. Chemoinformatics, edited by *Jürgen Bajorath, 2004*

274. Photosynthesis Research Protocols, edited by *Robert Carpentier, 2004*

273. Platelets and Megakaryocytes, Volume 2: *Perspectives and Techniques*, edited by *Jonathan M. Gibbins and Martyn P. Mahaut- Smith, 2004*

272. Platelets and Megakaryocytes, Volume 1: *Functional Assays*, edited by *Jonathan M. Gibbins and Martyn P. Mahaut-Smith, 2004*

271. B Cell Protocols, edited by *Hua Gu and Klaus Rajewsky, 2004*

270. Parasite Genomics Protocols, edited by *Sara E. Melville, 2004*

269. Vaccina Virus and Poxvirology: *Methods and Protocols*,edited by *Stuart N. Isaacs, 2004*

268. Public Health Microbiology: *Methods and Protocols*, edited by *John F. T. Spencer and Alicia L. Ragout de Spencer, 2004*

267. Recombinant Gene Expression: *Reviews and Protocols, Second Edition*, edited by *Paulina Balbas and Argelia Johnson, 2004*

266. Genomics, Proteomics, and Clinical Bacteriology: *Methods and Reviews*, edited by *Neil Woodford and Alan Johnson, 2004*

265. RNA Interference, Editing, and Modification: *Methods and Protocols*, edited by *Jonatha M. Gott, 2004*

264. Protein Arrays: Methods and Protocols, edited by *Eric Fung, 2004*

263. Flow Cytometry, *Second Edition*, edited by *Teresa S. Hawley and Robert G. Hawley, 2004*

262. Genetic Recombination Protocols, edited by *Alan S. Waldman, 2004*

261. Protein–Protein Interactions: *Methods and Applications*, edited by *Haian Fu, 2004*

260. Mobile Genetic Elements: *Protocols and Genomic Applications*, edited by *Wolfgang J. Miller and Pierre Capy, 2004*

259. Receptor Signal Transduction Protocols, *Second Edition*, edited by *Gary B. Willars and R. A. John Challiss, 2004*

258. Gene Expression Profiling: *Methods and Protocols*, edited by *Richard A. Shimkets, 2004*

257. mRNA Processing and Metabolism: *Methods and Protocols*, edited by *Daniel R. Schoenberg, 2004*

256. Bacterial Artifical Chromosomes, Volume 2: *Functional Studies*, edited by *Shaying Zhao and Marvin Stodolsky, 2004*

255. Bacterial Artifical Chromosomes, Volume 1: *Library Construction, Physical Mapping, and Sequencing*, edited by *Shaying Zhao and Marvin Stodolsky, 2004*

254. Germ Cell Protocols, Volume 2: *Molecular Embryo Analysis, Live Imaging, Transgenesis, and Cloning*, edited by *Heide Schatten, 2004*

253. Germ Cell Protocols, Volume 1: *Sperm and Oocyte Analysis*, edited by *Heide Schatten, 2004*

252. Ribozymes and siRNA Protocols, *Second Edition*, edited by *Mouldy Sioud, 2004*

251. HPLC of Peptides and Proteins: *Methods and Protocols*, edited by *Marie-Isabel Aguilar, 2004*

250. MAP Kinase Signaling Protocols, edited by *Rony Seger, 2004*

249. Cytokine Protocols, edited by *Marc De Ley, 2004*

248. Antibody Engineering: Methods and Protocols, edited by *Benny K. C. Lo, 2004*

247. Drosophila Cytogenetics Protocols, edited by *Daryl S. Henderson, 2004*

246. Gene Delivery to Mammalian Cells: *Volume 2: Viral Gene Transfer Techniques*, edited by *William C. Heiser, 2004*

245. Gene Delivery to Mammalian Cells: Volume 1: *Nonviral Gene Transfer Techniques*, edited by *William C. Heiser, 2004*

244. Protein Purification Protocols, *Second Edition*, edited by *Paul Cutler, 2004*

243. Chiral Separations: Methods and Protocols, edited by *Gerald Gübitz and Martin G. Schmid, 2004*

METHODS IN MOLECULAR BIOLOGY™

Patch-Clamp Methods and Protocols

Edited by

Peter Molnar

and

James J. Hickman

HUMANA PRESS TOTOWA, NEW JERSEY

999 Riverview Drive, Suite 208
Totowa, New Jersey 07512

www.humanapress.com

This publication is printed on acid-free paper. ∞
ANSI Z39.48-1984 (American Standards Institute) Permanence of Paper for Printed Library Materials

Cover illustration:

Production Editor: Jennifer Hackworth
Cover design by

For additional copies, pricing for bulk purchases, and/or information about other Humana titles, contact Humana at the above address or at any of the following numbers: Tel.: 973-256-1699; Fax: 973-256-8341; E-mail: humana@humanapr.com; or visit our Website: www.humanapress.com

Printed in the United States of America. 10 9 8 7 6 5 4 3 2 1

ISBN 978-1-61737-724-2 eISBN 978-1-59745-529-9

Preface

Patch-clamp electrophysiology became commonly used since its development by Bert Sakmann and Erwin Neher in 1981. The past two decades brought major advances not only in the patch-clamp technique but in the method's extensive applications to solve diverse scientific problems. In addition to an increase in the number of people utilizing this technique, there has been a tremendous improvement in the range of sophisticated equipment, elevating this method from the level of "wizardry" to routine usage. Recently, with the emergence of automated/parallel patch clamp, a major reorganization has taken place in the pharmacological applications of this methodology.

The goal in this volume was to summarize the typical patch-clamp applications and to assist scientists in identifying problems, which could be best addressed by this technique. In addition, this volume will provide the step-by-step procedures to perform the experiments that will answer those questions. The experiments described in this book will require a basic level of electrophysiological training. Our intention was not to teach the reader patch-clamp instrumentation and technique, as several basic and advanced level books have already been published, but to help an experimenter safely and effectively venture into new areas of electrophysiology and to routinely use the patch-clamp technique to answer scientific questions.

To enhance the enthusiasm of non-electrophysiologists toward this method, we have introduced several scientific problems as examples where combined techniques (e.g., electrophysiology + imaging or molecular biology) are most effective. Thus, encouraging scientists with diverse backgrounds to attempt to utilize patch-clamp electrophysiology.

This book was organized by the major patch-clamp application areas, namely pharmacology, physiology, and biophysics. As there are strong overlaps in the techniques utilized in these application areas, we have discussed individual techniques where they are most commonly used. Thus, avoiding any unnecessary repetition.

The first chapter provides examples and step-by-step instructions on how to use whole-cell and single-channel patch-clamp methods for testing drugs in industrial settings. The reader will learn how to perform the electrophysiological method as well as gain insight into the drug-screening process and the standard procedures in using it for this application. A reader will also learn about how electrophysiology can be best utilized in drug research and development. The second part of this chapter provides an opportunity to compare the current automated patch-clamp systems for typical applications and thus filling a gap in this literature.

The second chapter provides a wide selection of patch-clamp applications in physiological studies. Emphasis was given to techniques where patch clamp was combined with other methodologies such as photostimulation, force measurement, polymerase chain reaction (PCR), cell patterning, or computer modeling. Also, in this chapter, two typical applications of the relatively new dynamic-clamp method are introduced. In addition, emphasis was given to the high diversity of cell types as targets of electrophysiological studies. Through examples and detailed protocols, the reader can learn how to dissect/handle/culture cell lines, primary neurons, stem cells, cardiac myocytes, and skeletal muscle cells.

The last chapter focuses on the biophysical applications of the patch-clamp method using single-channel recordings or statistical analysis of whole-cell currents to obtain parameters that describe ion channel properties or transmitter release. In this chapter, we have also included an extensive theoretical treatise concerning single-channel kinetic analysis.

We hope that many electrophysiologists and non-electrophysiologists will find this book useful in designing and performing a wide variety of patch-clamp experiments in conjunction with other state-of-the-art methodologies.

Peter Molnar

Contents

Part III: Biophysics

Contributors

BRUNO ALLARD • *Physiologie Intégrative Cellulaire et Moléculaire, Université Claude Bernard Lyon 1, France*
GÉZA BERECKI • *Department of Experimental Cardiology, Academic Medical Center, University of Amsterdam, The Netherlands*
JONATHAN BETTENCOURT • *Department of Biomedical Engineering, Center for BioDynamics, Center for Memory and Brain, Boston University, Boston, MA, and Department of Physiology and Biophysics, Weill Medical College of Cornell University, New York, NY*
NEELIMA BHARGAVA • *Nanoscience Technology Center, University of Central Florida, Orlando, FL*
THIERRY CENS • *CRBM, CNRS, France*
PIERRE CHARNET • *CRBM, CNRS, France*
CLAUDE COLLET • *Laboratoire de Toxicologie Environnementale, Ecologie des Invertébrés, France*
UWE CZUBAYKO • *Flyion GmbH, Germany*
MAINAK DAS • *Nanoscience Technology Center, University of Central Florida, Orlando, FL*
HANS-ULRICH DODT • *Max-Planck-Institute of Psychiatry, Germany*
ALAN D. DORVAL II • *Department of Biomedical Engineering, Center for BioDynamics, Center for Memory and Brain, Boston University, Boston, MA, and Department of Biomedical Engineering, Duke University, Durham, NC*
MATHIAS EDER • *Max-Planck-Institute of Psychiatry, Germany*
MICHAEL FEJTL • *Flyion GmbH, Germany*
ARNOUD C. FIJNVANDRAAT • *Department of Experimental Cardiology and Experimental and Molecular Cardiology Group, Academic Medical Center, University of Amsterdam, The Netherlands*
LÁSZLÓ FODOR • *Pharmacology and Drug Safety Research, Gedeon Richter Ltd., Hungary*
ANDREA GHETTI • *AVIVA Biosciences Corp, San Diego, CA*
KATE GILLING • *Preclinical Research & Development, Merz Pharmaceuticals GmbH, Germany*

ANTONI C. G. VAN GINNEKEN • *Department of Experimental Cardiology and Experimental and Molecular Cardiology Group, Academic Medical Center, University of Amsterdam, The Netherlands*
ANTÓNIO GUIA • *AVIVA Biosciences, San Diego, CA*
ROMAIN GUINAMARD • *CNRS, UMR 6187, Université de Poitiers, France*
JAMES J. HICKMAN • *Nanoscience Technology Center, University of Central Florida, Orlando, FL*
ALEXANDER HÜMMER • *Flyion GmbH, Germany*
VINCENT JACQUEMOND • *Physiologie Intégrative Cellulaire et Moléculaire, Université Claude Bernard Lyon 1, France*
JUNG-FONG KANG • *Nanoscience Technology Center, University of Central Florida, Orlando, FL*
TOBIAS KRAUTER • *Flyion GmbH, Germany*
MIKE KUESTER • *Bayer Technology Services GmbH, Germany*
MATTHIAS G. LANGER • *University of Ulm, Germany*
CHRISTINE LEISGEN • *Multi Channel Systems MCS GmbH, Germany*
ALBRECHT LEPPLE-WIENHUES • *Flyion GmbH, Germany*
HENRY MARKRAM • *Brain and Mind Institute, EPFL, Switzerland*
CHRISTOPH METHFESSEL • *Bayer Technology Services GmbH, Germany*
PETER MOLNAR • *Nanoscience Technology Center, University of Central Florida, Orlando, FL*
JÓZSEF NAGY • *Pharmacology and Drug Safety Research, Gedeon Richter Ltd., Hungary*
THEODEN I. NETOFF • *Department of Biomedical Engineering, University of Minnesota Minneapolis, MN*
ERIC S. NISENBAUM • *Neuroscience Division, Lilly Research Laboratories, Eli Lilly and Company, Indianapolis, IN*
CHRIS G. PARSONS • *Head In Vitro Pharmacology, Preclinical Research & Development, Merz Pharmaceuticals GmbH, Germany*
SANDRINE POUVREAU • *Physiologie Intégrative Cellulaire et Moléculaire, Université Claude Bernard Lyon 1, France*
FENG QIN • *Department of Physiology and Biophysics, State University of New York at Buffalo, Buffalo, NY*
JENNIFER C. QUIRK • *Neuroscience Division, Lilly Research Laboratories, Eli Lilly and Company, Indianapolis, IN*
GERHARD RAMMES • *Max-Planck-Institute of Psychiatry and Technical University, Klinikum Rechts der Isar, Germany*
CHIARA SAVIANE • *University College London, London, UK*
ROBIN ANGUS SILVER • *University College London, London, UK*

EDWARD R. SIUDA • *Neuroscience Division, Lilly Research Laboratories, Eli Lilly and Company, Indianapolis, IN*
MARIA TOLEDO-RODRIGUEZ • *Brain and Mind Institute, EPFL, Switzerland*
JOHN A. WHITE • *Department of Biomedical Engineering, Center for BioDynamics, Center for Memory and Brain, Boston University, Boston, MA*
RONALD WILDERS • *Department of Physiology, Academic Medical Center, University of Amsterdam, The Netherlands*
JIA XU • *AVIVA Biosciences, San Diego, CA*
JAN G. ZEGERS • *Department of Physiology, Academic Medical Center, University of Amsterdam, The Netherlands*
WALTER ZIEGLGÄNSBERGER • *Max-Planck-Institute of Psychiatry, Germany*

I

Pharmacology

1

Pharmacological Analysis of Recombinant NR1a/2A and NR1a/2B NMDA Receptors Using the Whole-Cell Patch-Clamp Method

László Fodor and József Nagy

Summary

N-methyl-D-aspartate receptors (NMDARs) are ligand-gated ion channels belonging to the family of ionotropic glutamate receptors. Functional NMDARs are heterotetrameric assemblies of NR1 subunits with at least one type of NR2 subunits. Various combinations of these subunits form distinct NMDAR subtypes involved in a variety of physiological and pathological processes. Several pharmaceutical companies search subunit-selective drugs for curing various neurological diseases and having favorable side-effect profile. We applied the whole-cell patch-clamp technique for testing NR2B subunit-specific drugs in HEK cells transiently or stably expressing different types of NMDAR subunits. In stable cell lines, we applied an inducible mammalian expression system; cDNAs of NR1 and either NR2A or NR2B subunits were inserted into an ecdyson-inducible mammalian expression vector and were introduced into HEK293 cells. These expression systems proved to be suitable to analyze precisely the subtype selectivity of newly synthesized NR2B-selective NMDAR antagonists by using whole-cell patch-clamp technique.

Key Words: Whole-cell patch clamp; NMDA receptor; NR2B subunit; ecdysone receptor; HEK293; inducible gene expression; subtype-selective; drug discovery.

1. Introduction

Patch-clamp technique is still the golden standard for studying interaction of drug molecules and ion channel receptors. Patch experiments can provide detailed characterization of drug effect on ion channel function more reliably

From: *Methods in Molecular Biology, vol. 288: Patch-Clamp Methods and Protocols*
Edited by: P. Molnar and J. J. Hickman © Humana Press Inc., Totowa, NJ

than other indirect methods. Despite its low throughput regarding the number of compounds tested, patch clamp is considered as inevitable in drug discovery. The technique is very commonly used with human embryonic kidney (HEK)-293 cell-based recombinant systems for testing a wide range of expressed target molecules, including among others ligand- and voltage-gated ion channels, G protein-coupled receptors, and postsynaptic density proteins ***(1)***. Here, we show an example of application of the patch-clamp technique in studying *N*-methyl-D-aspartate receptor (NMDAR) subunit-specific drugs in HEK cells transiently or stably expressing different types of NMDAR subunits.

NMDARs are ligand-gated ion channels belonging to the family of glutamate receptors. They are widely expressed in the nervous system and implicated in a variety of neuropathological processes such as ischemia-related brain damage, epilepsy, neuropathic pain, and several neurodegenerative disorders ***(2–4)***. They are permeable for Na^+ and Ca^{2+} ions; hence, their activation results in depolarization of the cell membrane.

NMDARs have a tetrameric structure containing different types of subunits ***(5)***. Multiple genes that encode NMDAR subunits have been identified. These include the NR1 subunit with eight distinct mRNA splice variants (NR1a–h), the NR2 subunits that are coded by four different genes (NR2A, NR2B, NR2C, and NR2D), and in addition, two recently cloned subunits named NR3A and NR3B ***(6)***.

Expression studies with various combinations of these subunits in heterologous systems suggest that heteromeric assemblies of the NR1 subunit with at least one type of the NR2 subunits produce functional NMDARs with electrophysiological and pharmacological characteristics similar to those observed in brain tissue ***(7)***. According to subunit compositions, at least four heteromeric receptors can be formed, which have distinctive properties with respect to ligand binding, channel function, and varying sensitivity to pharmacological agents ***(8–11)***.

NMDARs containing NR2B subunit are involved in serious diseases such as neuropathic pain or neurodegenerative disorders ***(3,4)***. Furthermore, selective blockade of NR2B subtype of NMDARs offers a therapeutic action with less risk of adverse effects than that of NMDAR inhibition using non-subtype-selective agents. During lead optimization and candidate selection processes of drug discovery, a major issue is to ensure that the chosen compound has a selective effect on its target thereby minimizing its potential off-target effects. Subunit selectivity of a compound can be tested by evaluating its activity in cells purely expressing recombinant NMDARs with different subunit compositions. Here, we show the application of such a selectivity screening strategy by a pharmaceutical company, Gedeon Richter Ple. (Hungary). We show the use of two non-neural cell lines expressing recombinant NMDAR subunits for whole-cell current measurement. One of the cell lines is transiently transfected with NR1a and NR2A subunits, whereas the other one is stably transfected with NR1a plus NR2B subunits.

2. Materials

2.1. Preparation of Cells

1. Phosphate-buffered saline (PBS): 137 m*M* NaCl, 2.7 m*M* KCl, 10 m*M* Na_2HPO_4, and 1.8 m*M* KH_2PO_4 (all from Sigma, USA) dissolved in deionized water. Filter-sterilize and store at 4°C.
2. Dulbecco's modified Eagle's medium (D-MEM) supplemented with 10% fetal bovine serum (FBS, Gibco/BRL, USA). Filter-sterilize and store at 4°C.
3. pIND(SP1)Neo, pIND(SP1)Hygro, and pIND(SP1)/green fluorescent protein (GFP) ecdysone-inducible eukaryotic expression vectors (Invitrogen, the Netherlands).
4. cDNAs encoding the rat NR1a, NR2A, and NR2B subunits were obtained from Prof. P. Seeburg *(7)*.
5. Pfx-7 PerFect lipid transfection agent: 1:1 mix of a cationic lipid and L-alpha-Dioleoyl-Phosphatidylethanolamine (DOPE), 1.56×10^{-9} *M* of positive charge/μg lipid mix, average molecular weight: 1011 g/mol (Invitrogen).
6. Muristerone A (MuA), inducing agent (Invitrogen).
7. G418 and hygromycin, selecting agents (Invitrogen).
8. Poly-D-lysine (PDL) cell-adherence-agent (Sigma).
9. Ketamine, NMDAR-blocking agent (Sigma).
10. Ecdysone receptor (EcR)293 cells: HEK293 cells stably expressing EcR and retinoid X receptor (RXR) (Invitrogen).

2.2. Whole-Cell Patch Clamp on Transfected Cells

1. Extracellular solution: 140 m*M* NaCl, 5 m*M* KCl, 2 m*M* $CaCl_2$, 5 m*M* 5m*M* N-(2-hydoxyethyl) piperazine-N'-(2-ethane sulfonic acid) (HEPES), 5 m*M* HEPES-Na, and 20 m*M* glucose (all from Sigma). HCl or NaOH was used to adjust pH to 7.35. Osmolarity was 310 mOsm, adjusted with sucrose or distilled water.
2. Pipette filling solution: 140 m*M* CsCl, 10 m*M* HEPES, and 10 m*M* 10 m*M* ethylene glycol-bis (β- aminoethylether)-N,N,N′,N′- tetraacetic acid (EGTA) (Sigma). CsOH was used to adjust pH to 7.25, and osmolarity was 290 mOsm. Both solutions can be kept at −20°C up to a few weeks.

3. Methods

In neurons, overactivation of NMDARs results in cell death. NMDAR-mediated death correlates with an increased cytosolic calcium concentration *(12)*. Therefore, co-expression of NR1 with NR2A and/or NR2B subunits in non-neuronal cells may cause cell death unless cells are protected by inhibitors of the NMDAR ion channel ***(13)***. Another possibility to avoid excitotoxicity is that the expression of the receptor proteins is kept under control by an inducible gene-regulation system. For this reason, we have applied an ecdysone-inducible mammalian expression system: pIND from Invitrogen. Ecdysone is an insect steroid hormone that triggers metamorphosis of *Drosophila melanogaster*. Its effect is mediated by a heterodimer of the Ecdyson receptor (EcR) and the product of the ultraspiracle (USP) gene ***(14)***. Responsiveness of mammalian cells to the synthetic ecdysone analogs MuA or ponasterone A can be engineered by co-expressing the EcR with the mammalian homolog of the USP gene product, the RXR ***(15)***. Using this system, the expression of a chosen cDNA product can be kept under the control of the EcR/RXR complex that allows gene expression only in presence of the exogenous hormone. There are several advantages of the ecdysone system over the widely used tetracyclin-based gene induction strategies making them suitable for both in vitro and in vivo studies: ecdysone analogs are lipophilic, are not toxic or teratogenic, and have short half-life and favorable pharmacokinetic profile ***(16)***.

In addition to applying this inducible system in our present experiments, the culture medium was supplemented with ketamine following the induction of expression of NMDARs in order to completely rule out a possible NMDAR-mediated cellular damage.

cDNAs of different types of rat NMDAR subunits inserted into the pIND ecdyson-inducible mammalian expression vectors were introduced into HEK293 cells stably expressing EcR and RXR (EcR293 cells). NR1/NR2x receptor expression was performed either transiently (NR1/NR2A) or stably (NR1/NR2B). When transient expression was used, cDNA of the GFP in the pIND vector was also added to the cells to mark cells, which had taken up DNA successfully, and to test the effectiveness of induction. Because cell lines stably expressing recombinant NMDARs are more appropriate for robust assays desirable during compound screening, such kind of cell lines were also developed.

Patch clamp was used primarily for testing drug effect on NMDARs in transfected cells but also for selecting the most useful clonal cell lines. Clones that proved to show sustained responsiveness over several passages and high amplitude current responses to NMDAR activation were selected as tools for drug effect studies ***(17,18)***. Besides the specific agonist NMDA, glycine, a high-affinity endogenous co-agonist on NMDARs, was also necessary for activation.

3.1. Transient Expression of Rat NR1a/NR2A Subunits

1. EcR293 cells in D-MEM supplemented with 10% FBS were grown to 50–60% confluence in four-well plates that contained sterilized glass coverslips previously coated with PDL (*see* **Note 1**).
2. Stock solution (approximately 1 mg/ml) of cDNAs encoding the rat NR1a, NR2A subunits, and GFP subcloned into pIND(SP1) vectors were prepared in sterile water.
3. Vector solution was mixed with Pfx-7 PerFect lipid (supplied as a 2-mg/ml solution) in a 6:1 (w/w) lipid to DNA ratio (*see* **Note 2**).
4. Serum-free medium (D-MEM) was added to this mixture to reach 1–5 μg/ml DNA concentration (*see* **Note 3**). The mixture was incubated for 10 min at room temperature to allow lipid–DNA complexes to be formed.
5. Cells were washed by aspirating medium from the well, adding 500 μl sterile PBS/well, and aspirating the PBS. Washing removes serum that may interfere with transfection (*see* **Note 4**).
6. Hundred microliters of lipid–DNA mixture was added to each well, then cells were incubated in a CO_2 incubator (37°C) for 4 h (*see* **Note 5**).
7. After incubation, the lipid–DNA mixture was aspirated, and fresh serum-supplemented medium (1 ml/well) was added to the cells.
8. One day after transfection, the medium was replaced with fresh medium supplemented with the inducing agent MuA (3 μ*M*), and in order to prevent excitotoxicity, 500 μ*M* ketamine was also added. Until cells were used for patch-clamp experiments (24–96 h), cultures were held at 37°C in 5% CO_2/95% air atmosphere.

3.2. Generation of Stably Transfected Cell Lines Co-expressing Rat NR1a and NR2B Subunits

1. EcR293 cells were grown in 96-well plates with D-MEM supplemented with 10% FBS until they reached 50–60% confluence.
2. Stock solution (approximately 1 mg/ml) of cDNAs encoding the rat NR1a and NR2B subunits subcloned into the ecdysone-inducible eukaryotic expression vectors pIND(SP1)Hygro and pIND(SP1)Neo, respectively, were prepared in sterile water.
3. The NR1a- and NR2B-containing vectors were mixed with Pfx-7 PerFect lipid (supplied as a 2-mg/ml solution) in a 6:1 (w/w) lipid to DNA ratio.
4. Serum-free medium (D-MEM) was added to this mixture to reach 5 μg/ml DNA concentration. The mixture was incubated for 10 min at room temperature to allow lipid–DNA complexes to form.
5. Cells were washed by aspirating medium from the well, adding 100 μl sterile PBS/well, and aspirating the PBS.
6. Thirty microliters of lipid–DNA mixture was added to each well, then cells were incubated in a CO_2 incubator (37°C) for 4 h.
7. Then the lipid–DNA mixture was aspirated, and fresh serum-supplemented medium (100 μl/well) was added to the cells.

8. Twenty-four hours after transfection, G418 (600 μg/ml) and hygromycin (100 μg/ml) were added to the cultures in fresh serum-supplemented medium (100 μl/well).
9. After 2-week selection (*see* **Note 6**), clonal cell lines resistant to both G418 and hygromycin were picked and tested for functionally active NMDARs. Cells from each clone were seeded into four-well plates that contained sterilized glass coverslips coated with PDL and then were treated with the inducing agent MuA (1 μ*M*) for 24–96 h (*see* **Note 7**). The medium was also supplemented with 500 μ*M* ketamine. Until the experiments, cells were held at 37°C in 5% CO_2 atmosphere.

3.3. Whole-Cell Patch Clamp on Transfected Cells

Conventional whole-cell patch-clamp recordings *(19)* were made from transiently transfected cells or from cell lines 1–4 days after MuA induction.

1. A coverslip with attached cells was transferred to a microscope-mounted recording chamber (RC-25, Warner Instrument Corp., USA) and was constantly superfused with the extracellular solution at room temperature (23–25°C).
2. Patch electrodes (resistances: 3–7 MΩ) pulled with a P-87 micropipette puller (Sutter Instrument Co., USA) from borosilicate capillary glass (GC120F-10, Harvard Apparatus Edenbridge, UK) were filled with the pipette filling solution.
3. Cell selection and patch-clamping protocol was done under an inverted microscope (Eclipse TE2000, Nikon Instech Co. Ltd., Japan). GFP fluorescence was used for highlighting successfully transfected cells (*see* **Fig. 1**). Wild-type GFP can

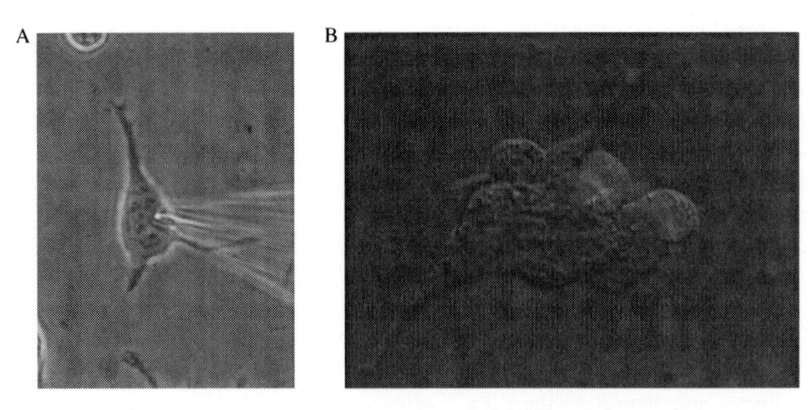

Fig. 1. (**A**) Patch clamping a HEK293 cell. The patch pipette (left) is touching the cell surface. (**B**) Green fluorescent protein (GFP) fluorescence. This group of cells contains well-expressing (light) and poorly expressing (dark) cells. The intensity of GFP fluorescence indicates the level of NMDAR expression.

be characterized with excitation peak at 395 nm and emission intensity peak at 510 nm; therefore, the microscope was equipped with a mercury lamp and also with a filter set (Sapphire/UV GFP, Chroma Technology Corp., USA) specific for wild-type GFP fluorescence. Piezoelectric micromanipulation system (PCS-5000, Burleigh Instruments Inc., USA) was used for positioning the patch pipette (*see* **Note 8**).

4. Most of the test drugs were poorly soluble in water. Therefore, dimethyl-sulfoxide (DMSO, Sigma), solutions were used to prepare stock solutions, so that after dilution in the extracellular solution, the final concentration of DMSO was 0.1% (v/v) in each case (including the control solution).

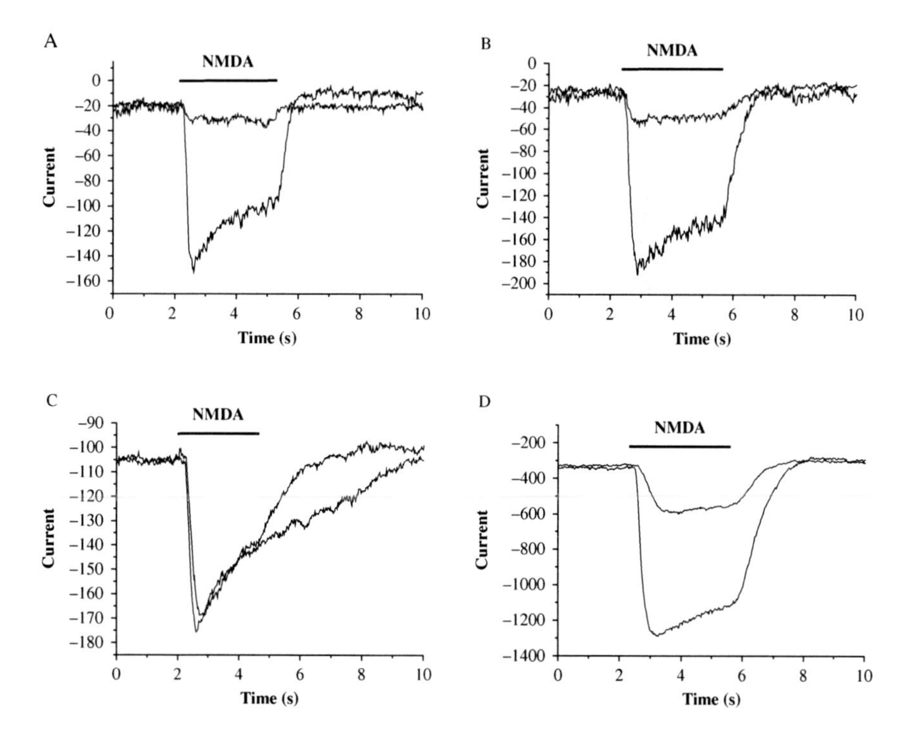

Fig. 2. Drug effect on 100 μ*M* *N*-methyl-D-aspartate (NMDA) and 10 μ*M* evoked current, individual recordings. NMDA was applied for 3 s in the absence or presence of the test compound. The upper traces were recorded in the presence of the drug. The effect of non-subtype-specific, competitive NMDA receptor blocker APV (100 μ*M*) on NR1/2A-expressing (**A**) and NR1/2B-expressing (**B**) cells is nearly equal. The NR2B-selective inhibitor CP101,606 did not show significant effect on NR1/2A-mediated current (**C**) at high concentration (10 μ*M*), whereas strongly inhibited the current of an NR1/2B-expressing cell (**D**) at much lower concentration (80 n*M*).

5. The solutions of the test compounds and *N*-methyl-D-aspartate (NMDA) were applied onto the cells through multi-barreled (10–20 in. water or $2.5–5 \times 10^3$ Pa) pressure-driven ejection pipettes (*see* **Note 9**) controlled by electromagnetic valves. The common output tip was positioned at a distance of 300–500 μM from the cell.
6. First a solution containing 100 μM NMDA (Sigma) and 10 μM glycine (Sigma) was repeatedly administered for 3 s at 30-s intervals, and after stabilization of the responses, the agonists were given in the presence of the test compound (*see* **Note 10**).
7. The inward current elicited by NMDA (*see* **Fig. 2**) was recorded at a holding potential of −76 mV using Axoclamp 200A amplifier (Molecular Devices Corp., USA) (*see* **Note 11**). Current signals were lowpass filtered at 0.025 kHz, digitized at a sampling rate of 50 Hz (Digidata 1320A interface, Molecular Devices Corp.), and captured and analyzed using pClamp 8.0 software (Molecular Devices Corp.) (for a summary of currents recorded from stably expressing cells *see* **Note 12**).
8. Data are presented as means ± SEM. Inhibition was calculated from the current peak amplitude evoked by NMDA compared with the baseline current in the presence and absence of the test compounds. The concentration-dependent effects of NMDA and inhibitory compounds were fitted to the Hill equation using Origin 6.0 (OriginLab Corp., USA). The concentration causing half-maximal inhibition (IC_{50}) was used to characterize the potency of the drug (*see* **Fig. 3**).

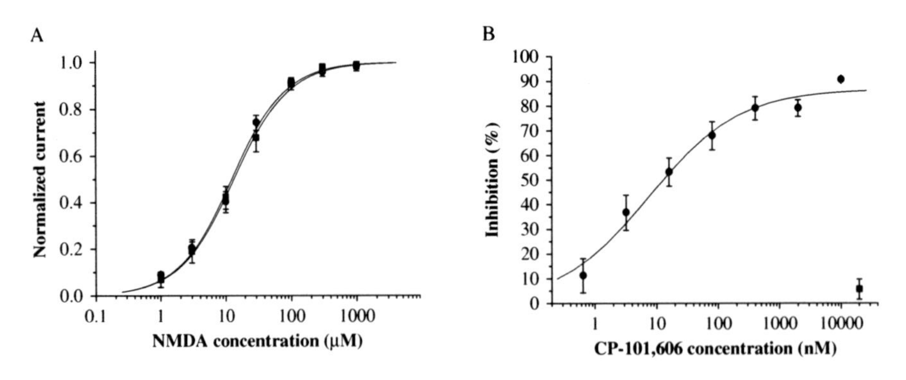

Fig. 3. (**A**) Concentration-response curves for NMDA (in the presence of 10 μM glycine) in NR1/2A- and NR1/2B-expressing cells. EC_{50} values were $12.6 \pm 1.2\,\mu M$ ($n = 7$) for 2B (•) and $14.9 \pm 3.6\,\mu M$ ($n = 6$) for 2 A (▪), whereas Hill coefficients were 1.2 ± 0.1 and 1.1 ± 0.2, respectively. (**B**) Concentration-dependent effect of CP101,606 on peak current was evoked by 100 μM NMDA on two NMDAR subtypes. In NR1/2B-expressing cells (•), the IC_{50} value was 7.6 ± 2.5 nM ($n = 5–8$) with a maximum current blockade of 87%, whereas 20 μM CP101,606 caused only a $5.6 \pm 4\%$ ($n = 4$) inhibition in the current of NR1/2A cells (▪).

4. Notes

1. The use of poly-lysine (D or L)-coated plates improves cell adherence and transfection results. The inner diameter of a well in a four-well plate is 16 mm (0.63 in.).
2. Avoid pipetting DNA stock solution! Always add DNA-free components to the DNA-containing vial and gently mix it with tipping or vortexing the vial.
3. Optimization of transfection conditions is essential for the highest-efficiency transfections and the lowest toxicity. The conditions that should be optimized include lipid composition (that is set, if a commercially available lipid mixture like Pfx-7 PerFect lipid is used), lipid and DNA concentrations, cell number, and incubation time with the DNA–lipid complex. To optimize the amount of the lipid, start with cells at approximately 50% confluency and 1–2 μg/ml DNA. With cell number and DNA concentration held constant, vary the amount of the lipid to determine its optimal concentration.
4. In general, serum decreases the efficiency of transfection. Grow the cells in complete medium (with serum) and perform transfections with serum-free medium, if possible. If your cell line cannot survive 4–24 h without serum, then try with adding serum to the transfection solution.
5. Longer time of incubation with the lipid–DNA mixture can increase transfection efficiency. Depending on cell line, it can be even 8–12 h.
6. To speed up clonal growth after the 2-week selection period, selecting agents (G418 and hygromycin) can be removed. Best clones are usually grown up within 1 week. Clones that need more than 2 weeks to grow up after ceasing the cytotoxic treatment are usually useless. To ensure the resistance of the clones, another 2-week selection period is required, or the clones must be maintained in presence of the selecting agents.
7. The extent of expression and NMDA responsiveness of transfected cells is sensitive to the applied concentration of MuA ***(18)***. For patch-clamp experiments, using 1 μ*M* MuA gives strong current responses. Responsiveness is also a function of the time of incubation with MuA. Lengthening time results in larger amplitudes using low concentrations (0.1 μ*M*) of MuA, but at high MuA, the strong responses that appear already 1–2 days after induction became weaker at later culturing times than that with using low MuA ***(17)***.
8. It is recommended to patch solitary cells or at least cells that are members of a small group of cells. As neighboring HEK cells can be electrically coupled through gap junctions, patching a cell in a large group could lead to space clamp errors ***(1)***.
9. We used several types of manifolds for drug ejection. The formerly used homemade ones consisted of plastic pipette tips converging to a common output tip that is also made from the tip of a pipette. All these parts were fixed by a two-component (epoxy based) glue. Sometimes we had problems with this type of manifold, mainly concerning their flow issues, namely, we experienced different flow rates among different input tubes. In order to improve flow performance,

now we work with commercially available manifolds. These are Perfusion Pencil Body for eight channels with removable tip (internal diameter: 250 μM) by World Precision Instruments, Germany and BPS-QMM-8WT Quartz Micromanifold for eight channels (output tube diameter: 200 μM) by ALA Scientific Instruments, Inc., USA. Both manifolds consist of polyimide-coated quartz capillary tubing.

10. In a few cases, the disappearance or gradual decrease of the agonist-evoked responses could be seen. This phenomenon might be because of the dilution of the agonist in the tip of the given pipette or tubing. In the case of such a decrease of agonist responses, we applied a long (1–5 min) non-stop flow in the given channel, which in most cases restores the originally recorded or assumed current amplitude elicited by the agonist. Leaving this point out of consideration may lead to serious overestimation or underestimation of the potency of an NMDAR-blocking drug.
11. Axon Instruments, the manufacturer of electrophysiology instruments, is part of Molecular Devices Corp. since 2004.
12. In stably expressing NR1a/2B cells, the average peak amplitude of inward current elicited by 100 μM NMDA and 10 μM glycine was 370 ± 36 pA ($n = 86$). The responsiveness (the ratio of cells that responded with at least 50 pA current in amplitude) was 60%. Current recorded from cells patched in our experiments was generally stable for 15 min (or even 80 min) (at least 15-min recording time in 82% of patched cells), although often showed moderate rundown.

Acknowledgments

The authors would like to thank Dr. István Tarnawa for valuable discussions and for critically reading the manuscript. The excellent technical assistance of Anikó Dóbiász and Anikó Kossár is also acknowledged.

References

1. Thomas P. and Smart T. G. (2005) HEK293 cell line: a vehicle for the expression of recombinant proteins. *J. Pharmacol. Toxicol. Methods* **51**, 187–200.
2. Hynd M. R., Scott H. L. and Dodd P. R. (2004) Glutamate-mediated excitotoxicity and neurodegeneration in Alzheimer's disease. *Neurochem. Int.* **45**, 583–595.
3. Loftis J. M. and Janowsky A. (2003) The N-methyl-D-aspartate receptor subunit NR2B: localization, functional properties, regulation, and clinical implications. *Pharmacol. Ther.* **97**, 55–85.
4. Parsons C. G. (2001) NMDA receptors as targets for drug action in neuropathic pain. *Eur. J. Pharmacol.* **429**, 71–78.
5. Furukawa H., Singh S. K., Mancusso R. and Gouaux E. (2005) Subunit arrangement and function in NMDA receptors. *Nature* **438**, 185–192.
6. Chatterton J. E., Awobuluyi M., Premkumar L. S., Takahashi H., Talantova M., Shin Y., Cui J., Tu S., Sevarino K. A., Nakanishi N., Tong G., Lipton S. A. and Zhang D. (2002) Excitatory glycine receptors containing the NR3 family of NMDA receptor subunits. *Nature* **415**, 793–798.

7. Monyer H., Sprengel R., Schoepfer R., Herb A., Higuchi M., Lomeli H., Burnashev N., Sakmann B. and Seeburg P. H. (1992) Heteromeric NMDA receptors: molecular and functional distinction of subtypes. *Science* **256**, 1217–1221.
8. Cull-Candy S., Brickley S. and Farrant M. (2001) NMDA receptor subunits: diversity, development and disease. *Curr. Opin. Neurobiol.* **11**, 327–335.
9. Qian A., Buller A. L. and Johnson J. W. (2005) NR2 subunit-dependence of NMDA receptor channel block by external Mg^{2+}. *J. Physiol. (Lond.)* **562**, 319–331.
10. Vicini S., Wang J. F., Li J. H., Zhu W. J., Wang Y. H., Luo J. H., Wolfe B. B. and Grayson D. R. (1998) Functional and pharmacological differences between recombinant N-methyl-D-aspartate receptors. *J. Neurophysiol.* **79**, 555–566.
11. Williams K. (1993) Ifenprodil discriminates subtypes of the N-methyl-D-aspartate receptor: selectivity and mechanisms at recombinant heteromeric receptors. *Mol. Pharmacol.* **44**, 851–859.
12. Raymond L. A., Moshaver A., Tingley W. G., Shalaby I. and Huganir R. L. (1996) Glutamate receptor ion channel properties predict vulnerability to cytotoxicity in a transfected nonneuronal cell line. *Mol. Cell. Neurosci.* **7**, 102–115.
13. Anegawa N. J., Guttmann R. P., Grant E. R., Anand R., Lindstrom J. and Lynch D. R. (2000) N-methyl-D-aspartate receptor mediated toxicity in nonneuronal cell lines: characterization using fluorescent measures of cell viability and reactive oxygen species production. *Mol. Brain Res.* **77**, 163–175.
14. Yao T. P., Forman B. M., Jiang Z., Cherbas L., Chen J. D., McKeown M., Cherbas P. and Evans R. M. (1993) Functional ecdysone receptor is the product of EcR and ultraspiracle genes. *Nature* **366**, 476–479.
15. No D., Yao T. P. and Evans R. M. (1996) Ecdysone-inducible gene expression in mammalian cells and transgenic mice. *Proc. Natl. Acad. Sci. U* **93**, 3346–3351.
16. Saez E., No D., West A. and Evans R. M. (1997) Inducible gene expression in mammalian cells and transgenic mice. *Curr. Opin. Biotechnol.* **8**, 608–616.
17. Kurkó D., Dezsö P., Boros A., Kolok S., Fodor L., Nagy J. and Szombathelyi Z. (2005) Inducible expression and pharmacological characterization of recombinant rat NR1a/NR2A NMDA receptors. *Neurochem. Int.* **46**, 369–379.
18. Nagy J., Boros A., Dezsö P., Kolok S. and Fodor L. (2003) Inducible expression and pharmacology of recombinant NMDA receptors, composed of rat NR1a/NR2B subunits. *Neurochem. Int.* **43**, 19–29.
19. Hamill O. P., Marty A., Neher E., Sakmann B. and Sigworth F. J. (1981) Improved patch clamp techniques for high resolution current recording from cells and cell-free membrane patches. *Pflügers. Arch.* **391**, 85–100.

2

Memantine as an Example of a Fast, Voltage-Dependent, Open Channel *N*-Methyl-D-Aspartate Receptor Blocker

Chris G. Parsons and Kate Gilling

Summary

Electrophysiological techniques can be used to great effect to help determine the mechanism of action of a compound. However, many factors can compromise the resulting data and their analysis, such as the speed of solution exchange, expression of additional ion channel populations including other ligand-gated receptors and voltage-gated channels, compounds having multiple binding sites, and current desensitization and rundown. In this chapter, such problems and their solutions are discussed and illustrated using data from experiments involving the uncompetitive *N*-methyl-D-aspartate (NMDA) receptor antagonist memantine. Memantine differs from many other NMDA receptor channel blockers in that it is well tolerated and does not cause psychotomimetic effects at therapeutic doses. Various electrophysiological parameters of NMDA-induced current blockade by memantine have been proposed to be important in determining therapeutic tolerability, potency, onset and offset kinetics, and voltage dependency. These were all measured using whole cell patch-clamp techniques using hippocampal neurons. Full results are shown here for memantine, and these are summarized and compared with those from similar experiments with other NMDA channel blockers. The interpretation of these results is discussed, as are theories concerning the tolerability of NMDA channel blockers, with the aim of illustrating how electrophysiological data can be used to form and support a physiological hypothesis.

Key Words: NMDA; uncompetitive; concentration dependence; concentration clamp; voltage dependence; kinetics.

From: *Methods in Molecular Biology, vol. 288: Patch-Clamp Methods and Protocols*
Edited by: P. Molnar and J. J. Hickman © Humana Press Inc., Totowa, NJ

1. Introduction

Memantine is an uncompetitive *N*-methyl-D-aspartate (NMDA) receptor antagonist that is registered in Europe and the United States for the treatment of moderate to severe Alzheimer's disease (AD). It has clear symptomatic effects in both AD patients ***(1,2)*** and animal models of AD ***(3)*** and, on the basis of its mechanism of action, is also likely to show neuroprotective activity in AD ***(3,4)***. This compound blocks the channel in an use-dependent manner, meaning that it can only gain access to the channel in the presence of agonist, and remains trapped in the channel following removal of agonist ***(3,5,6)***. Both the clinical tolerability and the symptomatic effects of memantine have been attributed to its fast blocking kinetics and strong voltage dependency ***(3,5–7)***. These properties have been characterized by numerous groups using whole cell patch-clamp recordings from primary cultures of hippocampal and cortical neurons ***(6,8–15)***. However, there are several factors that must be taken into account when performing such experiments to ensure the quality of the recordings and their analysis.

- Fast blocking kinetics can only be measured accurately when fast concentration-clamp techniques are used to apply antagonists. This is particularly problematic with primary cultures of hippocampal/cortical neurons due to their large dendritic arborization, the inability to lift such cells from the bottom of the dish, and resulting problems of buffered diffusion. Fluid-in-fluid fast concentration-clamp systems with relatively large application diameters, for example, theta glass stepping motor systems, are preferred over systems such as U-type application tubes with which offset kinetics cannot be addressed because there is no real "wash off" of compounds with the latter technique.
- The native cells used probably have mixed receptor populations (for example, N1a/2A and NR1a/2B), and differences in the potency of the antagonist at each of the receptor subtypes may exist. Uncompetitive antagonists such as memantine may also bind to multiple sites within the NMDA receptor channel ***(14,15)***. Both of these aspects can lead to double exponential blocking/unblocking kinetics that first become apparent when fast concentration-clamp techniques are used. When present, these double kinetics must be measured accurately and subsequently analyzed. In addition, the unblocking kinetics are voltage dependent, and this aspect cannot be addressed by using ramping protocols in voltage-dependency experiments (*see* point iii).
- Voltage-dependency experiments are often hampered by the presence of additional voltage-gated ion channels, and their contribution to the currents recorded must be minimized and/or accounted for in the analysis. One way around this problem is to use tetrodotoxin (TTX) to block voltage-activated sodium channels (VASCs) and replace K^+ with Cs^+ in the recording solutions to reduce the effects of

voltage-activated potassium channels (VAKCs). The contribution of voltage-activated calcium channels on the currents measured can be reduced by lowering Ca^{2+} in the extracellular solution (*see* point iv). The contribution of all voltage-activated channels (VACs) can also be reduced by avoiding the use of relatively fast ramping protocols and rather recording individual NMDA-induced currents after holding at different holding potentials and allowing the VACs to desensitize/inactivate for several tens of seconds following each incremental depolarizing step. For example, shifting the holding potential from −90 to +70 mV in steps of 10 mV, allowing at least 30 s at each new holding potential before applying agonist. Residual VAC currents can then be subtracted from the agonist-induced currents mediated through ligand-gated channels such as the NMDA receptor channel. However, such protocols are long and are associated with other potential problems (*see* points iv and v).

- NMDA receptors show various forms of Ca^{2+}-dependent desensitization (with time constants of 500 ms to several seconds) ***(16–18)***, the presence of which can interfere with experiments assessing the blocking kinetics of open channel blockers such as memantine. This problem can be minimized by decreasing Ca^{2+} concentrations in the solutions used (*see also* point iii), but Ca^{2+} is an important cation for membrane stability. One way around this additional problem is to reduce Ca^{2+} concentrations in the presence of agonist (0.2 m*M*) but maintain normal Ca^{2+} concentrations (1.5 m*M*) between agonist applications to allow the cell membrane to "recover" between agonist applications.
- Glycine is a co-agonist for NMDA receptors (at the glycine$_B$ site) with an EC_{50} of around 1 μ*M*, and its presence is a prerequisite for receptor channel activation by glutamate or NMDA ***(19)***. At non-saturating glycine concentrations, NMDA receptors show strong glycine-dependent desensitization, with quite fast kinetics ($\tau = 100$–400 ms), which can impede the analysis of antagonist blocking kinetics ***(20)***. Moreover, glycine concentrations in the solutions can be dynamically altered by the presence of microbial organisms in the perfusion setup as glycine is involved in metabolism. As such, the magnitude and rate of glycine-sensitive desensitization can change during long recordings, and it is essential to keep the whole perfusion system very clean. D-Serine is also a co-agonist for the glycine site ***(21)***, but it is not metabolized so easily and is the preferred co-agonist for such experiments, used at saturating concentrations. Clean perfusion systems are nonetheless very important for such experiments, for example, contamination with previously used "sticky" (most often lipophyllic) compounds should be avoided.
- Aside from the various forms of receptor desensitization detailed in points iv and v, NMDA receptors also show moderate rundown that should be minimized by the choice of appropriate intracellular (adenosine 5′-triphosphate (ATP) regenerating) and extracellular (low Ca^{2+}) solutions. However, some form of run-down compensation is essential in the analysis to account for such dynamic changes, especially when assessing the potency of antagonists with several concentrations being tested sequentially over time.

Most scientific papers on patch-clamp experiments tend to minimize the description of the methods used, and minor details that could be very important for the final outcome are often not apparent from such descriptions. The aim of this chapter is to describe, in detail, the methods used to address such aspects for memantine as an example of a fast, voltage-dependent, NMDA receptor channel blocker.

2. Materials

2.1. Cell Culture

1. Mg^{2+}-free Hanks' buffered salt solution (Gibco BRL, Germany) stored at 2–5°C and warmed to approximately 35°C before use.
2. Solution of 0.05% DNase and 0.3% ovomucoid (Sigma Aldrich, Germany) in phosphate-buffered saline (PBS, Gibco BRL) stored at −20°C and warmed to approximately 35°C before use.
3. Solution of 0.66% Trypsin and 0.1% DNase (Sigma Aldrich) in PBS stored at −20°C and warmed to approximately 35°C before use.
4. Minimum essential medium (Gibco BRL) stored at 2–5°C and warmed to approximately 35°C before use.
5. Poly-D-L-ornithine (500 μg/ml) dissolved in 0.5 *M* boric acid (both from Sigma Aldrich), stored at −20°C, and warmed to approximately 35°C before use.
6. Laminin (Sigma Aldrich) dissolved in PBS to a concentration of 10 μg/ml, stored at −20°C, and warmed to approximately 35°C before use.
7. $NaHCO_3$/HEPES-buffered minimum essential medium supplemented with 5% fetal calf serum and 5% horse serum (all from Gibco BRL), stored at 2–5°C, and warmed to approximately 35°C before use.
8. Cytosine-β-D-arabinofuranoside (Sigma Aldrich) stored at 2–5°C.
9. Plasticware including flasks, Petri dishes, and pipettes (Corning Incorporated, Germany).

2.2. Patch Clamp

1. Borosilicate glass for recording pipettes with an outer diameter of 1.5 mm and an inner diameter of 1.275 mm (Hilgenberg GmbH, Germany; cat. no. 1408411).
2. The P-97 horizontal pipette puller (Sutter Instruments, USA).
3. Square-walled application pipette glass with a wall-to-wall measurement of 700 μ*M* (Warner Instruments LLC, USA; cat. no. P/N 3SG700-5).
4. The SF-77B Perfusion Fast-Step: a stepper motor-driven, double-barreled theta glass application pipette delivery system (Warner Instruments LLC).
5. Silicon-glass tubing with an external diameter of 0.43 mm and an internal diameter of 0.32 mm (S.G.E. GmbH, Germany).
6. Polyethylene tubing of the PE-10 size (Clay Adams, USA).
7. Low-volume manifolds (MM series six-port manifolds; Warner Instruments LLC).

8. Manifold valves (Lee Hydraulische Miniaturkomponenten GmbH, Germany; cat. no. LFAA1201718H).
9. TIB 14S digital output trigger interface (HEKA, Germany).
10. EPC-9 or EPC-10 amplifier (HEKA).
11. Axiovert 35 inverted microscope (Carl Zeiss, Germany).
12. A valve driver, similar to those produced by Lee (Lee Hydraulische Miniaturkomponenten GmbH; cat. no. IECX0501500A).
13. Software for data acquisition and analysis, such as TIDA 5.0 (HEKA), Excel 2000 (Microsoft, USA), and GraFit 5.0 (Erithacus Software Ltd., UK), and suitable computer hardware.
14. Intracellular solution used for recording NMDA receptor-mediated currents from hippocampal neurons, consisting of 120 m*M* CsCl, 10 m*M* ethylene glycol-tris(2-aminoethylether)-N,N,N′,N′-tetraacetic acid (EGTA), 1 m*M* $MgCl_2$, 0.2 m*M* $CaCl_2$, 10 m*M* glucose, 20 m*M* tetraethyl ammonium chloride, 2 m*M* ATP, and 0.2 m*M* adenosine 3′,5′-cyclic monophosphate (cAMP). All these components were purchased from Sigma Aldrich and stored according to manufacturer's instructions.
15. Extracellular bath solution used for recording NMDA receptor-mediated currents from hippocampal neurons, consisting of 140 m*M* NaCl, 3 m*M* CsCl, 10 m*M* glucose, 10 m*M* HEPES, 0.2 m*M* $CaCl_2$, and 4.5 m*M* sucrose, and supplemented with 0.35 μ*M* TTX. All these components were purchased from Sigma Aldrich and stored according to manufacturer's instructions.
16. NMDA (Sigma Aldrich) stored as a 100-m*M* stock solution in distilled water at −20°C. The stock solution of the co-agonist D-serine (Sigma Aldrich) prepared and stored under the same conditions but at a concentration of 10 m*M*.
17. The NMDA receptor antagonists memantine, neramexane (Merz Pharmaceuticals GmbH, Germany), ketamine, phencyclidine (PCP), dextromethorphan, and dextrorphan (Sigma Aldrich) all stored in distilled water at 2–5°C as stock solutions of 10 m*M*.

3. Methods

3.1. Cell Culture

Hippocampal tissue was obtained from rat embryos (E20–E21) and was then transferred to Ca^{2+}- and Mg^{2+}-free Hanks' buffered salt solution on ice. Cells were mechanically dissociated in 0.05% DNase/0.3% ovomucoid solution following an 8-min pre-incubation with 0.66% Trypsin/0.1% DNAase solution. The dissociated cells were then centrifuged at 18 *g* for 10 min, re-suspended in minimum essential medium, and plated at a density of 150,000 cells/cm^2 onto poly-D-L-ornithine/laminin-precoated plastic Petri dishes. These dishes were precoated by treating dishes overnight at 37°C with poly-D-L-ornithine, washing twice with PBS, and then incubating with laminin solution overnight at 37°C.

Excess solution was aspirated and dishes washed with PBS followed by the cell medium before cell plating. The cells were nourished with $NaHCO_3$/HEPES-buffered minimum essential medium supplemented with 5% fetal calf serum and 5% horse serum and incubated at 37°C with 5% CO_2 at 95% humidity. The medium was exchanged completely following the inhibition of further glial mitosis with cytosine-β-D-arabinofuranoside after about 5 days in vitro (DIV). Patch-clamp recordings were made after 12–15 DIV.

3.2. Patch Clamp

3.2.1. Recording

Voltage-clamp recordings were made in the whole cell configuration of the patch-clamp technique at a holding potential of −70 mV, unless otherwise stated. All recordings were made at room temperature (20–23°C) (Note: The kinetics of drug/receptor interactions are highly dependent on temperature.) Pyramidal cells were visualized using an inverted microscope under phase contrast and selected for patching based upon their position and morphology. The cells were opened by suction after the formation of a giga seal between the pipette and cell membrane and were allowed to stabilize for 1–2 min before recordings were made. Patch-clamp pipettes were pulled from borosilicate glass using a horizontal puller and, when filled with intracellular solution, had resistances of 1–3 MΩ. Currents were recorded using an EPC-9/10 amplifier, and TIDA 5.0. software was used for the collection and storage of data. For full details of the functions, measurements, and compensations performed by the EPC-9/10 amplifier and related software *see* **refs.** ***22*** and ***23***. Briefly, offset compensation was performed for each open pipette in order to ensure that the command potential is equal to the membrane potential. The liquid junction potential was measured for each set of solutions using an agar bridge in place of the usual silver chloride pellet as the ground electrode. Liquid junction potential was measured by filling one perfusion chamber with intracellular solution as the reference and the other chamber with extracellular solution. For the solutions used for recording from the hippocampal neurons, the liquid junction potential was measured to be 3.4 mV.

Series resistance was measured (mean value of 4.22 ± 0.14 MΩ) and accordingly compensated for in conjunction with capacitance. Fast capacitive currents were corrected for by the EPC-9/10 upon formation of the giga seal, and whole cell capacitance correction was performed after the cell was opened. These procedures were performed semi-automatically using the amplifier and TIDA 5.0 software.

The current signal was filtered by the EPC-9/10 amplifier using the three-pole prefilter with Bessel 10 kHz bandwidth and the four-pole filter set to 2.9 kHz with Bessel characteristic. Current measurements were acquired at a rate of 10 kHz to avoid potential problems of aliasing.

3.2.2. Perfusion System

Test substances were applied by switching channels of a modified stepper motor-driven, double-barreled theta glass application pipette. The openings of the square-walled application pipette glass were reduced to 200–250 μ*M* by pulling these glass capillaries, by hand, over a Bunsen burner and then cleanly separating "two" new perfusion pipettes by cutting them with a diamond cutter. Furthermore, the internal dead volume of such application pipettes was reduced to a minimum by the following procedure. Silicon-glass tubing was inserted as far as possible toward the tips of the theta glass application "pipettes" (around 2–3 mm from the tip). The non-tip, open ends of the silicon-glass tubing was blocked with acrylic glue, and the pipettes were then reverse filled with molten wax almost to the open tips of the silicon-glass tubing. After this procedure, wax was cleaned from the outside of the perfusion pipettes, and the glued ends of the silicon-glass tubing were cut free and attached through conventional polyethylene tubing to the low volume six-port manifolds. These manifolds were then connected to an automatic perfusion array gated through manifold valves to gravity-fed syringes containing the solutions of interest. Perfusion was controlled using the TIB 14S digital output trigger interface in conjunction with the EPC-9/10 and the TIDA data acquisition system. Valves were driven by a custom-made valve driver to provide the necessary power (490 mW per valve) and spiked voltage jumps. Twenty volts for 10 ms then held at 10 V.

Optimal positioning of the pipette was practiced using solutions of different osmolarity to visualize the interface between solutions under phase-contrast microscopy. The best angle was found to be 45%, and care was taken to keep the lower edge of the application pipette tip parallel to, and as close as possible to, the bottom of the dish without scratching the plastic during the stepping motor movement. The lower edge of the application pipette was positioned with the start channel centered some 150–250 μ*M* from the neuron of interest. The solution exchange time of this perfusion system as measured using small, lifted cells was approximately 20 ms. Complete exchange of the perfused solutions to be applied through the application pipette was of the order of 1–2 s. The level of bath solution was kept constant using a vacuum-driven glass suction pipette. This is important to avoid changes in recording pipette capacitive characteristics.

3.2.3. Solutions

The composition of the intracellular solution used when recording NMDA receptor mediated from hippocampal neurons is given in **Subheading 2.2 point 14**. The absence of intracellular K^+ and the presence of intracellular TEA should block VAKCs. ATP and cAMP were included to decrease rundown although more elaborate ATP regenerating systems can be used for more problematic receptors such as neuronal nicotinic receptors *(24)*. The corresponding extracellular bath solution composition is given in **Subheading 2.2 point 15**. TTX was included at 0.35 μ*M* in order to block VASCs, and D-serine was present at 10 μ*M* in all extracellular solutions—this concentration being sufficient to saturate the glycine$_B$ site, which should remain stable during the course of the experiments.

3.2.4. Cumulative Protocols

Cumulative protocols can be used for faster determination of concentration dependency of blockade and were shown to produce very similar IC_{50} determinations as kinetic protocols (*see* **Table 1**). In these protocols, five to six sequentially increasing concentrations (in a log 3 progression, for example, 0.3, 1, 3, 10, and 30 μ*M*) of memantine or other standard uncompetitive NMDA receptor antagonists are applied in a cumulative regime, each for 10–30 s, in the continuous presence of NMDA (200 μ*M*) for 100–200 s, and recovery is only recorded after the last and highest concentration. With such protocols, cells do not have to remain stable for such long recording durations. However, they give little useful information on the kinetics of block, and desensitization/rundown is more of a problem.

3.2.5. Kinetic Experiments

Kinetic experiments were performed by applying various single concentrations of memantine or standard uncompetitive NMDA receptor antagonists for 10–30 s in the continuous presence of NMDA (200 μ*M*) for 30–120 s. When using this protocol, currents are allowed to recover after each application of antagonist, and onset and offset kinetics can be measured.

3.2.6. Voltage Dependency

Fractional block of currents by memantine (10 μ*M*) at various holding potentials was used to determine the voltage dependency of this effect. The holding potential was changed every 120 s from −80 to +60 mV in 10 mV increments, and NMDA (200 μ*M*) was applied for 41 s at each holding potential. Memantine

Table 1

Summary of the Results From Experiments Measuring the Antagonistic Properties of Various Known *N*-Methyl-D-Aspartate (NMDA) Receptor Antagonists Against NMDA-Induced Currents Recorded From *Xenopus* Oocytes and Cultured Hippocampal Neurons

Substance	Cumulative IC_{50} (μM)	Cumulative Hill	Kinetic IC_{50} (μM)	Kinetic Hill	K_{on} ($10^4\ M/s$)	K_{off} (per second)	K_d (μM)	IC_{50} (0 mV) (μM)	δ	β
Memantine	1.87 ± 0.17	0.80 ± 0.04	1.27 ± 0.08	0.93 ± 0.06	5.94 ± 0.35	0.125 ± 0.019	2.100 ± 0.450	17.35 ± 1.78	0.83 ± 0.04	0.08 ± 0.02
Neramexane	1.21 ± 0.04	0.88 ± 0.02	0.85 ± 0.05	0.81 ± 0.03	9.30 ± 0.90	0.120 ± 0.008	1.330 ± 0.200	16.99 ± 2.09	0.96 ± 0.04	0.10 ± 0.03
Ketamine	1.47 ± 0.22	0.82 ± 0.08	0.98 ± 0.09	1.01 ± 0.08	3.95 ± 0.50	0.070 ± 0.005	1.670 ± 0.320	13.16 ± 0.90	0.85 ± 0.05	0.08 ± 0.02
Phencyclidine (PCP)	0.13 ± 0.01	0.78 ± 0.08	0.11 ± 0.03	0.70 ± 0.05	7.00 ± 0.19	0.010 ± 0.003	0.143 ± 0.004	1.26 ± 0.11	0.65 ± 0.04	0.10 ± 0.17
Dextromethorphan	1.95 ± 0.17	1.37 ± 0.15	2.11 ± 0.17	1.16 ± 0.10	1.74 ± 0.16	0.127 ± 0.027	7.470 ± 1.820	15.34 ± 1.51	0.84 ± 0.05	0.14 ± 0.02
Dextrorphan	0.34 ± 0.05	1.02 ± 0.10	0.36 ± 0.08	1.00 ± 0.17	19.90 ± 0.78	0.063 ± 0.004	0.317 ± 0.023	3.39 ± 0.70	0.61 ± 0.06	−0.01 ± 0.14
Mg^{2+}	NT	NT	37.62 ± 1.41	0.82 ± 0.02	11.28 ± 0.75	21.61 ± 0.60	191.6 ± 13.8	5521 ± 261	0.95 ± 0.02	−0.32 ± 0.01

NT, not tested.

Only compounds produced by Merz Pharmaceuticals GmbH were tested using *Xenopus* oocytes. The hippocampal IC_{50} and Hill coefficient values and onset and offset kinetics are calculated from recordings in which the NMDA currents were allowed to recover between the application of various concentrations on antagonist. All data are shown in the form of value ± SE.

(10 μ*M*) was applied for 11 s during each NMDA application period. During the recovery period, 15 s following the removal of memantine, neurons were clamped to +70 mV for 5 s in the continuing presence of NMDA to facilitate complete recovery from antagonism. Similar experiments were performed with PCP, dextrorphan, and (+)MK-801 except that the application and recovery times had to be increased for these very slow channel blockers.

In order to subtract any residual VAC currents, mirror voltage-clamp (P5) protocols with smaller (20%) voltage steps in the opposite direction were run between agonist and antagonist applications at each holding potential, for example, the equivalent for a step from −90 to +70 mV (difference of −160 mV) was −90 to −122 mV (difference of −32 mV).

3.2.7. Analysis of Data

TIDA 5.0 software was used for the quantification of individual current amplitudes and kinetics. Excel 2000 was used to pool these data and GraFit 5.0 software used to fit pharmacological equations and pooled kinetic K_{on} and K_{off} values. For all data points, the value given is the mean of results from four to eight cells per concentration.

Rundown was usually not extreme for these NMDA receptor currents recorded from cultured hippocampal neurons (normally less than 10% over a 1-min period in the continuous presence of agonist). However, in order to produce the most accurate assessment of potency, the analysis nevertheless corrected for extrapolated linear current rundown—this was increasingly important for high-affinity compounds where the duration of the agonist exposure was prolonged to assure that low concentrations of antagonist reached steady-state blockade. Antagonism of NMDA receptor-mediated currents was measured as the magnitude of the steady-state blocked current as a percentage of the control current. For non-kinetic antagonistic protocols with five cumulatively increasing concentrations of antagonist, the control current for each antagonist concentration was calculated by a linear projection from the steady-state current before and after antagonist application, that is, for the first antagonist concentration, control current = 0.85 × current before antagonist + 0.15 × recovery current. For the second antagonist concentration, these same currents were multiplied by 0.7 and 0.3, respectively, and so on for all concentrations. For kinetic protocols, single concentrations of antagonist were applied in the continuing presence of agonist, and the control current was taken as the mean of the steady-state current before and after antagonist application.

Potency of compounds was assessed by plotting the mean percentage current magnitude, calculated with standard error against antagonist concentration, and a curve was fit using the logistic equation for which the variable parameters IC_{50} and Hill coefficient (n) were free and the range and background values were normally set to 100 and 0, respectively;

$$\% \text{ current} = \frac{\text{range}}{1 + \left(\text{antagonist}/IC_{50}\right)^{n}} + \text{background}. \tag{1}$$

For kinetic protocols, exponential fits to both onset and offset kinetics were made using TIDA 5.0 software, and most onset and offset responses were often well fit by a single exponential function, where a and b are current amplitudes and c represents the time constant, τ:

$$\text{amplitude} = a + b \cdot e^{\text{time}/c}. \tag{2}$$

When double exponential fits described the data better, these were fit with the following equation:

$$\text{amplitude} = a + b \cdot e^{\text{time}/c} + d \cdot e^{\text{time}/f}, \tag{3}$$

where a, b, and d are current amplitudes and c and f are time constants (c represents τ_{fast} and f is τ_{slow}). These were then integrated to produce a single time constant ($t_{combined}$) according to **Eq. 4**.

$$t_{combined} = \frac{\left[c \cdot b/(b+d)\right] + \left[f \cdot d/(b+d)\right]}{\left[b/(b+d)\right] + \left[d/(b+d)\right]}. \tag{4}$$

The rational for this weighting procedure was to allow simple comparison between the rate kinetics produced by this investigation and also with previous data for memantine and other channel blockers for which similar techniques or simple single exponentials were used to fit the data. Another reason for this weighting was to allow a simple calculation of calculated K_d (*see* **Eq. 6**).

Mean values of the time constant (τ) were calculated for each concentration of antagonist, and $1/\tau$ was plot against concentration. These data were fit using linear equations, where m is the gradient of the line, which corresponds to K_{on} for the onset rate fit, and g is the intercept at the y-axis, which corresponds to K_{off} for the offset rate fit:

$$y = m \cdot x + g. \tag{5}$$

All compounds tested showed concentration-dependent open channel blocking kinetics, whereas the offset rate was essentially concentration independent. As expected, the fit of τ_{on} against concentration intercepted the

y-axis at similar values to the fit of τ_{off} against concentration. Calculation of the ratio K_{off}/K_{on} was used to reveal an apparent K_d that was then compared with the IC_{50} calculated at equilibrium:

$$K_d = \frac{K_{off}}{K_{on}}. \tag{6}$$

To assess voltage dependency, single concentrations of antagonist were applied to currents in the plateau phase and then the cell was allowed to recover before stimulation was repeated at a more positive holding potential (holding potential was increased by increments of +10 mV each time). Blockade was expressed as a percentage of the mean control current recorded before and after the application of antagonist, as in the kinetic experiments as described earlier in this section. The pooled data were then fit by the following equation, where $IC_{50}(0)$ is the IC_{50} at 0 mV, β is the fraction of voltage-independent sites, δ is the fraction of the electric field sensed by the voltage-dependent site, and all other parameters have their normal meaning (z, valency; F, Faraday's constant; R, universal gas constant; and T, absolute temperature):

$$\text{fractional current} = \frac{1-\beta}{[1+\text{antagonist}]/[IC_{50}(0)\cdot e^{-z\delta FV/RT}]}. \tag{7}$$

4. Results

The application of NMDA (200 μ*M*) in the continuous presence of D-serine (10 μ*M*) to hippocampal neurons evoked currents that were attenuated by all the uncompetitive NMDA receptor antagonists used in this study. The data are summarized in **Table 1**.

4.1. Potency

Initial experiments involved the application of five sequentially increasing concentrations of antagonist. However, kinetic measures of both onset and offset rates for each concentration of antagonist cannot be determined using this application protocol, so the kinetic protocol was employed that involved the application of antagonist in the continuous presence of NMDA and D-serine with a period of current recovery between each antagonist concentration. This protocol could also be considered to give more accurate measure of potency, as the magnitude of blockade by each concentration is assessed compared with the current before and after each application of antagonist, but the difference between values determined from each set of results is clearly only slight (*see* **Table 1**).

Memantine antagonized NMDA currents concentration-dependently with an IC_{50} value of $1.87 \pm 0.17\,\mu M$ and a Hill coefficient of 0.80 ± 0.04 when the cumulative protocol was used (*see* **Fig. 1**). From the data produced using the kinetic protocol, as shown in **Fig. 1B,** a new concentration–response curve was produced, and the IC_{50} value calculated from this curve was only marginally lower ($1.27 \pm 0.08\,\mu M$) although the difference does reach a level of significance ($p = 0.019$, $t = 3.19$; Student's t-test using raw data). The Hill coefficient is slightly higher (0.93 ± 0.06) than that derived from the earlier data, but values derived using these two protocols are not significantly different ($p > 0.05$; Student's t-test using raw data).

4.2. Kinetics

Simple analysis of the kinetic data shows memantine to have concentration-dependent blocking and concentration-independent unblocking kinetics ($K_{on} = 5.94 \pm 0.35 \times 10^4\, M/s$ and $K_{off} = 0.125 \pm 0.019$ per second). From these values, K_d can be calculated as $2.10 \pm 0.45\,\mu M$ according to the equation $K_d = K_{off}/K_{on}$, which correlates well with the IC_{50} values calculated by both of the concentration–response curves.

However, memantine did indeed sometimes show double exponential kinetics (*see* **Fig. 2**). The onset and offset kinetics of blockade following concentration jumps with memantine ($10\,\mu M$ at -70 mV) showed double exponential kinetics: τ_{on} fast 86.9 ± 6.3 ms(64.7%); τ_{on} slow 1383 ± 122 ms; τ_{off} fast 834 ± 321 ms(22.9%); and τ_{off} slow 4795 ± 921 ms. It should be noted that the kinetics of blockade by memantine in these experiments with D-serine ($10\,\mu M$ at -70 mV) were somewhat faster than those using glycine at a subsaturating concentration ($1\,\mu M$ at -70 mV) ***(12)***. This is not surprising as channel gating kinetics also influence the kinetics of open channel blockade. The relief of blockade following voltage jumps to $+70$ mV in the continuous presence of memantine ($10\,\mu M$) also showed rapid, double exponential kinetics: τ_{off} fast 98.7 ± 38.1 ms(43.9%) and τ_{off} slow 725 ± 122 ms. Reblock by memantine following jumps back to -70 mV was extremely fast: τ_{on} fast 7.7 ± 2.2 ms(41.6%) and τ_{on} slow 285 ± 41 ms. The extremely rapid reblock by memantine and additional minor effects of such voltage steps on the kinetics of NMDA currents in the absence of antagonist necessitated fitting responses with memantine as the ratio between responses in the presence/absence of memantine. The difference in onset kinetics between the concentration-clamp and voltage-step protocols may be explained by a second, extracellular holding site for memantine on the NMDA receptor ***(3)***. Alternatively, it may just indicate that the concentration clamp was not fast enough.

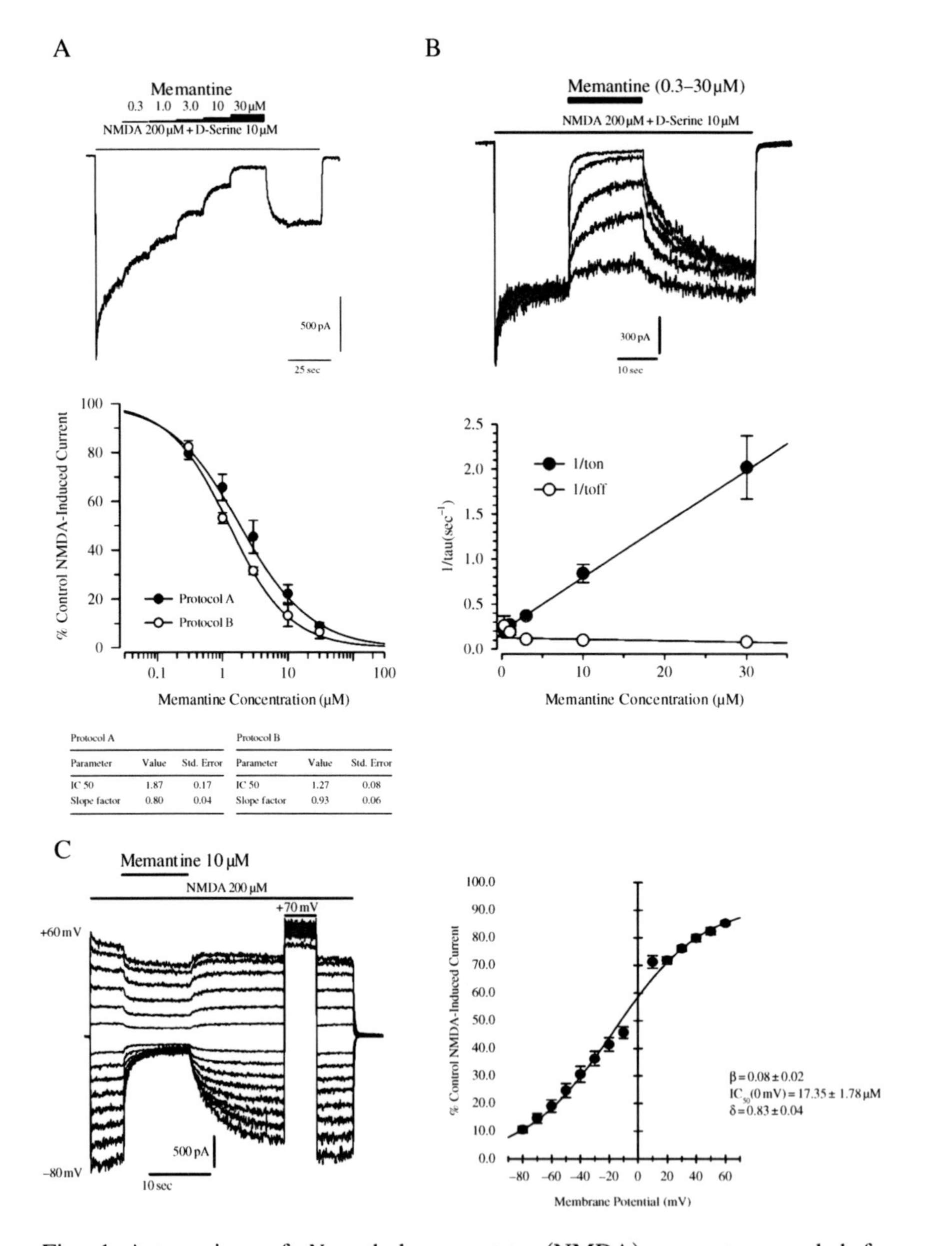

Protocol A			Protocol B		
Parameter	Value	Std. Error	Parameter	Value	Std. Error
IC 50	1.87	0.17	IC 50	1.27	0.08
Slope factor	0.80	0.04	Slope factor	0.93	0.06

Fig. 1. Antagonism of *N*-methyl-D-aspartate (NMDA) currents recorded from cultured hippocampal neurons by memantine. **(A)** Stepwise application of increasing concentrations of antagonist during constant application of agonists. Mean rundown-corrected percentage blockade (±SE) was plotted against antagonist concentration

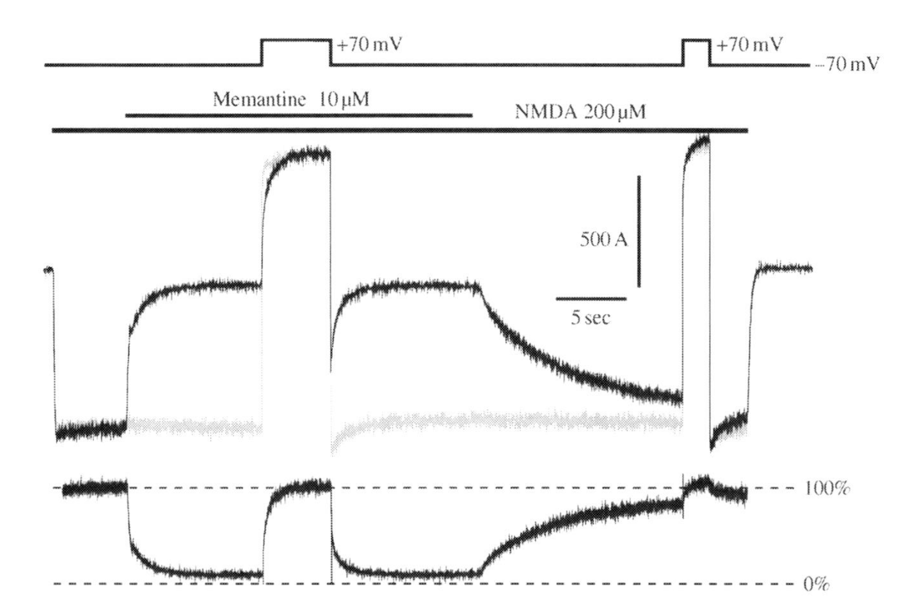

Fig. 2. Double exponential kinetics of blockade by memantine in cultured hippocampal neurons. Traces are averages of 10 recordings, and residual capacitive artifacts were subtracted. The bottom trace was constructed by basing recordings in the presence of memantine (black) as a percentage of those in the absence of memantine (gray). This was then used to fit the kinetics of the data. NB: Neurons were shortly clamped to +70 mV during the memantine washout phase to facilitate recovery.

Fig. 1. on a logarithmic scale and the curve fit using **Eq. 1** (filled circles). **(B)** Attenuation of currents by memantine (0.3–30 μ*M*) with current recovery between antagonist applications. Mean control-corrected percentage blockade (±SE) was plotted against antagonist concentration on a logarithmic scale and the curve fit using **Eq. 1** (open circles). The onset and offset kinetics followed a single or double exponential time course. Mean $1/t$ values (±SE) were plotted against antagonist concentration and fit according to **Eq. 5** from which K_{on} and K_{off} were read. These two values were then used to calculate K_d. **(C)** Voltage dependence of antagonism of the NMDA currents by a single concentration of memantine (10 μ*M*). The initial holding potential of −80 mV was increased for each stimulation in increments of 10 mV until it reached +60 mV. In the example recording shown, it can be seen that a voltage step to +70 mV was added during current recovery, but this was not deemed necessary for further experiments. The mean percentage blockade of control currents (±SE) was plotted against holding potential and fit using **Eq. 7**.

4.3. Voltage Dependency

Blockade of NMDA receptor-mediated currents by memantine was voltage dependent, as illustrated by the δ value of 0.83 ± 0.04, and an approximately 10-fold higher IC_{50} value at 0 mV compared with $-70\,mV[IC_{50}(0mV) = 17.35 \pm 1.78\,\mu M]$ (*see* **Fig. 1C**). The proportion of the voltage-independent sites, β, was very small (0.08 ± 0.02). To ensure that memantine had completely left the channel and the current was fully recovered, a 5-s voltage step to +70 mV was added during the current recovery, as can been seen in the trace shown, but this was not considered necessary for subsequent experiments with moderate affinity blockers that show similar fast kinetics. However, such procedures can be very useful when determining the voltage dependency of more potent, slower blockers such as (+)MK-801 *(3)*.

5. Notes: The Interpretation and Application of Electrophysiological Data

The reason for the better therapeutic safety of memantine compared with other channel blockers such as (+)MK-801 and phencyclidine is still a matter of debate, and data such as those presented in this chapter (summarized in **Table 1**) have been utilized to support several theories. The interpretation of these electrophysiological data and how they have been used to form hypotheses concerning actions of compounds in vivo are described in the following notes.

1. Memantine and other well-tolerated open channel blockers such as amantadine, dextromethorphan, (+-)-5-aminocarbonyl-10, 11-dihydro-5H-dibenzo[a,d] [a,d] cyclohepten-5, 10-imine (ARL 15896AR), and (*s*)-*alpha*-phenyl-2-pyridine-ethanamine dihydrochloride (ADCI) show much faster open channel blocking/unblocking kinetics than compounds burdened with negative psychotropic effects such as (+)MK-801 or phencyclidine ***(3,6,7,10,25,26)***. The kinetics of (+)MK-801 and phencyclidine are too slow to allow them to leave the channel upon depolarization, which is reflected in apparently weaker voltage dependency. These two parameters are directly related to affinity, with lower affinity compounds such as memantine showing faster kinetics and apparently stronger voltage dependency, as reflected in estimated δ value ***(13)***. The δ value describes the percentage of the transmembrane field the drug experiences when blocking the NMDA receptor channel *(3)*. The unblocking rate of memantine in the continuous presence of this antagonist following depolarizing voltage steps is very rapid and well within the time course of an NMDA receptor-mediated excitatory postsynaptic potential (EPSP).
2. Memantine blocks and unblocks open NMDA receptor channels with double exponential kinetics. The amplitude and speed of the fast component of block increases with memantine concentration. In contrast, the speed of fast unblock

remains constant, but the amplitude decreases with memantine concentration ***(8,9,14,15,27)***. Moreover, the predominant effect of depolarization is to increase dramatically the weight of the faster recovery time constant ***(9,11,27)***. These data indicate that memantine binds to at least two sites within the channel ***(14,15)***.

3. Both Lipton's and Rogawski's groups have proposed that the ability of low-affinity open channel blockers to gain rapid access to the NMDA receptor channel is important in determining their therapeutic safety in ischemia and epilepsy ***(7,26,28)***. However, this hypothesis alone cannot explain the better therapeutic profile of memantine as, even if receptors are only blocked following pathological activation, they would then remain blocked in the continuous presence of memantine, and therefore be unavailable for subsequent physiological activation. Physiologically, NMDA receptors are transiently activated by millimolar concentrations of glutamate ***(29)*** following strong depolarization of the postsynaptic membrane that rapidly relieves their voltage-dependent blockade by Mg^{2+} ***(30)***, whereas during pathological activation, NMDA receptors are activated by lower concentrations of glutamate but for much longer periods of time ***(31–36)***. Unfortunately, the voltage dependency of the divalent cation Mg^{2+} is so pronounced that it also leaves the NMDA channel upon moderate depolarization under pathological conditions. Although uncompetitive antagonists also block the NMDA receptor channel, high-affinity compounds such as (+)MK-801 have much slower unblocking kinetics than Mg^{2+} and less pronounced functional voltage dependency and are therefore unable to leave the channel within the time course of a normal NMDA receptor-mediated excitatory postsynaptic potential. As a result, (+)MK-801 blocks both the pathological and the physiological activation of NMDA receptors ***(3)***.
4. We were the first to suggest that the combination of fast offset kinetics and strong voltage dependency allows memantine to rapidly leave the NMDA channel upon transient physiological activation by millimolar concentrations of synaptic glutamate but blocks the sustained activation by micromolar concentrations of glutamate under moderate pathological conditions ***(6,12,13)***. This hypothesis is further supported by the fact that although the predominant component of offset kinetics at near resting membrane potentials is still too slow to allow synaptic activation—that is, around 5 s—the relief of blockade in the continuous presence of memantine upon depolarization is much faster due to an increase in the weight of the faster recovery time constant ***(9,27,37,38)***. These kinetics are likely to be even faster in vivo owing to higher temperatures ***(39)***. Furthermore, the rate of recovery from memantine blockade is dependent on the open probability of NMDA channels ***(10)*** and therefore would be faster in the presence of higher, synaptic concentrations of glutamate ***(29)***. Given the crucial role of NMDA receptors in neuronal plasticity, the fact that memantine improves cognition and neuronal plasticity seems paradoxical at the first glance. It should be realized, however, that Mg^{2+} is an endogenous NMDA channel blocker, and its removal from the channel leads both to an impairment in neuronal plasticity ***(40,41)*** and a neuronal death ***(42)***. Any dysfunction of postsynaptic

neurons leading to weakened blockade by Mg^{2+}, for example, because of partial depolarization as a consequence of an energy deficit, may trigger such functional (plasticity) and structural (neuronal loss) deficits ***(4,43)***. Because memantine is more potent and slightly less voltage-dependent than Mg^{2+}, it may thus serve as a more effective surrogate for Mg^{2+} ***(6)***. As a result of its somewhat less pronounced voltage dependency, memantine is more effective than Mg^{2+} in blocking tonic pathological activation of NMDA receptors at moderately depolarized membrane potentials. However, following strong synaptic activation, memantine like Mg^{2+} can leave the NMDA receptor channel with voltage-dependent, fast unblocking kinetics. In turn, memantine suppresses synaptic noise but allows the relevant physiological synaptic signal to be detected. This provides both neuroprotection and symptomatic restoration of synaptic plasticity by one and the same mechanism ***(3,4)***. Antagonists that have "too high" affinity for the channel or "too little" voltage dependence, such as dizocilpine ([+]MK-801), thus produce numerous side effects as they essentially act as an irreversible plug of the NMDA receptor channel and block both pathological and physiological function.

5. A moderate potentiation of NMDA-induced outward currents by memantine at positive potentials in hippocampal neurons has also been reported ***(11)*** (data not shown here). This could be related to the finding that Mg^{2+} and ketamine increased NMDA receptor-mediated currents in cultured mouse hippocampal neurons and HEK-293 cells expressing NMDA $\xi 1/\varepsilon 2$ receptors by increasing the affinity of the $glycine_B$ site for the co-agonist ***(44)***. Such a facilitation would be predicted to be more pronounced with lower concentrations of glycine. This could have important functional implications as the differentiation between the block of NMDA receptors at near resting membrane potentials and the lesser block following strong synaptic depolarization to around $-20\,mV$ would be enhanced by such a mechanism and would facilitate the ability of memantine to differentiate between pathological and physiological activation of NMDA receptors. Such a potentiation was not seen in this study, most likely because of the use of saturating concentrations of D-serine.
6. It should also be noted that a third theory was recently proposed in an excellent paper by Blanpied et al. ***(8)*** (*see* **ref. 5** for review) and supported by data from Sobolevsky et al. ***(14)*** (not shown here). The data indicate that memantine and amantadine appear to have a lesser tendency to be trapped in NMDA receptor channels than do phencyclidine or (+)MK-801. This difference was attributed to the ability of channel blockers to increase the affinity of NMDA receptors for agonist and the faster kinetics of the amino adamantanes. Receptors blocked by memantine retain agonist and thereby open and release memantine following removal of both agonist and memantine from the extracellular solution ***(10)***. This partial trapping is less pronounced for higher affinity compounds as their slower unblocking kinetics do not allow them to leave the channel quickly enough following agonist removal. The relief of block in the absence of agonist was greater in the experiments of Sobolevsky et al. ***(14)***. This, however, may have been due to the use of higher

concentrations of aspartate that would have increased the proportion of liganded receptors at the time of agonist/antagonist removal. Blanpied et al. *(8)* proposed that partial trapping may underlie the better therapeutic profile of memantine as a proportion of channels—around 15–20%—would always unblock in the absence of agonist and thereby be available for subsequent physiological activation. In other words, the antagonism by memantine is like that of a low intrinsic activity partial agonist in that it does not cause 100% blockade of NMDA receptors. Although this theory is attractive, it is only relevant for the therapeutic situation if partial trapping also occurs in the continuous presence of memantine. This point had not been addressed previously. This prompted us to perform experiments on partial trapping in the continuing presence of memantine, and the results of these studies were very similar to those reported by Blanpied et al. *(8)*, that is, around 15% of channels released memantine following agonist removal *(3)*. However, although this theory can be used to explain the therapeutic tolerability of memantine, it provides no mechanism of action for the symptomatic effects observed in AD patients.

References

1. Reisberg, B., R. Doody, A. Stoffler, F. Schmitt, S. Ferris, and H.J. Motrius (2003) Memantine in moderate-to-sexere Alzheimer's disease. *M. Engl. J. Med.* **348,** 1333–1341.
2. Tariot, P.N., M.R. Farlow, G.T. Grossberg, S.M. Graham, S. McDonald, and I. Gergel (2004) Memantine treatment in patients with moderate to severe Alzheimer disease already receiving donepezil – a randomized controlled trial. *JAMA* **291**, 317–324.
3. Parsons, C.G., W. Danysz, and G. Quack (1999) Memantine is a clinically well tolerated N-methyl-D-aspartate (NMDA) receptor antagonist – a review of preclinical data. *Neuropharmacology* **38**, 735–767.
4. Danysz, W. and C.G. Parsons (2003) The NMDA receptor antagonist memantine as a symptomatological and neuroprotective treatment for Alzheimer's disease: preclinical evidence. *Int. J. Geriatr. Psychiatry* **18**, S23–S32.
5. Johnson, J.W. and S.E. Kotermanski (2006) Mechanism of action of memantine. *Curr. Opin. Pharmacol.* **6**, 61–67.
6. Parsons, C.G., R. Gruner, J. Rozental, J. Millar, and D. Lodge (1993) Patch-clamp studies on the kinetics and selectivity of N-methyl-D-aspartate receptor antagonism by memantine (1-amino-3,5-dimethyladamantan). *Neuropharmacology* **32**, 1337–1350.
7. Rogawski, M.A. (1993) Therapeutic potential of excitatory amino-acid antagonists – channel blockers and 2,3-benzodiazepines. *Trends Pharmacol. Sci.* **14**, 325–331.
8. Blanpied, T.A., F.A. Boeckman, E. Aizenman, and J.W. Johnson (1997) Trapping channel block of NMDA-activated responses by amantadine and memantine. *J. Neurophysiol.* **77**, 309–323.

9. Bresink, I., T.A. Benke, V.J. Collett, A.J. Seal, C.G. Parsons, J.M. Henley, and G.L. Collingridge (1996) Effects of memantine on recombinant rat NMDA receptors expressed in HEK 293 cells. *Br. J. Pharmacol.* **119**, 195–204.
10. Chen, H.S.V. and S.A. Lipton (1997) Mechanism of memantine block of NMDA-activated channels in rat retinal ganglion cells: uncompetitive antagonism. *J. Physiol. (Lond.)* **499**, 27–46.
11. Parsons, C.G., S. Hartmann, and P. Spielmanns (1998) Budipine is a low affinity, N-methyl-D-aspartate receptor antagonist: patch clamp studies in cultured striatal, hippocampal, cortical and superior colliculus neurones. *Neuropharmacology* **37**, 719–727.
12. Parsons, C.G., V.A. Panchenko, V.O. Pinchenko, A.Y. Tsyndrenko, and O.A. Krishtal (1996) Comparative patch-clamp studies with freshly dissociated rat hippocampal and striatal neurons on the NMDA receptor antagonistic effects of amantadine and memantine. *Eur. J. Neurosci.* **8**, 446–454.
13. Parsons, C.G., G. Quack, I. Bresink, L. Baran, E. Przegalinski, W. Kostowski, P. Krzascik, S. Hartmann, and W. Danysz (1995) Comparison of the potency, kinetics and voltage-dependency of a series of uncompetitive NMDA receptor antagonists in-vitro with anticonvulsive and motor impairment activity in-vivo. *Neuropharmacology* **34**, 1239–1258.
14. Sobolevsky, A. and S. Koshelev (1998) Two blocking sites of amino-adamantane derivatives in open N-methyl-d-aspartate channels. *Biophys. J.* **74**, 1305–1319.
15. Sobolevsky, A.I., S.G. Koshelev, and B.I. Khodorov (1998) Interaction of memantine and amantadine with agonist-unbound NMDA-receptor channels in acutely isolated rat hippocampal neurons. *J. Physiol. (Lond.)* **512**, 47–60.
16. Clark, G.D., D.B. Clifford, and C.F. Zorumski (1990) The effect of agonist concentration, membrane voltage and calcium on N-methyl-D-aspartate receptor desensitization. *Neuroscience* **39**, 787–797.
17. Grantyn, R. and H.D. Lux (1988) Similarity and mutual exclusion of NMDA-activated and proton-activated transient Na+-currents in rat tectal neurons. *Neurosci. Lett.* **89**, 198–203.
18. Zilberter, Y., V. Uteshev, S. Sokolova, and B. Khodorov (1991) Desensitization of N-methyl-D-aspartate receptors in neurons dissociated from adult-rat hippocampus. *Mol. Pharmacol.* **40**, 337–341.
19. Johnson, J.W. and P. Ascher (1987) Glycine potentiates the NMDA response in cultured mouse-brain neurons. *Nature* **325**, 529–531.
20. Parsons, C.G., X.G. Zong, and H.D. Lux (1993) Whole-cell and single-channel analysis of the kinetics of glycine-sensitive N-methyl-D-aspartate receptor desensitization. *Br. J. Pharmacol.* **109**, 213–221.
21. Hashimoto, A. and T. Oka (1997) Free D-aspartate and D-serine in the mammalian brain and periphery. *Prog. Neurobiol.* **52**, 325–353.

22. Hamill, O.P., A. Marty, E. Neher, B. Sakmann, and F.J. Sigworth (1981) Improved patch-clamp techniques for high resolution current recording from cells and cell-free membrane patches. *Pflugers Arch.* **391**, 85–100.
23. Sigworth, F.J., H. Affolter, and E. Neher (1995) Design of the Epc-9, a computer-controlled patch-clamp amplifier. 2. Software. *J. Neurosci. Methods* **56**, 203–215.
24. Albuquerque, E.X., E.F.R. Pereira, N.G. Castro, M. Alkondon, S. Reinhardt, H. Schroder, and A. Maelicke (1995) Nicotinic receptor function in the mammalian central-nervous-system, in *Diversity of Interacting Receptors*. New York Acad Sciences: New York, pp. 48–72.
25. Black, M., T. Lanthorn, D. Small, G. Mealing, V. Lam, and P. Morley (1996) Study of potency, kinetics of block and toxicity of NMDA receptor antagonists using fura-2. *Eur. J. Pharmacol.* **317**, 377–381.
26. Rogawski, M.A., S.I. Yamaguchi, S.M. Jones, K.C. Rice, A. Thurkauf, and J.A. Monn (1991) Anticonvulsant activity of the low-affinity uncompetitive N-methyl-D-aspartate antagonist (+/−)-5-aminocarbonyl-10,11-dihydro-5h-dibenzo[a,D]cyclohepten-5, 10-imine (Adci) – comparison with the structural analogs dizocilpine (Mk-801) and carbamazepine. *J. Pharmacol. and Exp. Ther.* **259**, 30–37.
27. Frankiewicz, T., B. Potier, Z.I. Bashir, G.L. Collingridge, and C.G. Parsons (1996) Effects of memantine and MK-801 on NMDA-induced currents in cultured neurones and on synaptic transmission and LTP in area CA1 of rat hippocampal slices. *Br. J. Pharmacol.* **117**, 689–697.
28. Chen, H.S.V., J.W. Pellegrini, S.K. Aggarwal, S.Z. Lei, S. Warach, F.E. Jensen, and S.A. Lipton (1992) Open-channel block of N-methyl-D-aspartate (NMDA) responses by memantine – therapeutic advantage against NMDA receptor-mediated neurotoxicity. *Journal of Neuroscience* **12**, 4427–4436.
29. Clements, J.D., R.A.J. Lester, G. Tong, C.E. Jahr, and G.L. Westbrook (1992) The time course of glutamate in the synaptic cleft. *Science* **258**, 1498–1501.
30. Nowak, L., P. Bregestovski, P. Ascher, A. Herbet, and A. Prochiantz (1984) Magnesium gates glutamate-activated channels in mouse central neurons. *Nature* **307**, 462–465.
31. Andine, P., M. Sandberg, R. Bagenholm, A. Lehmann, and H. Hagberg (1991) Intracellular and extracellular changes of amino-acids in the cerebral-cortex of the neonatal rat during hypoxic-ischemia. *Brain Res. Dev. Brain Res.* **64**, 115–120.
32. Benveniste, H., J. Drejer, A. Schousboe, and N.H. Diemer (1984) Elevation of the extracellular concentrations of glutamate and aspartate in rat hippocampus during transient cerebral-ischemia monitored by intracerebral microdialysis. *J. Neurochem.* **43**, 1369–1374.
33. Buisson, A., J. Callebert, E. Mathieu, M. Plotkine, and R.G. Boulu (1992) Striatal protection induced by lesioning the substantia-nigra of rats subjected to focal ischemia. *J. Neurochem.* **59**, 1153–1157.

34. Globus, M.Y.T., R. Busto, E. Martinez, I. Valdes, W.D. Dietrich, and M.D. Ginsberg (1991) Comparative effect of transient global-ischemia on extracellular levels of glutamate, glycine, and gamma-aminobutyric-acid in vulnerable and nonvulnerable brain-regions in the rat. *J. Neurochem.* **57**, 470–478.
35. Globus, M.Y.T., M.D. Ginsberg, and R. Busto (1991) Excitotoxic index – a biochemical marker of selective vulnerability. *Neurosci. Lett.* **127**, 39–42.
36. Mitani, A., Y. Andou, and K. Kataoka (1992) Selective vulnerability of hippocampal Ca1 neurons cannot be explained in terms of an increase in glutamate concentration during ischemia in the gerbil – brain microdialysis study. *Neuroscience* **48**, 307–313.
37. Sobkowicz, H.M. and S.M. Slapnick (1992) Neuronal sprouting and synapse formation in response to injury in the mouse organ of corti in culture. *Int. J. Dev. Neurosci* **10**, 545–566.
38. Sobkowicz, H.M., S.M. Slapnick, and B.K. August (1993) Presynaptic fibers of spiral neurons and reciprocal synapses in the organ of corti in culture. *J. Neurocytol.* **22**, 979–993.
39. Davies, S.N., D. Martin, J.D. Millar, J.A. Aram, J. Church, and D. Lodge (1988) Differences in results from invivo and invitro studies on the use-dependency of N-methylaspartate antagonism by Mk-801 and other phencyclidine receptor ligands. *Eur. J. Pharmacol.* **145**, 141–151.
40. Coan, E.J., A.J. Irving, and G.L. Collingridge (1989) Low-frequency activation of the NMDA receptor system can prevent the induction of Ltp. *Neurosci. Lett.* **105**, 205–210.
41. Frankiewicz, T. and C.G. Parsons (1999) Memantine restores long term potentiation impaired by tonic N-methyl-D-aspartate (NMDA) receptor activation following reduction of Mg2+ in hippocampal slices. *Neuropharmacology* **38**, 1253–1259.
42. Furukawa, Y., M. Okada, N. Akaike, T. Hayashi, and J. Nabekura (2000) Reduction of voltage-dependent magnesium block of N-methyl-D-aspartate receptor-mediated current by in vivo axonal injury. *Neuroscience* **96**, 385–392.
43. Rogawski, M.A. and G.L. Wenk (2003) The neuropharmacological basis for the use of memantine in the treatment of Alzheimer's disease. *CNS Drug Rev.* **9**, 275–308.
44. Wang, L.Y. and J.F. Macdonald (1995) Modulation by magnesium of the affinity of NMDA receptors for glycine in murine hippocampal-neurons. *J. Physiol. (Lond.)* **486**, 83–95.

3

Methods for Evaluation of Positive Allosteric Modulators of Glutamate AMPA Receptors

Edward R. Siuda, Jennifer C. Quirk, and Eric S. Nisenbaum

Summary

Hypofunctioning of glutamate synaptic transmission in the central nervous system (CNS) has been proposed as a factor that may contribute to cognitive deficits associated with various neurological and psychiatric disorders. Positive allosteric modulation of the α-amino-3-hydroxy-5-methyl-4-isoxazoleproprionic acid (AMPA) subtype of glutamate receptors has been proposed as a novel therapeutic approach, because these receptors mediate the majority of rapid excitatory neurotransmission and are intimately involved in long-term changes in synaptic plasticity thought to underlie mnemonic processing. By definition, positive allosteric modulators do not affect AMPA receptor activity alone but can markedly enhance ion flux through the ion channel pore in the presence of bound agonist. Despite this commonality, positive allosteric modulators can be segregated on the basis of the preferential effects on AMPA receptor subunits, their alternatively spliced variants and/or their biophysical mechanism of action. This chapter provides a detailed description of the methodologies used to evaluate the potency/efficacy and biophysical mechanism of action of positive allosteric modulators of AMPA receptors.

Key Words: AMPA; glutamate; desensitization; deactivation; LY503430; cyclothiazide; GluR2; positive allosteric modulator.

1. Introduction

Glutamate is the neurotransmitter primarily responsible for excitatory synaptic transmission in the central nervous system (CNS). An evolving collection of evidence has indicated that dysfunction of glutamatergic signaling

From: *Methods in Molecular Biology, vol. 288: Patch-Clamp Methods and Protocols*
Edited by: P. Molnar and J. J. Hickman © Humana Press Inc., Totowa, NJ

in the CNS may contribute to cognitive deficits associated with various neurological and psychiatric disorders ***(1–7)***. In an effort to explore novel therapies for cognitive dysfunction in these disorders, several approaches are under investigation, all of which share the common goal of enhancing glutamatergic synaptic transmission. One strategy has focused on compounds that allosterically modulate glutamate α-amino-3-hydroxy-5-methyl-4-isoxazoleproprionic acid (AMPA) receptors. AMPA receptors are tetrameric complexes assembled in various stoichiometries from four distinct subunits (termed GluR1-4 or GluRA-D), each having two splice variants (flip [i] and flop [o]) giving rise to a wide array of functional receptors ***(8–15)***. These receptors mediate the majority of rapid excitatory neurotransmission and are intimately involved in long-term changes in synaptic plasticity (e.g., long-term potentiation) thought to represent a neural substrate for certain forms of memory encoding ***(16)***.

The functional impact of positive allosteric modulation of AMPA receptors is to enhance glutamatergic postsynaptic depolarization by increasing ion flux through AMPA receptors/channels in the presence of agonist ***(17,18)***. An important consequence of the enhanced depolarization is a decrease in the threshold for induction of long-term potentiation (LTP), as well as an increase in the degree of potentiation ***(19,20)***. Consistent with these modifications in long-term synaptic efficacy, numerous experimental studies have reported that positive allosteric modulators can enhance performance of animals in multiple models of cognitive function ***(6,19,21–27)***. More importantly, clinical studies with AMPA receptor allosteric modulators have shown promising results in treating cognitive deficits associated with Parkinson's disease, Alzheimer's disease, schizophrenia, and aging ***(28–31)***.

Many structurally different positive allosteric AMPA receptor modulators have been described, all of which lack any intrinsic activity at AMPA receptors when applied alone but on binding of glutamate (or other agonists) can enhance ion flux through the ion pore of the receptor/channel complex. Despite this common feature, AMPA receptor modulators can be segregated along several functional dimensions, including their potency/efficacy for modulation of specific subunits or splice variants, as well as their biophysical mechanism of action, which can involve suppression of the desensitization (decay of channel conductance in the presence of agonist) and/or the deactivation (decay of channel conductance on removal of agonist) process of the receptor ***(32,33)***. Efforts to develop novel AMPA receptor modulators as therapies for cognitive dysfunction have focused on identification of molecules with selectivity for receptor subtype and/or biophysical mechanism of action to determine whether such preferential activity confers greater efficacy in preclinical models and ultimately in the clinic. This chapter provides a detailed methodology for evaluating the potency/efficacy and biophysical mechanism of action of positive allosteric modulators of AMPA receptors. The studies focus on the nature of modulation of the flip splice variant of recombinant homomeric GluR2 receptors (GluR2i) by LY503430 and cyclothiazide.

2. Materials (*see* Notes 1 and 2)

2.1. Human Embryonic Kidney 293 Cells

1. Molecular biology and cell production.
   a. Mammalian expression vector pcDNA5/FRT (Invitrogen, cat. no. V6010-20) containing the GluR2i subunit gene.
   b. Flp-In™ pcDNA™5/FRT Core Kit (Invitrogen K6010-02).
2. Cell-culture media.
   a. Dulbecco's modified Eagle's medium (DMEM; GIBCO-BRL, cat. no. 11965-092). Store at 4 °C and once opened dispose after 6 months.
   b. Ten percent fetal bovine serum (FBS), certified, and heat inactivated (GIBCO-BRL, cat. no. 10082-147). Aliquot when first thawed and freeze 50-ml aliquots (for 500 ml media) at −35 °C (*see* **Note 3**).
   c. Hygromyocin B (100 μg/ml) [Hygromyocin B in phosphate-buffered saline (PBS) 50 mg/ml; Invitrogen, cat. no. 10687-010]. Store at 4 °C and once opened dispose after 6 months.
   d. Once cell-culture media is made, sterile filter, store at 4 °C, and dispose after a month.
   e. Cell-culture media is warmed in a 37 °C water bath prior to use.
3. Splitting cells.
   a. Trypsin-ethylenediaminetetraacetic acid (EDTA) (0.05%; GIBCO-BRL, cat. no. 25300–054). Store at 4 °C.
   b. Cell-culture flasks (Fisher, cat. nos. 10-126-28, 10-126-37, and 10-126-34).
   c. Dulbecco's PBS (DPBS; GIBCO-BRL 14190-144). Store at room temperature.
4. Plating cells.
   a. Coverslips (Fisher, cat. no. 12-545-101).
   b. Collagen (Sigma, cat. no. C8919) or poly-D-lysine (Sigma, cat. no. P7280) (*see* **Note 4**).
   c. BD Falcon six-well tissue culture plates (Fisher, cat. no. 08-772-1B).

2.2. Whole-Cell Patch-Clamp Recording

1. Internal electrode solution: 165 m*M* *N*-methyl-D-glucamine, 4 m*M* $MgCl_2$, and 40 m*M* HEPES. Solution is aliquotted and stored at −30 °C. On day of recording, thaw aliquot and add 3 m*M* 1,2-bis(2-aminophenoxy)ethane-N, N, N′, N′-tetraacetic

acid (BAPTA), 12 mM phosphocreatine, 2 mM Na_2ATP, 0.2 mM guanosine-5′-triphosphate; pH = 7.2 ± .02 (adjusted with NaOH) and osmolarity = 280 ± 2 mOsm/l (adjusted with phosphocreatine or dH_2O). Filter (0.2 μm) to remove impurities and store on ice while recording.

2. Extracellular solution: 120 mM NaCl, 5 mM $BaCl_2$, 1 mM $MgCl_2$, 20 mM CsCl, 10 mM glucose, and 10 mM HEPES; pH = 7.4 ± 0.03 (adjusted with 1N NaOH) and osmolarity = 300 ± 3 mOsm/l (adjusted with glucose or dH_2O). Store at 4 °C for up to 7 days. Add 500 nM tetrodotoxin (TTX) on the day of recording (*see* **Note 5**).
3. Patch-clamp capillary glass electrodes (PG52165-4, WPI Inc.) are pulled using a multistage puller (P-97; Sutter Instruments Inc.) and fire-polished using a microforge (Narishige MF-830) to a tip diameter of approximately 1–2 μm. Electrode resistances range between 2 and 5 MΩ.
4. Recording electrodes are inserted into a Plexiglas electrode holder with side port (HL-U; Molecular Devices) which is attached to an amplifier headstage (CV203BU, Molecular Devices). The headstage is secured to a pivoting mounting plate attached to a micromanipulator (PCS-5000 Series; EXFO Life Sciences Group) and positioned using a translational stage (433 Series; Newport Inc.). The translational stage and micromanipulator are in turn attached to a two-sided microscope mount (PCS-500-15; EXFO Life Sciences Group) secured to an inverted microscope (Eclipse TE300; Nikon).
5. Voltage-clamp recordings are made using an Axopatch 200B amplifier (Molecular Devices). Currents are amplified and subsequently digitized at 20 kHz using an A/D board. Currents are initially filtered using an 8-pole Bessel filter (5 kHz cutoff frequency) and then passed through a Humbug 50/60 Hz noise eliminator (Quest Scientific) and monitored with pCLAMP software (v8.1 or higher; Molecular Devices).

2.3. Outside-Out Patch-Clamp Recording

1. Thin-wall glass filament-filled electrodes (TW150F-3; WPI Inc.) are pulled using a multistage puller. Electrodes are fire-polished using a microforge to a final tip diameter of approximately 0.5–1 μm.
2. Internal electrode solution: 135 mM CsCl, 10 mM CsF, 10 mM HEPES, 1 mM $MgCl_2$, and 0.5 mM $CaCl_2$; pH = 7.2 ± 0.03 (adjusted with CsOH) and osmolarity = 295 ± 0.03 mOsm/l (adjusted with sucrose or dH_2O) (*see* **Note 6**).
3. Extracellular recording solution: 145 mM NaCl, 3 mM KCl, 5 mM HEPES, 1 mM $MgCl_2$, and 1.8 mM $CaCl_2$; pH = 7.3 ± 0.03 (adjusted with NaOH) and osmolarity = 295 ± 0.03 mOsm/l (adjusted with sucrose or dH_2O).
4. An additional sucrose-based extracellular solution, containing 250 mM sucrose, 3 mM KCl, 5 mM HEPES, 1 mM $MgCl_2$, and 1.8 mM $CaCl_2$; pH = 7.3 ± 0.03 (adjusted with NaOH); and osmolarity = 295 ± 0.03 mOsm/l (adjusted with sucrose or dH_2O), is mixed 1:4 with the extracellular recording solution and added to the recording dish (50-mm Petri dish; Pall 7232) creating a background bath.

This sucrose-based solution is used to aid in visualization of the interface between the two solutions flowing from the theta tube (*see* **Fig. 2C and 2D**).

5. High-performance VIDICON camera controller and camera head (C2400 series; Hamamatsu Photonics) and black-and-white video monitor (PVM-137; Sony Corporation) are used for visualization and alignment of recording electrode with theta tube (*see* **Subheading 2.5.**).

2.4. Sixteen Barrel Solution Delivery System

1. Sixteen square glass microcells (VitroCom 8260) are secured into a plastic holder (manufactured in-house), aligned under a dissecting microscope, and glued (14420, Ted Pella Inc.) together. The plastic holder has a metal post used to attach the holder to an actuator for proper positioning (*see* **Fig. 1B**, **C**).
2. P10 polyethylene tubing is inserted into each glass microcell and glued into place. P50 polyethylene tubing is then inserted over the PE10 tubing and glued, creating a single continuous tube connected to each microcell (*see* **Fig. 1B**) (*see* **Note 7**).
3. The plastic holder for the 16-barrel array is attached to a manual linear stage with XYZ axes (461 series; Newport Corporation). The manual liner stage is mounted onto another translation stage (460A series; Newport Corporation) controlled by a motorized linear actuator (850G series; Newport Corporation) and a Universal Motion Controller/Driver (ESP300; Newport Corporation) (*see* **Fig. 1A**).
4. Sixteen 10-ml syringes are suspended approximately 12 inches above the stage of the inverted microscope (*see* **Fig. 1A**). Attached to the end of each syringe is a polypropylene female luer fitting (06359-25, Cole-Palmer Instrument Company). C-flex tubing (2.4 mm outer diameter; 064240-60, Cole-Palmer Instrument Company) is used to connect the end of each luer fitting to the input of a two-way solenoid valve (LFAA0501518H, The Lee Company). The output of each solenoid valve is then connected using the C-flex tubing to the P50 polyethylene tubing attached to each barrel of the drug delivery system (*see* **Note 8**). Opening/closing of the solenoid valves is controlled by a switch panel that can supply each valve with a −5 V input independently.
6. L-Glutamate (0218, Tocris) was dissolved in dH_2O and equimolar NaOH to create a 40-m*M* stock solution that was aliquotted and stored at −30 °C. Aliquots were stable over the long term. Drugs were prepared and stored according to their specific requirements (*see* **Note 9**).

2.5. Rapid Solution Delivery System

1. Double-barrel borosilicate glass theta tube (BT 150-10; Sutter Instruments Inc.) is pulled using a multistage electrode puller. An electrode polisher (Narishige Inc.) is used to slowly and carefully polish the tip until final tip diameter is approximately 50–200 μm (*see* **Fig. 2A**) (*see* **Note 10**). Larger tip diameters are used for rapid

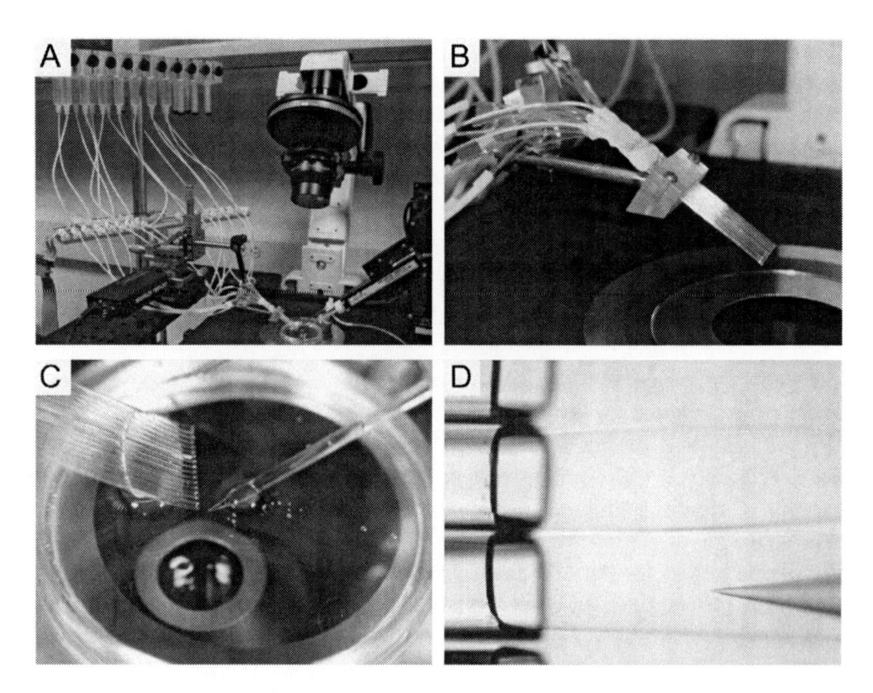

Fig. 1. Sixteen-barrel solution delivery system. (**A**) Overview of the solution delivery system showing the 16-syringe solution holders with tubing connecting to the input port of individual solenoid valves and the output port of each valve connected with additional tubing to individual glass barrels. (**B**) Close-up view of the 16-barrel glass tubing array. (**C**) At the beginning of each concentration-response experiment, the glass barrels are positioned at approximately 45° to the bottom of the recording dish. While viewing through the microscope, the cell is positioned in front of the first glass barrel by moving the microscope stage. (**D**) High-magnification image showing the relationship between the recording electrode tip and the solution barrels. Solution is flowing from the two center barrels in this image.

application onto whole cells, and smaller tip diameters are required for outside-out patches. Quartz tubing (160-2255-5, Agilent Technologies) is inserted into each barrel for additional stability. The unpolished end of each theta tube barrel is attached to a 1–2 in. length of P10 polyethylene tubing using epoxy glue. A 1–2 in. length of P50 polyethylene tubing is inserted around the P10 polyethylene tubing and secured using epoxy glue to create one continuous tube (*see* **Fig. 2A**). A four-to-one manifold (MPP-4; Warner Instruments LLC) is attached to the input of each barrel of the theta tubing to permit multiple solutions to be applied to an individual cell.

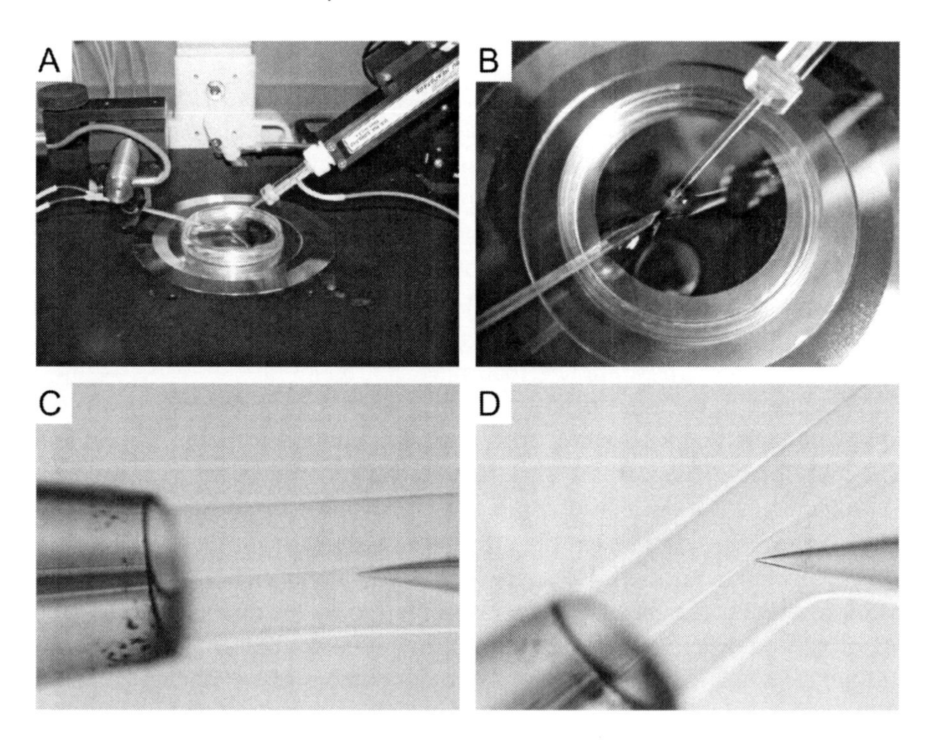

Fig. 2. Rapid solution delivery system. (**A**) Photograph of the rapid solution delivery system showing the recording electrode and associated headstage and the theta tube assembly. (**B**) Photograph illustrating the orientation of the recording electrode with the theta tube positioned in the recording chamber. (**C**) High magnification image of solution flowing through the theta tube barrels. The recording electrode is positioned adjacent to the solution interface. This orientation is amenable for measuring desensitization of α-amino-3-hydroxy-5-methyl-4-isoxazoleproprionic acid (AMPA) receptors recorded using the whole-cell configuration. (**D**) High-magnification image of the preferred orientation of the recording electrode and theta tube for measuring deactivation of AMPA receptors from outside-out patches of membrane.

2. The theta tube is inserted into the pipette holder of a piezoelectric microstage (PZS-200; EXFO Life Sciences Group). The position of the theta tube assembly is shifted between barrels using an amplifier driver (PZ-150M; EXFO Life Sciences Group) (*see* **Fig. 2A**).
3. Solutions are delivered to the theta tubing using a multi-barrel syringe pump (702103, Harvard Apparatus) set to a delivery rate of 0.2–0.3 ml/min. The end of the each syringe pump is connected to an input of the four-to-one manifold using C-flex tubing (*see* **Note 11**).

3. Methods

The methods described next provide instructions on how to measure the potency, efficacy, and biophysical mechanism(s) of action of positive allosteric modulators of AMPA receptors. Although the specific description focuses on evaluation of homomeric GluR2i receptors transfected into human embryonic kidney 293 (HEK293) cells, the methodologies can be generalized to measurements from other AMPA receptor subunits and splice variants transiently or stably expressed in other immortalized cells, as well as heteromeric receptors expressed in native neurons isolated from different regions of the CNS. In these studies, two electrophysiological recording configurations are used, including whole-cell and outside-out patch methods, and two drug delivery systems are employed, including a 16-barrel array for concentration-response studies and a theta-tube for biophysical analyses. Whole-cell recordings are very amenable to measurements of the concentration responsiveness of allosteric modulators and also can be used to assess the effects of modulators on the rate and extent of receptor desensitization. However, because of the insufficient temporal and spatial resolution of agonist application conferred by the diameter of the whole cell, effects on deactivation cannot be accurately measured using this recording configuration and instead require outside-out patches of membrane containing AMPA receptors (*see* **Note 12**). A detailed description of the different types of recording configurations is reviewed in Hille ***(39)***.

3.1. Preparation of HEK293 Cells Stably Expressing GluR2i Receptors

1. Cell line production.
   a. GluR2i was first subcloned into the mammalian expression vector pcDNA5/FRT. Native GluR2 subunits are modified in the pore region (TM2), so that a glutamine residue (Q; CGA) is edited to an arginine (R; CGG). However, these edited subunits have a significantly lower single-channel conductance ***(40)***. Therefore, in the present studies, the unedited or Q form of the GluR2 subunit is used. Stable cell lines are then generated using the Flp-In System. Cell lines are established on the basis of Hygromyocin B resistance according to the manufacturer's instructions.
   b. Cells are grown in cell-culture flasks, maintained in an incubator at 37 °C and 95% O_2 monitored periodically and split when confluent.
2. Splitting cells.
   a. Use aseptic techniques for the following protocol.
   b. Preheat cell-culture media and Trypsin-EDTA to 37 °C.
   c. Aspirate media out of flask with a sterile aspirating pipette and rinse three times with PBS.

d. Add Trypsin-EDTA to flask and let sit for 1 min.
e. Tap the flask firmly with your hand to loosen the attached cells.
f. Add the proper volume of cell-culture media to this mixture.
g. Centrifuge the cell suspension for approximately 8 min at just under 2000 rpm (914 g) at room temperature and aspirate off the supernatant. Add 1 ml fresh media and resuspend the pellet with gentle trituration.
h. Determine the concentration of cells suspended in media with a hemocytometer and either seed into a new flask or plate out for electrophysiology experiments.

3. Plating cells.

a. In a sterile laminar flow hood, place coverslips in each well of a Falcon six-well tissue culture plate.
b. Coat coverslips with approximately 100–400 μl collagen or poly-D-lysine (40 μg/ml).
c. Let the coverslips sit under ultraviolet light for at least 30 min.
d. Rinse the coverslips three times with PBS.
e. Coverslips can be used immediately after coating or may be stored in a sterile laminar flow hood. Do not use coverslips if over 1 month old.
f. Add 3 ml of cell-culture media to each well.
g. Plate out approximately 1×10^5 cells/dish (*see* **Note 13**).
h. Plate cells out the day before electrophysiological recording and store in an incubator overnight at 37 °C.

3.2. Concentration-Response Measurements of Positive Allosteric Modulators on GluR2i Receptors

1. The whole-cell variant of the patch-clamp technique is used for measuring the concentration responsiveness of positive allosteric modulators in homomeric GluR2i receptors stably transfected into HEK293 cells.
2. Coverslips are removed from the incubator and placed in a recording dish filled with extracellular recording solution. The recording dish is then secured onto the X-Y stage of the inverted microscope.
3. The 16-barrel solution delivery array is inserted into the dish, and a suction line is also added to maintain the fluid level when solutions are applied to the cell (*see* **Fig. 1A**, **C**).
4. A cell is identified for recording, and the first barrel of the 16-barrel solution delivery array is positioned approximately 50 μ*M* from the cell. The array should be parallel to the cell and the motorized steps of the actuator adjusted, so that each step will move the array to position successive barrels in front of the cell (*see* **Fig. 1C**, **D**).
5. A glass recording electrode is pulled, and the tip is fire-polished prior to use. The recording electrode is filled with the internal recording solution making sure

to remove all bubbles from the electrode tip by gently tapping the shank of the electrode with forefinger.

6. The recording electrode is inserted into the Plexiglas electrode holder, and the holder is attached to the amplifier headstage. A small amount of constant positive pressure (2–3 cm H_2O) is applied to the side port of the holder using a 10-cc syringe connected to a manometer with a three-way stopcock. The positive pressure is applied prior to immersing the electrode into the extracellular bath and maintained as the electrode is advanced through the bath to prevent debris from clogging the electrode tip.
7. Upon placing the recording electrode into the extracellular bath, offset potentials are corrected and electrode capacitance is compensated.
8. After identifying a cell for recording, the electrode is initially advanced to within approximately 50 μm of the cell using the coarse controls of the micromanipulator and increasing levels of magnification through the inverted microscope. The solenoid valve controlling solution flow through the first barrel is turned on, so that the cell is in a stream of extracellular fluid.
9. The recording electrode is then slowly advanced onto the cell using the fine controls of the micromanipulator, so that the membrane of the cell dimples slightly, prior to releasing the positive pressure on the electrode. The stopcock attached to the manometer is quickly switched, so that negative pressure is applied to the recording electrode rupturing the membrane at the tip of the electrode and achieving a whole-cell recording configuration with a gigaohm "seal" resistance between the tip of the electrode and the cell membrane.
10. The membrane potential of the cell is clamped to −80 mV. Series resistance and whole-cell capacitance are compensated (70–85%) and monitored periodically.
11. The cell is lifted off the recording dish, and solution through the first and second barrels is permitted to flow (*see* **Fig. 1D**). The cell is initially exposed to the control extracellular solution from the first barrel for 30 s. A continuous (chart-recorder mode in pCLAMP) recording of current is begun. The 16-barrel solution delivery array is then shifted, so that the solution from the second barrel, 100 μ*M* glutamate (or other agonist), is bathing the cell for 10 s and then returned to the control solution to recover. The current evoked by application of glutamate is recorded. This procedure can be repeated two to three times to obtain an estimate of the effects of glutamate alone. It should be noted that because of the relatively slow switching speed of the actuator, the current recorded in these experiments is the desensitized AMPA receptor current (*see* **Fig. 3A**, **B**).
12. After the effects of glutamate alone have been measured, solution to the first two barrels is turned off and the solution from barrels 3 and 4 is turned on. The cell is exposed to the lowest concentration (e.g., 30 n*M*) of allosteric modulator alone for 30 s and then bathed for 10 s in the presence of allosteric modulator and glutamate. Subsequent evaluations of increasing concentrations of the positive

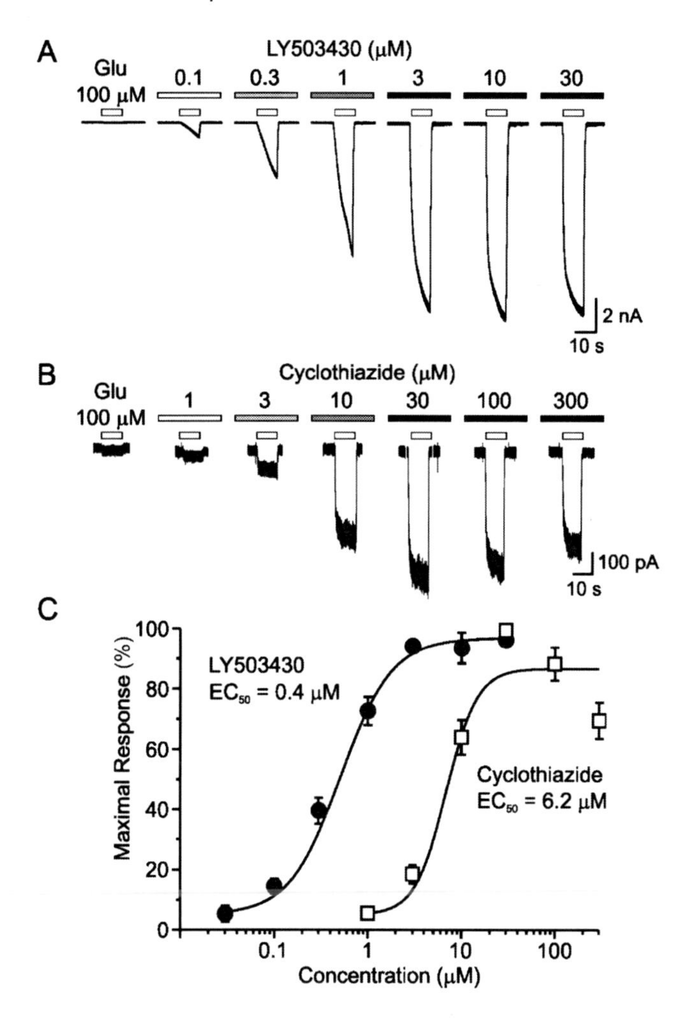

Fig. 3. Concentration-response measurements of positive allosteric modulators on α-amino-3-hydroxy-5-methyl-4-isoxazoleproprionic acid (AMPA) receptors. (**A**) Representative responses of a human embryonic kidney (HEK) cell stably transfected with GluR2i receptors to application of glutamate alone and in combination with LY503430. Individual responses represent epochs from a longer continuous recording. Application of glutamate (100 μ*M*) evoked a small response that reflects the desensitized or steady-state current. Subsequent application of LY503430 (0.03–30 μ*M*) alone does not alter GluR2i receptor activity but markedly enhances the desensitized current when co-administered with glutamate. Note that the responses to LY503430 do not achieve a steady state during the 10-s glutamate application period. (**B**) Representative responses of a HEK cell stably transfected with GluR2i receptors to application of glutamate

allosteric modulator are assessed in the same manner (*see* **Fig. 3A, B**) (*see* **Note 14**).

13. Concentration-response profiles can be constructed by measuring the current amplitude during a 10-s co-application of compound and glutamate (100 μ*M*). The data are then normalized to the maximum peak current obtained from co-application of compound and glutamate and plotted as a function of potentiator concentration. The plotted points then were fit with points averaged at each concentration curve using a four-parameter logistic curve: $Y = \text{min} + [(\text{max} - \text{min})/1 + 10(\log EC_{50} - x)\ \text{Hillslope}]$, where EC_{50} is the concentration equal to 50% of the maximally effective concentration and *n* is the Hill coefficient. Average EC_{50} values can be determined and reported as mean ± SEM (*see* **Fig. 3C**).

3.3. Biophysical Evaluation of Positive Allosteric Modulators on GluR2i Receptors

1. The methods for achieving an outside-out patch configuration are similar to those described in **Subheading 3.2.** for whole-cell recordings.
2. Coverslips are removed from incubator and placed in a recording dish filled with extracellular recording solution. The recording dish is then secured onto the X-Y stage of the inverted microscope.
3. The theta tube is inserted into the dish, and a suction line is also added to maintain the fluid level when solutions are applied to the cell.
4. Outside-out patch recording electrodes are fabricated and filled with internal recording solution (*see* **Subheading 2.3.**) according to procedures described in **Subheading 3.2.**
5. The recording electrode is inserted into the Plexiglas electrode holder, and the holder is attached to the amplifier headstage as described in **Subheading 3.2.**
6. On placing the recording electrode into the extracellular bath, offset potentials are corrected, and electrode capacitance is compensated.

Fig. 3. *(Continued)* alone and in combination with cyclothiazide. Similar to responses to LY503430, application of cyclothiazide (1–300 μ*M*) alone does not alter GluR2i receptor activity but markedly enhances the desensitized current when co-administered with glutamate. In contrast to LY503430, responses to cyclothiazide do achieve a steady state during the 10-s glutamate application. (**C**) Concentration-response plots for LY503430 (•) and cyclothiazide (□). Data points represent the mean (±SEM) response of GluR2i receptors to each concentration of positive allosteric modulator when co-applied with glutamate. The data points are plotted as a function of modulator concentration and fitted with a logarithmic function. The average EC_{50} values obtained from the fitted data were 410 n*M* for LY503430 ($n = 12$) and 6.1 μ*M* for cyclothiazide ($n = 10$). Glu, glutamate.

7. Identify a cell for recording. The cell does not have to be isolated as in whole-cell recording. Only a small portion of the membrane will be removed. Occasionally, multiple patches of membrane can be pulled from the same cell for separate recordings.
8. The initial steps in establishing an outside-out patch recording require that a prior whole-cell configuration be achieved using the procedures described in **Subheading 3.2**. However, before attempting to establish the whole-cell configuration and pulling a patch of membrane, make sure there is no solution flow as it may disturb the patch during pulling.
9. After achieving the whole-cell configuration, clamp the membrane potential to −80 mV.
10. Adjust electrode capacitance compensation, but do not compensate for series resistance or whole-cell capacitance (*see* **Note 15**).
11. Wait at least for 10 s before attempting to pull a patch.
12. Slowly pull the micromanipulator directly upward, not backward or at an angle, stretching the membrane attached to the electrode (*see* **Note 16**).
13. Once the patch is pulled, the cell capacitance transients will reverse. At this point, do not make any further compensations or adjustments.
14. Proper alignment of recording electrode with pulled patch of membrane with the theta tube is essential (*see* **Fig. 2B**, **C**). To facilitate this alignment, the recording electrode and theta tube should be positioned at the center of the field of view on the video monitor that is connected to the tube camera (a CCD camera may be substituted) attached to the camera port of the inverted microscope.
15. Turn on the flow of control solution from the theta tube, ensuring that it is positioned away from the electrode, so that no air bubbles coming out of the tube will disrupt the patch.
16. Subsequently, position the electrode tip approximately 50 μm from the end of the theta tube at about a 45° angle. When the electrode is positioned correctly, the solution interface will bend around the electrode (*see* **Fig. 2D**).
17. An episodic recording mode is used to measure deactivation and desensitization of GluR2i receptors during application of glutamate alone and in the presence of positive allosteric modulators. For outside-out patch recordings, ensure a constant flow of the sucrose solution into the bath to allow for the visibility of the solution interface and to remove debris and excess compound.
18. Currents evoked by rapid application of glutamate are initially measured by shifting the theta tube from the control solution to a solution containing 10 m*M* glutamate (*see* **Fig. 4A**, **B**).This procedure can be repeated three to four times, and subsequent averaging of the evoked currents can be conducted to obtain an accurate measure of the effects of glutamate alone. The same procedures are used to measure deactivation and desensitization of GluR2i receptors except the length of the glutamate exposure is less than or equal to 1 ms for deactivation and typically greater than or equal to 100 ms for desensitization (*see* **Figs. 4** and **5**).

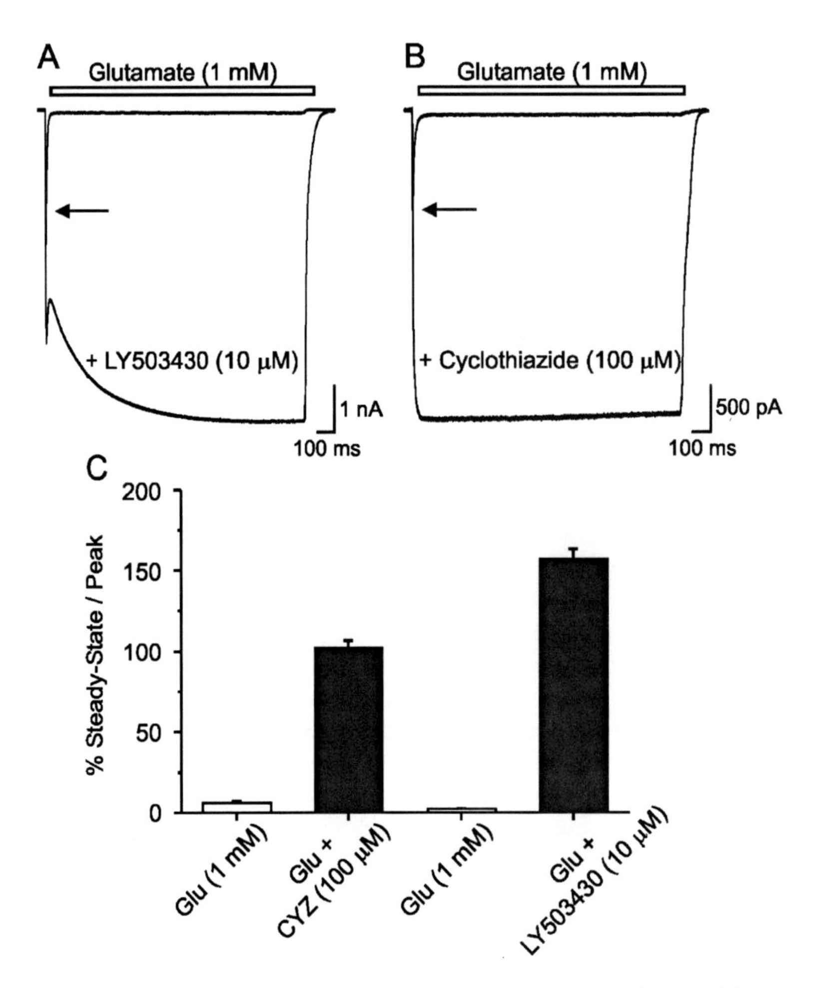

Fig. 4. Effects of positive allosteric modulators on the extent of desensitization of α-amino-3-hydroxy-5-methyl-4-isoxazoleproprionic acid (AMPA) receptors. (**A**) Representative responses of GluR2i receptors to rapid application of glutamate (1 m*M*, 1.5 s duration) alone and in the presence of saturating concentrations, either (**A**) LY503430 (10 μ*M*) or (**B**) cyclothiazide (100 μ*M*). The arrows in **A** and **B** represent the peak current elicited by glutamate alone. Both LY503430 and cyclothiazide suppress desensitization of GluR2i receptors. (**C**) The degree to which positive allosteric modulators attenuate desensitization can be assessed by measuring the percent of the steady-state current amplitude relative to the peak current amplitude before and after application of the modulator. The bar graph shows that the percent steady-state/peak current was less than 10% during administration of glutamate alone, indicating near-complete desensitization. In contrast, co-application of saturating concentrations of LY503430 (10 μ*M*; $n = 24$) or cyclothiazide (100 μ*M*; $n = 8$) increased the percent

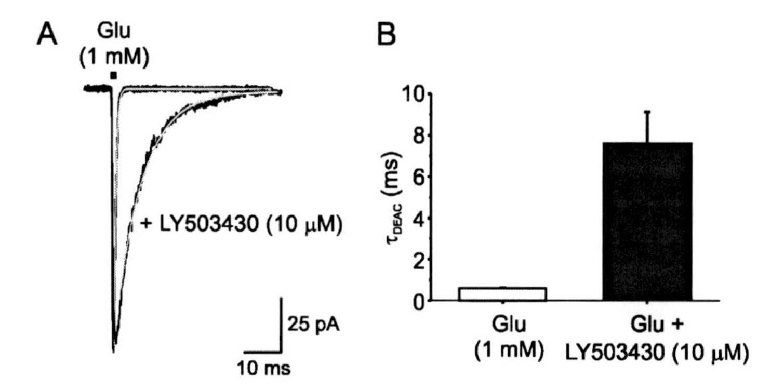

Fig. 5. Effects of positive allosteric modulators on the deactivation kinetics of α-amino-3-hydroxy-5-methyl-4-isoxazoleproprionic acid (AMPA) receptors. (**A**) Representative responses of an outside-out patch of membrane containing GluR2i receptors to brief (1 ms duration) application of glutamate (1 mm.) alone and in combination with a saturating concentration of LY503430 (10 μ*M*). Exponential fits of the decay phase of each response were used to measure the time constant of deactivation [τ_{DEAC} (ms)] (gray traces). The τ_{DEAC} of the response in the presence of LY503430 ($n = 7$) was significantly slower than that for the response during control conditions. (**B**) Bar graph of the average (±SEM) τ_{DEAC} of responses to glutamate alone and in the presence of LY503430. Glu, glutamate.

19. Flow from the syringes containing control and glutamate alone solutions are stopped, whereas solution flow is initiated from two additional syringes containing a saturating concentration of positive allosteric modulator alone and with glutamate (10 m*M*). Again, currents evoked by rapid co-application of glutamate and modulator are measured as described in **step 18** (*see* **Figs. 4** and **5**).
20. When recording is completed, apply positive pressure to the recording electrode to remove the patch of membrane from the tip. Perform a junction potential measurement (*see* **Notes 17** and **18**).
21. Time constants of deactivation (τ_{DEAC}) can be obtained by fitting the decay of the digitized trace in the presence of glutamate from 75–95% of the peak current to steady state with a single exponential function in pCLAMP. Time constants of deactivation are presented as mean ± SEM (*see* **Fig. 5B**).

Fig. 4. *(Continued)* steady-state/peak current to 100% or greater, indicating that desensitization was almost completely eliminated. Glu, Glutamate; CYZ, cyclothiazide.

4. Notes

1. Vendor information are as follows:

   a. Agilent Technologies, USA.
   b. Cole-Palmer Instrument Company, USA.
   c. EXFO Life Sciences Group, Canada (formerly Burleigh Instruments).
   d. Fisher Scientific, USA.
   e. GIBCO-BRL, USA.
   f. Hamamatsu, Japan.
   g. Harvard Apparatus, USA.
   h. Invitrogen, USA.
   i. The Lee Company, USA.
   j. Narishige Inc., Japan.
   k. Newport Corporation, USA.
   l. Nikon, USA.
   m. Molecular Devices, USA (formerly Axon Instruments).
   n. Pall Corporation, USA.
   o. Quest Scientific, Canada.
   p. Sigma, USA.
   q. Sutter Instruments Inc., USA.
   r. Sony Corporation, Japan.
   s. Ted Pella, Inc., USA.
   t. Tocris, USA.
   u. VitroCom Inc., USA.
   v. Warner Instruments LLC, USA.
   w. WPI Inc., USA.

2. All experiments are performed at room temperature and chemicals purchased from commercially available sources, except for LY503430 that was synthesized at Eli Lilly and Company (Indianapolis, IN).
3. FBS should be thawed a maximum of two times before using in media.
4. Collagen is used to plate cells for the whole-cell patch-clamping experiments shown here, because it allows a stable substrate to which cells can adhere yet does not anchor them to the coverslip. This permits individual, isolated cells to be pulled up from the bottom of the recording dish. Poly-D-lysine is used for the outside-out patch experiments as it tends to more securely adhere the cells to the coverslip when pulling a patch.
5. Ba^{+2} in the extracellular recording solution can facilitate formation of a "seal" between the cell and the recording electrode.
6. The presence of CsF in the internal recording solution for outside-out patches is crucial for pulling a patch off the cell membrane.

7. In all steps involving glue, it is essential that enough time is given for the glue to cure and form a water proof barrier. ProBond Epoxy (Elmer's E-609) works well; however, other adhesives could be substituted.
8. Before initiating any type of electrophysiological recording, it is critical to eliminate all bubbles from the solution tubing by running the recording solutions through the tubing lines. Bubbles leaving either the 16-barrel solution delivery system or the theta tube can dislodge the cell (whole-cell) or the piece of cell membrane (outside-out) from the recording electrode, thereby prematurely terminating the experiment.
9. Although L-glutamate (0218, Tocris) was used for the majority of these experiments, mono-sodium glutamate (G2834, Sigma) dissolves more readily and may be substituted. Storage conditions and compound stability are similar to that of L-glutamate. LY503430 (Eli Lilly and Company) and cyclothiazide (0713, Tocris) were dissolved in 100% dimethyl sulfoxide (DMSO) to create 20 m*M* stock solutions and maintained at −30 °C when not in use. Solution aliquots are stable for several months at −30 °C.
10. Using a microgrinder to polish the theta tube tip can be quite difficult. It is not uncommon for the glass to break. When polishing is complete, make sure debris is removed by gently forcing either air or water through the tube. These impurities will clog the theta tube and distort the flow of solution. In addition, ensure that the tip of the theta tube is not cracked to prevent distortion of the solution flow. Solution flow can be visualized by running distilled water through the theta tube into a dish containing a highly concentrated NaCl solution to check for any irregularities. The solution interface can be visualized and tested by performing a junction potential measurement. After each day's experiment, be sure to thoroughly rinse the theta tube and store dry. With proper care, a theta tube can last indefinitely.
11. A multi-barrel syringe pump may be substituted for the gravity-induced flow method for the 16-barrel delivery system. However, depending on the number of concentrations being tested, it may be difficult to find a commercially available pump that can accommodate all syringes.
12. Studies examining channel kinetics typically use outside-out patch clamp electrophysiology because of better temporal and spatial resolution, especially for deactivation. Previously published data ***(34)*** examining desensitization kinetics of GluR2i using whole-cell electrophysiology showed no difference from data obtained with outside-out patches of GluR2i receptors ***(35,36)***. However, it is important to note that other reports measuring desensitization time constants using whole-cell recordings tend to be slower than those obtained from outside-out patches ***(37,38)***. Individual differences between laboratories most likely account for these discrepancies.
13. Plating approximately 1×10^5 cells/dish was based on the cell line used in these experiments (GluR2i expressed in HEK293 cells). This value may be adjusted depending on how rapidly the cell line grows. For the whole-cell patch-clamp

recordings, it is important to find isolated cells for recording. Therefore, do not plate the cells too densely or they will become confluent by the start of the experiment.

14. The current amplitude evoked by 100 μ*M* glutamate alone can be very low. Therefore, when analyzing these data, the peak of each response was normalized to the maximum response obtained when LY503430 was co-applied with glutamate (regardless of concentration) to construct a concentration–response curve.
15. Unlike whole-cell recording conditions, outside-out patch recording does not require isolated cells. Membrane patches can be pulled from large clumps of cells. Because the whole-cell configuration is just a step in achieving an outside-out patch recording, it is not necessary to compensate for series resistance or whole-cell capacitance.
16. Successful completion of this step in achieving an outside-out configuration can be quite difficult. It is essential to very carefully pull the electrode directly upward off the cell. For this reason, it is important for the cells be well-adhered to the recording dish or they will be pulled up with the electrode. Furthermore, pulling patches from some cell lines can be quite difficult. With these cell lines, the membrane does not pinch off and instead forms long strings from the electrode to the cell. These cells are not amenable to pulling outside-out patches.
17. Before initiating any experiments, while making the recording solutions for the day, add 2–5% dH_2O to the control solution to offset the osmolarity. This will allow you to be able to perform a junction potential at the end of the experiment without having to switch to other solutions.
18. Theta tube position was activated through an episodic stimulation square pulse protocol established in pCLAMP and shifted using a piezoelectric actuator having a charging time of 0.3 ms. Onset-to-offset time of junction potential (0.1 *M* and 1.0 *M* NaCl) was approximately 700 μs.

Acknowledgements

The authors thank Mark Fleck PhD and Elizabeth Cornell for assistance with the outside-out patch recording technique.

References

1. Yamada KA (2000) Therapeutic potential of positive AMPA receptor modulators in the treatment of neurological disease. *Expert Opin Investig Drugs* 9:765–778.
2. Staubli U, Rogers G, Lynch G (1994) Facilitation of glutamate receptors enhances memory. *Proc Natl Acad Sci USA* 91:777–781.
3. Sekiguchi M, Yamada K, Jin J, Hachitanda M, Murata Y, Namura S, Kamichi S, Kimura I, Wada K (2001) The AMPA receptor allosteric potentiator PEPA ameliorates post-ischemic memory impairment. *NeuroReport* 12:2947–2950.

4. Oepen G, Eisele K, Thoden U, Birg W (1985) Piracetam improves visuomotor and cognitive deficits in early Parkinsonism–a pilot study. *Pharmacopsychiatry* 18:343–346.
5. Ingvar M, Ambros-Ingerson J, Davis M, Granger R, Kessler M, Rogers GA, Schehr RS, Lynch G (1997) Enhancement by an ampakine of memory encoding in humans. *Exp Neurol* 146:553–559.
6. Dimond SJ, Scammell RE, Pryce IG, Huws D, Gray C (1979) Some effects of piracetam (UCB 6215 Nootropyl) on chronic schizophrenia. *Psychopharmacology* 64:341–348.
7. Hampson RE, Rogers G, Lynch G, Deadwyler SA (1998) Facilitative effects of the ampakine CX516 on short-term memory in rats: correlations with hippocampal neuronal activity. *J Neurosci* 18:2748–2763.
8. Burnashev N, Monyer H, Seeburg PH, Sakmann B (1992) Divalent ion permeability of AMPA receptor channels is dominated by the edited form of a single subunit. *Neuron* 8(1): 189–198.
9. Jonas P, Sakmann B (1992) Glutamate receptor channels in isolated patches from CA1 and CA3 pyramidal cells of rat hippocampal slices. *J Physiol* 455: 143–171.
10. Raman IM, Trussell LO (1995) The mechanism of alpha-amino-3-hydroxy-5-methyl-4-isoxazolepropionate receptor desensitization after removal of glutamate. *Biophys J* 68:137–146.
11. Lambolez B, Audinat E, Bochet P, Crepel F, Rossier J (1992) AMPA receptor subunits expressed by single Purkinje cells. *Neuron* 9(2): 247–258.
12. Geiger JR, Melcher T, Koh DS, Sakmann B, Seeburg PH, Jonas P, Monyer H (1995) Relative abundance of subunit mRNAs determines gating and Ca^{2+} permeability of AMPA receptors in principal neurons and interneurons in rat CNS. *Neuron* 15(1):193–204.
13. Keinanen K, Wisden W, Sommer B, Werner P, Herb A, Verdoorn TA, Sakmann B, Seeburg PH (1990) A family of AMPA-selective glutamate receptors. *Science* 249:556–560.
14. Sommer B, Keinanen K, Verdoorn TA, Wisden W, Burnashev N, Herb A, Kohler M, Takagi T, Sakmann B, Seeburg PH (1990) Flip and flop: a cell-specific functional switch in glutamate-operated channels of the CNS. *Science* 249: 1580–1585.
15. Hollmann M, O'Shea-Greenfield A, Rogers SW, Heinemann S (1989) Cloning and functional expression of a member of the glutamate receptor family. *Nature* 342:643–648.
16. Staubli U, Perez Y, Xu F, Rogers G, Ingvar M, Stone-Elander S, Lynch G (1994) Centrally active modulators of glutamate receptors facilitate the induction of long-term potentiation *in vivo*. *Proc Natl Acad Sci USA* 91:11158–11162.
17. Armstrong N, Sun Y, Chen GQ, Gouaux E (1998) Structure of a glutamate-receptor ligand-binding core in complex with kainate. *Nature* 395(6705):913–917.

18. Baddeley A. (1992) Working memory. *Science* 255:556–559.
19. Dudkin KN, Kruchinin VK, Chueva IV (1997) Synchronization processes in the mechanisms of short-term memory in monkeys: the involvement of cholinergic and glutaminergic cortical structures. *Neurosci Behav Physiol* 27: 303–308.
20. Mayer M, Vyklicky L Jr. (1989) Concanavalin A selectively reduces desensitization of mammalian neuronal quisqualate receptors. *Proc Natl Acad Sci USA* 86:1411–1415.
21. Brown TH, Chapman PF, Kairiss EW, Keenan CL (1988) Long-term synaptic potentiation. *Science* 242(4879):724–728.
22. Carlsson A, Waters N, Holm-Waters S, Tedroff J, Nilsson M, Carlsson ML (2001) Interactions between monoamines, glutamate, and GABA in schizophrenia: new evidence. *Annu Rev Pharmacol Toxicol* 41:237–260.
23. Carlsson ML (2001) On the role of cortical glutamate in obsessive-compulsive disorder and attention-deficit hyperactivity disorder, two phenomenologically antithetical conditions. *Acta Psychiatr Scand* 102(6):401–413.
24. Chen Q, Flores-Hernandez JF, Jiao Y, Reiner A, Surmeier DJ (1998) Physiological and molecular properties of AMPA/kainate receptors expressed by striatal medium spiny neurons. *Develop Neurosci* 20:242–252.
25. Coyle JT (1996) The glutamatergic dysfunction hypothesis for schizophrenia. *Harv Rev Psychiatry* 3(5):241–253.
26. Dingledine R, Borges K, Bowie D, Traynelis SF (1999) The glutamate receptor ion channels. *Pharmacol Rev* 51(1):7–61.
27. Eastwood SL, Burnet PW, Harrison PJ (1997) GluR2 glutamate receptor subunit flip and flop isoforms are decreased in the hippocampal formation in schizophrenia: a reverse transcriptase-polymerase chain reaction (RT-PCR) study. *Brain Res Mol Brain Res* 44(1):92–98.
28. Barkley R, Grodzinsky G, DuPaul G (1992) Frontal lobe functions in attention deficity disorder with and without hyperactivity: a review and research report. *J Abnorm Child Psychol* 2;20:163–188.
29. Baumbarger PJ, Muhlhauser M, Zhai J, Yang CR, Nisenbaum ES (2001) Positive modulation of alpha-amino-3-hydroxy-5-methyl-4-isoxazole propionic acid (AMPA) receptors in prefrontal cortical pyramidal neurons by a novel allosteric potentiator. *J Pharmacol Exp Ther* 298(1):86–102.
30. Benson D (1991) The role of frontal dysfunction in attention deficit hyperactivity disorder. *J Child Neurol* 6(suppl):S9–S12.
31. Bleakman D, Gates MR, Ogden A, Ornstein PL, Zarrinmayeh H, Nisenbaum ES, Baumbarger P, Jarvie KR, Miu P, Ho K et al. (2000). Novel AMPA receptor potentiators LY392098 and LY404187: effects on recombinant human and rat neuronal AMPA receptors. *Soc Neurosci Abstr* 30:173.
32. Arai A, Lynch G (1992) Factors regulating the magnitude of long-term potentiation induced by theta pattern stimulation. *Brain Res* 598:173–184.

33. Armstrong N, Gouaux E (2000) Mechanisms for Activation and Antagonism of an AMPA-sensitive glutamate receptor: crystal structures of the GluR2 ligand binding core. *Neuron* 28:165–181.
34. Quirk JC, Siuda ER, Nisenbaum ES (2004) Molecular determinants responsible for differences in desensitization kinetics of AMPA receptor splice variants. *J Neurosci* 24(50): 11416–11420.
35. Yelshansky MV, Sobolevsky AI, Jatzke C, Wollmuth LP (2004) Block of AMPA receptor desensitization by a point mutation outside the ligand-binding domain. *J Neurosci* 24(20):4728–4736.
36. Koike M, Tsukada S, Tsuzuki K, Kijima H, Ozawa S. (2000) Regulation of kinetic properties of GluR2 AMPA receptor channels by alternative splicing. *J Neurosci* 20(6):2166–2174.
37. Grosskreutz J, Zoerner A, Schlesinger F, Krampfl K, Dengler R, Bufler J. (2003) Kinetic properties of human AMPA-type glutamate receptors expressed in HEK293 cells. *Eur J Neurosci* 17(6):1173–1178.
38. Partin KM, Fleck MW, Mayer ML (1996) AMPA Receptor flip/flop mutants affecting deactivation, desensitization, and modulation of cyclothiazide, aniracetam and thiocyanate. *J Neurosci* 16(21):6634–6647.
39. Hille B (2001) *Ion Channels of Excitable Membranes.* Sinaur Associates, Sunderland, MA.
40. Swanson GT, Kamboj SK, Cull-Candy SG (1997) Single-channel properties of recombinant AMPA receptors depend on RNA editing, splice variation, and subunit composition. *J Neurosci* 17:58–69.

4

Automated Voltage-Clamp Technique

Andrea Ghetti, António Guia, and Jia Xu

Summary

The voltage-clamp electrophysiology method is the gold standard for measuring the function of ion channels. In the past, this technique has had limited applicability in pharmaceutical drug discovery because of its low throughput, steep learning curve, and challenges in standardization of the experiments. Recently, new electrophysiology platforms have been developed, which are based on the use of planar electrodes. One key advantage of the new electrode geometry is that it makes the process of cell-to-electrode sealing amenable to automation, thus increasing the throughput and significantly reducing the skill-set needed to run the experiments. The further addition of computer-controlled fluidics, voltage-clamping electronics, and automated data handling makes it possible to perform multiple electrophysiology experiments in parallel with a high degree of consistency and in completely automated mode. Among the new offerings for automated voltage clamp, one of the systems, PatchXpress/Sealchip, is quickly becoming the new gold standard for the quantification of ion channel function. In this chapter, we provide an overview of the new planar patch-clamping platforms and describe how electrophysiology experiments are performed on the PatchXpress/Sealchip automated system.

Key Words: Parallel voltage clamp; automated electrophysiology; ion channel screening; ion channel drug discovery.

1. Introduction

The application of voltage-clamp technique *(1)* to membrane patches and whole cells *(2,3)* has enabled the high-resolution electrophysiological characterization of biological membranes and the properties of ion channels that reside in

From: *Methods in Molecular Biology, vol. 288: Patch-Clamp Methods and Protocols*
Edited by: P. Molnar and J. J. Hickman © Humana Press Inc., Totowa, NJ

these membranes. Despite its widespread application and continuous improvements, voltage clamp remains a challenging methodology to many researchers because of the requirement to operate fairly sophisticated electronic instruments, the labor intensive and low-throughput operation, and the lack of standardization. For ion channel pharmaceutical screening, voltage clamp has given way to technologies that produce far-less relevant data such as fluorescence-based screening in exchange for higher throughput and lower cost. In addition, the need for assessing cardiac safety of a large number of early-stage compounds has precipitated the needs for automating the voltage-clamp technology to keep up with the capacity demand of medicinal chemistry support.

Since 2002, two classes of automated voltage-clamp devices have become available. Based on planar substrates, AVIVA Biosciences introduced a high-quality 16-channel voltage-clamp recording product—Sealchip_{16}—capable of generating GΩ-resistance seals with over 75% success rate on PatchXpress, the automation system co-developed by AVIVA and Axon Instruments ***(4,5)***.

This was followed by Sophion Biosciences that have approximately 50% giga seal success rate. On the other hand, the IonWorks system invented by Essen Instruments provides low-quality seals with approximately 100 MΩ resistance but has a higher (48×) throughput ***(6)***. An alternative to automation by planar electrode approach is represented by Flyion, whose product, Flyscreen, forms GΩ-resistance seals inside a glass pipette.

In this chapter, we will take the AVIVA Sealchip as an example to discuss voltage-clamp automation, in the context of hERG testing. The objective is to provide an example for our readers of the scope and capabilities of automated voltage-clamp systems. We also provide procedural details and notes aimed at maximizing the productivity of current PatchXpress users.

2. Materials

2.1. Cell Culture

1. Growth medium: Dulbecco's modified Eagle's medium with nutrient Mix F-12 (D-MEM/F-12) containing 15 m*M* 4-(2-hydroxyethyl)-1-piperazineethanesulfonic acid (HEPES) buffer and L-glutamine (Invitrogen, USA), 10% fetal bovine serum (FBS) (Invitrogen), and penicillin–streptomycin (Invitrogen).
2. G-418 (Invitrogen).
3. Trypsin–ethylenediaminetetraacetate (EDTA) (0.05% Trypsin, with EDTA 4Na) (Invitrogen).
4. Dulbecco's phosphate-buffered saline (D-PBS) without calcium, magnesium, and phenol red (HyClone, USA).

2.2. Cell Isolation

1. D-PBS without calcium, magnesium, and phenol red.
2. Accumax (Innovative Cell Technologies, USA).
3. D-MEM/F-12 containing 15 m*M* HEPES buffer and L-glutamine.
4. FBS.
5. Trypan blue (Cambrex, USA).

2.3. Whole Cell Recording

1. External solution (ES): 1.8 m*M* $CaCl_2$; 1.0 m*M* $MgCl_2$; 4 m*M* KCl; 137 m*M* NaCl; 10 m*M* glucose; 10 m*M* HEPES; pH 7.4 with 1 *M* NaOH; osmolarity approximately 295 mOsm.
2. Internal solution (in m*M*): 130 m*M* KCl, 1 m*M* $MgCl_2$, 5 m*M* ethylene glycol tetraacetic acid (EGTA), 10 m*M* HEPES, 5 m*M* adenosine 5′-triphosphate (ATP); pH adjusted to 7.25 with KOH; osmolarity approximately 280 mOsm.

3. Method

Unintended inhibition of the cardiac potassium channel hERG is considered the main culprit in drug-induced arrhythmias known as *torsade de pointes*. Electrophysiology is considered to be the most reliable in vitro screening method for identifying potential cardiac hERG liabilities, but only the recent advent of planar electrode-based voltage-clamp electrophysiology provides sufficient throughput to support the drug-testing needs of most drug-discovery programs *(7–9)*. In PatchXpress™, the recording electrode consists of the AVIVA's Sealchip™ planar electrode, which contains 16 independent recording chambers each having a 1–2 μm hole at the bottom (*see* **Fig. 1**).

The cells are voltage clamped from their bottom surface in the Sealchip™ electrode and on their side or top surface with traditional voltage-clamp

Fig. 1. Sealchip, the 16 chambers are visible from the top.

electrodes. With the Sealchip™ electrode, compounds are delivered from a 96-well plate into the recording chambers by means of a disposable tip.

At the beginning of the experiment, dissociated cells are dispensed in suspension in each chamber, and a single cell, chosen randomly, landing on the hole at the bottom of the chamber can form a tight (GΩ) seal with the surface of the Sealchip™. Suction applied through the hole can then rupture the patch of membrane and establish electrical continuity between the inside of the cell and the underlying electrode chamber. The cell is then monitored for a brief period of time to assess its stability, and subsequently, recording is started by measuring hERG baseline current (which serves as the pre-drug control current) to determine the stability and quality of the ionic current. Cells with high-quality, stable recording configuration and exhibiting a stable baseline are then used to determine the effects of the test compounds.

3.1. hERG-CHO Cell Line Propagation

The cells are cultured in T75 flasks using standard culture medium D-MEM/F12 containing 10% FBS, 1% penicillin–streptomycin, and 500 μg/ml G-418. Cells are propagated using Trypsin (0.05%)/EDTA. Culture media are replaced 24 h following passage and then 2–3 days after.

1. Warm all media (37 °C) and reagents (room temperature) prior to splitting the cells.
2. Split cells from a T75 flask at approximately 90% confluency.
3. In a laminar flow hood aspirate media from culture, tilting flask to ensure that cells are not disturbed by suction.
4. Add 10 ml Dulbecco's Phosphate Balanced Salt Solution (DPBS) and tilt one or two times *gently* to wash the cell monolayer.
5. Aspirate the DPBS (repeat **steps 3** and **5** one more time).
6. Add 3 ml 0.05% Trypsin–EDTA mix.
7. Tilt once or twice to cover the cell monolayer.
8. Leave flask undisturbed for 1 min to trypsinize.
9. Aspirate Trypsin solution and leave cells on table for 2 min.
10. Dislodge cells from the flask by tapping the side of the flask a few times.
11. Add 10 ml growth medium.
12. Pipette up and down several times to mix and dissociate cell clumps until cells are separated.
13. Aliquot cell suspension into each new T75 flask with 10 ml growth media/flask.
14. Place flasks in 37 °C, 5% CO_2 incubator.

Seeding Density:

For 1 day split = 1:5 (2 ml cells each flask).
For 2 day split = 1:20 (0.5 ml cells each flask).
For 3 day split = 1:40 (0.25 ml cells each flask).

3.2. Cell Isolation

1. Cells should be about 70–85% confluence on the day of isolation. Note: 24–48 h prior to the electrophysiology experiments, the cells should be cultured in growth medium without G-418.
2. Before isolation warm media and reagents but do not warm up Accumax solution.
3. Remove media and wash twice with DPBS without calcium and magnesium.
4. Aspirate all remaining DPBS.
5. Mix 9 ml DPBS without calcium and magnesium with 3 ml Accumax and add to cells. Leave T75 flask with DPBS/Accumax mixture for 2 min at room temperature (upon microscope examination, cells should be turning round at this point).
6. Aspirate supernatant without disturbing the cells monolayer and repeat **step 5** and incubate T75 flask at 37 °C until 95% of cells dissociate from the flask (about 5–7 min).
   Note: Too short digestion time results in cells that may form GΩ seals less frequently but when they do tend to give longer lasting recordings. Too long digestion time increases seal formation efficiency but may reduce the lifespan of cells during the recordings.
7. Add 9 ml D-MEM/F12 containing 10% FCS into the flask to stop the digestion.
8. Gently (at a flow rate of approximately 1 ml/s) pipette up and down several times to mix and dissociate cell clumps.
9. Place cell suspension in 50-ml centrifuge tube.
10. Centrifuge 1000 rpm (300 g) for 2 min in a tabletop clinical centrifuge; carefully aspirate supernatant without disturbing the cell pellet.
    Resuspend cells into 5 ml D-MEM/F12 with 10% FCS and count the cells. Adjust cells dilution to approximately 0.5–1.0 million cells/ml and incubate cells in 37 °C, 5% CO_2 incubator.
11. Place isolated incubator to recover for at least 30 min before using for electrophysiology recording.

3.3. Cell Preparation Prior to Voltage-Clamp Recording

1. Take 1 ml isolated cells out of incubator, carefully re-suspend using a 1000 – μl pipette as to minimize air bubbles in cell suspension, and place in 1.25-ml mini-Eppendorf tube. Centrifuge at 1000 rpm (300 g) for 1 min and discard supernatant.
2. Re-suspend cells in 100 μl ES for recording. Avoid bubbles when re-suspending the cells.

3.4. Drug Preparation

Compound's dilutions (*see* **Fig. 1**) are prepared starting from a stock solution at 10 m*M* in pure dimethylsulphoxide (DMSO). All dilutions are performed using glass vials. Compounds should be prepared no longer than 20–30 min before they are tested.

1. Sonicate stock solution for about 2 min to insure proper mixing. The final DMSO concentration in the experiments should not exceed 0.3%. In typical experiments, the DMSO concentration in each sample is 0.1%.
2. Place compound plate in plate location 1 on the PatchXpress workstation.

3.5. Whole Cell Recording

1. Turn on PatchXpress, turn on vacuum and air pressure, and run the startup procedure.
2. Remove Sealchip (*see* **Fig. 2**) from container and rinse both sides with distilled water taking special care not to touch the chip's bottom side. Place in PatchXpress upside down and with the slanted part of Sealchip toward the front of the machine.
3. Place new cells tube (*see* **Subheading 2.3.**) in first slot on PatchXpress.
4. Cells are voltage clamped at −80 mV holding potential, and hERG current is activated by a depolarizing step first to −50 mV for 500 ms to serve as a reference

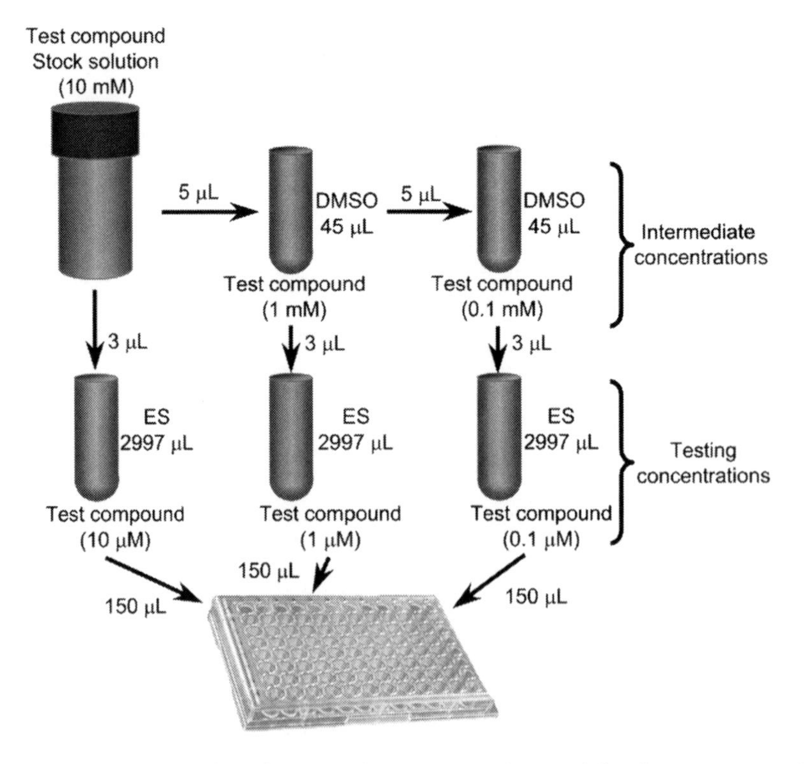

Fig. 2. Dilution procedure for preparing compounds tested for three concentrations: 10, 1, and 0.1 µ*M*.

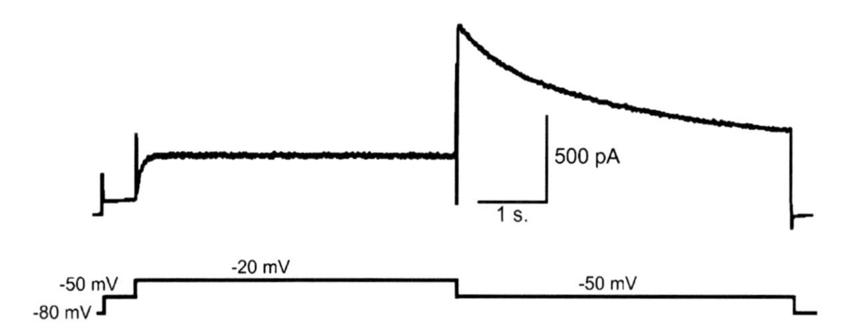

Fig. 3. hERG channel whole cell current (top) and waveform stimulation protocol (bottom).

point for the tail current, then to +20 mV for 2 s to activate the channels, then finally back to −50 mV for 2 s to remove the inactivation and observe the deactivating tail current (*see* **Fig. 3**).

3.6. PatchXpress Procedures

A standard procedure was designed to determine the IC_{50} of hERG blockers. In these studies, IC_{50} values are estimated from either a single concentration or three incremental concentrations of test compound (for example, 0.1, 1, and 10 μ*M*). (*see* **Notes 1–4**) In all studies, baseline hERG current amplitudes are established for 4 min before drug addition. There is a 300-s (maximum 500 s) wait period in between each drug addition. The same compound concentration is added three times with a 20- to 30-s wait period between each pair of drug addition in order to minimize any potential lost due to adsorption and due to dilution with resident buffer.

4. Conclusions

In recent years, the introduction of automated electrophysiology platform has created new opportunities for the identification of ion channel-targeting drugs to be used as therapeutic agents. It has also improved the efficiency of compound testing against ion channels known to have the potential for severe side effects. Although the efficient acquisition of high-quality electrophysiology data in automated mode has been made possible, significant challenges remain. The throughput offered by the current platforms, while much higher than the one achievable by conventional manual voltage-clamp methods, is still orders of magnitude below that of typical high-throughput instruments used in drug screening. Also, in some of the existing platforms, the higher throughput tends to come at the expense of data quality. The current trend toward higher throughput, increased data quality, and instrument flexibility is however a guarantee that ion channel drug discovery is entering a new and exciting age.

5. Notes

Electrophysiology is considered the gold standard method for ion channel studies because it delivers a direct and specific measurement of the activity of a membrane-bound protein, leaving little room for misinterpretation about the effect of a pharmaceutical agent on the membrane ion channel protein. There are nevertheless the usual issues that complicate dose–response studies, such as drug availability or steric effects of other components of the solutions, and a whole set of new issues that are specific to ion channels, such as voltage dependence of the protein, specificity for a substrate, presence of overlapping background signals, signal run-down or run-up, and so on. In the next few paragraphs, we will describe how some of the most common problems with high-throughput studies are addressed by the PatchXpress system and discuss how to deal with some of the electrophysiology-specific problems that often come up in experimental procedure design.

The goal of pharmaceutical screening is most often to test for inhibitory effect of compounds on a specific target, and as maximum inhibition is usually complete, potency is the only ranking system required to screen a library of compounds. In this case, the midpoint value of a Hill equation fit of the concentration-dependence data is referred to as a compound's IC_{50} value (concentration that produces 50% maximal inhibition). The measurement of an IC_{50} value makes a number of assumptions: (1) the concentration of the analyte must be well clamped during measurement, (2) the analyte must have access to the receptor (channel), (3) the assay must be capable of measuring the effect, and (4) conditions during the measurement, such as temperature and pH, can produce results that are predictive of in vivo effects.

1. The most frequent problem with high-throughput screens is a positive shift in the concentration-dependent effects of the analyte. This shift is the result of the removal of drug from the vicinity of its receptor, which is usually attributed to the insolubility of the drug and/or to drug absorption onto the various surfaces that come in contact with the compounds.

   On the PatchXpress, similarly to other systems, the compound plate was found to be a major contributor to drug adsorption. We can resolve this issue for specific families of compounds if we know their approximate cLogP (calculated water–oil partition coefficient) and can assume that the drug will not itself precipitate from solution either during the testing or anytime before the testing. For example, the drug Terfenadine can display an IC_{50} that is right-shifted by two orders of concentration when it is briefly stored on polystyrene multiwell compound plates before testing. Even polypropylene compound plates do not resolve this issue, and the most significant improvement in the drug potency measurement is only achieved with glass well compound plates *(7)*.

The remaining shift has been identified as drug dilution in the measurement chamber and possible drug consumption by other cells within the same chamber. Before drug is added to the test chamber, all but 8–12 μl of solution is withdrawn from the test chamber, and then 30–50 μl of drug solution is added to the resident buffer, diluting the final drug concentration by up to 20%, or a final test compound concentration that is 80% of the desired value. The usual fix is to add compound in solution while keeping the suction to waste open, resulting in flow-through solution exchange and producing a drug concentration that is better clamped. However, the relatively small number of molecules of analyte that are present in the small bath volume become more sensitive to drug absorption into other cells that are not the object of the measurements. Alternatively, the drug may be added multiple times, applying suction only immediately before the next drug addition, resulting in final concentrations of 80% after the first addition, 94% after the second addition, or 99% after the third addition of the test compound. The best resolution to the delivery problems would therefore be a hybrid approach of a first compound addition of 60 μl of drug-containing solution with suction during the addition (that is, exchange six fold the residual volume or > 98% exchange), followed by a second addition without suction (to increase the amount of molecules available for binding to cells and other surfaces within the test chamber).

Finally, drug concentration may also be decreased by the precipitation of compound from solution. In this context, to minimize the chances of aggregation and precipitation in aqueous buffer, it is important to delay mixing of compounds until immediately before the experiments and to sonicate the mixtures well before performing any dilutions.

2. Many drugs that affect ion channels may do so by interaction with a region of the protein that is responsible for ionic conductance. For this reason, many compounds may have to compete sterically for their binding position either because of the flow of ions that repel the analyte in the vicinity of the binding site or because a voltage-dependent protein conformational change shelters the interaction site, and hence these analytes display a voltage dependence of their activity. These complications are unique to ion channels, but they are easily dealt with by designing test protocols that span more than a single membrane potential (that is, determine voltage dependence of current and inhibition). The PatchXpress automated method provides a number of options to test voltage dependence of compound activity: the voltage protocol sweeps may include a voltage ramp, or a sequence of steps to various voltages, or a protocol with sweeps to multiple voltages may be applied repeatedly.
3. Although an ionic current may appear to be inhibited, there are a number of artifacts that may mimic inhibition. For example, channels may inactivate because of use dependence or be removed from the cell surface or modified by intracellular processes that may result in a reduction in ionic current that is not related to a drug

effect. Conversely, a true inhibition may be masked by recruitment of channels to the cell membrane or by intracellular modification of the channel to enhance ionic current.

To ensure that the experiment faithfully captured a result, the PatchXpress procedure includes the ability to continuously monitor ionic current during compound addition, then during compound removal (washout). This time course of ionic current should produce a discernable change in current amplitude that is correlated with the drug addition, then return to pre-drug levels when the drug is washed out (the classical pharmacological experiment), allowing extrapolation of onset and offset kinetics, and supporting theories for mechanism of action of the analytes.

Another factor that affects the fidelity of the measurements is the design of the test protocol. The protocol waveform and periodicity must be appropriate for the ionic current that is the object of the experiment. For example, the tetrodotoxin (TTX)-sensitive sodium current represents a large number of single-channel openings that occur almost simultaneously upon depolarizing the membrane. This current presents a challenge to assay as the desired effect is often an open-state inhibition (drug binds to an open channel but unbinds from a closed channel), and therefore the drug displays use-dependent inhibition. The degree of use dependence of a drug may therefore be assessed by either measuring drug effect at higher stimulation frequencies or maintaining the channels in their open state thereby promoting drug binding and therefore inhibition.

A further issue that usually confounds electrophysiology experiments is that the membrane must be voltage clamped very precisely to enable measurements of activation potentials or various other measures of ion channel function. Reproducibility of this type of data usually suffers from voltage-offset problems related to electrode drift, variations in solution preparation, restrictions to solution dialysis into the cytosol, and so on. For this purpose, the PatchXpress has the ability to learn key biophysical parameters during a preparatory section of a test procedure and then adjust subsequent test protocols. In one specific example, AVIVA assays sodium channels by first discovering the half-inactivation potential for each cell, then setting the holding potential for the drug assay to a voltage 7 mV hyperpolarized from the half-inactivation potential ***(10)***. This compensates at once for errors in solution liquid junction potential determination, for drift in electrode offset potentials, for differences in dialysis of cells, and for differences in physiological states of the cells.

4. Like any other experiment, the conditions of the experiment should allow the experimental results to be used to rank the compounds against each other but ideally should allow extrapolation of what the drug should do under physiological conditions. One of the most commonly ignored factors is temperature. PatchXpress allows all solutions to setting to room temperature, and it records into every data file the temperature in the immediate vicinity of the recording chambers. Temperature variations may then be used to flag experiments that may contain the temperature artifact.

In addition to temperature, a great many other experimental conditions, not unique to ion channels, can affect the outcome of an experiment. For example, pH is known to have a dominant effect on whether a compound can enter the cell membrane, and as many compounds inhibit the current only after permeating the membrane, this can have a significant effect on the experimental results.

References

1. Hodgkin, A.L. and Huxley, A.F. A quantitative description of membrane current and its application to conduction and excitation in nerve. *J Physiol* 1952; 117: 500–544.
2. Neher, E. and Sakmann, B. Single-channel currents recorded from the membrane of denervated frog muscle fibers. *Nature* 1976; 260: 799–802.
3. Hamill, O.P., Marty, A., Neher, E., Sakmann, B., and Sigworth, F.J. Improved patch-clamp techniques for high-resolution current recording from cells and cell-free membrane patches. *Pflugers Arch* 1981; 391: 85–100.
4. Xu, J., Guia, T., Rothwarf, D., Huang, M., Sithiphong, K., Ouyang, J., Tao, G., Wang, X., and Wu, L. A benchmark study with SealChip planar patch-clamp technology. *Assay Drug Dev* 2003; 1(5): 675–684.
5. Guia, A. and Xu, J. (2005) Planar electrodes: the future of voltage clamp. In *Ion Channels in the Pulmonary Vasculature*. Marcel Dekker, New York, 635–651.
6. Kiss, L., Bennett, P.B., Uebele, V.N., Koblan, K.S., Kane, S.A., Neagle, B., and Schroeder, K. High throughput ion-channel pharmacology: planar-array-based voltage clamp. *Assay Drug Dev Technol* 2003; 1: 127–135.
7. Tao, H., Santa Ana, D., Guia, A., Huang, M., Ligutti, J., Walker, G., Sithiphong, K., Chan, F., Guoliang, T., Zozulya, Z., Saya, S., Phimmachack, R., Sie, C., Yuan, J., Wu, L., Xu, J., and Ghetti, A. Automated tight seal electrophysiology for assessing the potential hERG liability of pharmaceutical compounds. *Assay Drug Dev Technol* 2004; 2: 497–506.
8. Dubin, A.E., Nasser, A., Rohrbacher, J., Hermans, A., Marrannes, R., Grantham, C., van Rossem, K., Cik, M., Chaplan, S.R., Gallacher, D., Xu, J., Guia, A., Byrne, N., and Mathes, C. Identifying modulators of hERG channel activity using the PatchXpress™ planar patch clamp. *J Biomol Screen* 2005; 10: 168–181.
9. Guo, L. and Guthrie, H. Automated electrophysiology in the preclinical evaluation of drugs for potential QT prolongation. *J Pharmacol Toxicol Methods* 2005; 52: 123–135.
10. Tao, H., Guia, A., Xie, B., Santa Ana, B., Manalo, G., Xu, J., and Ghetti, A. Efficient characterization of use-dependent ion channel blockers by real time monitoring of channel state. *Assay Drug Dev Technol* 2006; 4: 57–64.

5

Flip-the-Tip

Automated Patch Clamping Based on Glass Electrodes

Michael Fejtl, Uwe Czubayko, Alexander Hümmer, Tobias Krauter, and Albrecht Lepple-Wienhues

Summary

A conventional borosilicate glass patch pipette is glued into a plastic jacket, forming the entity of a FlipTip®. One or two three-channel modules of recording tip sockets are mounted on a liquid handler platform to take up FlipTips. The tip sockets are connected to preamplifiers (HEKA) and to a suction system. The inner chamber of the tip sockets is filled with intracellular solution (IS), and FlipTips are prefilled with extracellular solution (ES) immediately before use. The FlipTips are then inserted into the recording tip sockets, and suspended cells are taken out of a cell hotel and dispensed into the open back of the FlipTips. Simply by gravity, the cells move down toward the end of the pipette. Continuous gentle suction draws a single cell into the very end of the tip, forming a classical GigaSeal of 1–5 GΩ. Stronger suction pulses are then applied to establish the open whole-cell configuration. Under voltage-clamp conditions, the response of an ion channel is measured simultaneously in three to six recording sockets. Alternatively, the perforated patch method is used with Amphoterecine B as the pore-forming agent. Compound delivery is accomplished with a fused silica quartz pipette through the open back of the FlipTip directly onto the cell. All these tasks are performed automatically and completely unattended by the Flyscreen®8500 automated patch-clamp robot. A data throughput of several hundred data points per day can be accomplished with the system.

Key Words: Automated patch clamp; GigaSeal; Flip-the-Tip; Flyscreen; flyion; high-throughput screening; lead optimization; drug development; FlipTip®.

From: *Methods in Molecular Biology, vol. 288: Patch-Clamp Methods and Protocols*
Edited by: P. Molnar and J. J. Hickman © Humana Press Inc., Totowa, NJ

1. Introduction

The Human Genome Project has revealed approximately 30,000 genes, of which a population of several hundred encode for ion channels and/or receptor proteins. Research shows that diseases such as Alzheimer's or chronic pain are linked to a dysfunction of ion channels. Thus, the pharmaceutical industry is increasingly focusing on ion channels as targets to alleviate these illnesses by developing new drugs with a high specific potency and little side effects. Voltage-sensitive dyes or binding assays are indirect ways to address ion channel behavior and are therefore not ideal for screening ion channels in detail. In contrast, patch clamping has been acknowledged to be the "gold standard" to assess ion channel physiology. The method was introduced 1976 by Neher and Sakmann and has been successfully used in thousands of laboratories all over the world. However, the standard approach to patch clamping is time consuming, needs experienced academic personnel, and allows only a low throughput. Thus, flyion has developed the revolutionary Flip-the-Tip technology, providing full automation of the standard patch-clamp approach with glass electrodes, so that up to six cells can be recorded simultaneously. This technology allows higher throughput in the drug discovery process at reduced cost while retaining high data quality.

2. Materials

2.1. Cell Culture of Leukocyte Tyrosine Kinase (LTK) Kv1.5 Fibroblasts

1. The LTK mouse fibroblast cell line was obtained from Dr. Michael M. Tamkun, Colorado State University, CO.
2. Permanent culture growth medium: Dulbecco's modified Eagle's medium (DMEM; PAA Laboratories GmbH, Austria, cat. no. E15-009) without Na-pyruvate (PAA Laboratories GmbH, S11-003), 10% fetal calf serum (FCS) (PAA Laboratories GmbH, Austria, cat. no. A15-649), 2 m*M* L-glutamine (PAA Laboratories GmbH, Austria, cat. no. M11-004), 1× nonessential amino acids (NEAA; PAA Laboratories GmbH, Austria, cat. no. M11-003), 1× penicillin/streptomycin (P/S; PAA Laboratories GmbH, Austria, cat. no. P11-010), and 160 μg/ml geneticin (G418; PAA Laboratories GmbH, Austria, cat. no. P27-011).
3. Working culture growth medium: DMEM without Na-pyruvate, 10% FCS, 2 m*M* L-glutamine, 1× NEAA, 1× P/S, and dexamethasone (Sigma-Aldrich, Germany, cat. no. D-4902) for Kv1.5 induction (final concentration, 0.1 μ*M*).
4. Wash medium: Dulbecco's phosphate-buffered saline (DPBS without Ca and Mg; PAA Laboratories GmbH, Austria, cat. no. H15-002).
5. Medium for cell dissociation: Trypsin/ethylendiaminetetraacetic acid (EDTA) solution (0.25%/1 m*M* EDTA; Invitrogen GmbH, Germany, cat. no. 25200), Trypsin/EDTA solution (0.05%/0.53 m*M* EDTA; Invitrogen GmbH, Germany,

cat. no. 25300), Hank's balanced salt solution (HBSS) without Ca^{2+} or Mg^{2+} (PAA Laboratories GmbH, Austria, cat. no. H15-009).
6. Medium for Flyscreen®8500 robot: Leibovitz L-15 (PAA Laboratories GmbH, Austria, cat. no. E15-821) + glutamin + 10% FCS. FCS has to be added separately by hand. All other mixed solutions come ready-to-use.

2.2. Recording Solutions

1. Extracellular solution (ES) (m*M*): *N*-methyl-D-glutamine (NMDG) 145 m*M*; 5 m*M* KCl, 1 m*M* $MgCl_2$, 2 m*M* $CaCl_2$, 10 m*M* HEPES, and 10 m*M* glucose, osmolarity adjusted to 310 mOsm.
2. Intracellular solution (IS) (m*M*): 115 m*M* $KMeSO_3$, 3 m*M* $MgCl_2$, 0.7 m*M* $CaCl_2$, 10 m*M* HEPES-KOH, and 5 m*M* K_2ATP, osmolarity adjusted to 285 mOsm. All salts were purchased from Sigma-Aldrich.

3. Methods

3.1. Initiation of Culture

1. Prepare a 15-ml Falcon tube with 10 ml growth medium without geneticin. The vial containing the frozen cells has to be quickly thawed by gentle agitation in a 37 °C water bath. Immediately after defrosting the cell suspension, remove the vial from the water bath.
2. Wipe the vial with 70% alcohol and add the cell suspension to the prepared Falcon tube.
3. Gently mix the diluted cell suspension for 30 s to avoid local cell-toxic dimethyl sulfoxide (DMSO) concentrations.
4. Centrifuge the cells at 100 *g* for 5 min at room temperature.
5. Remove the supernatant by aspiration and add 12 ml of fresh growth medium.
6. Resuspend the cell pellet by gently pipeting up and down with a 10-ml serological pipette.
7. Transfer the cell suspension to a T-75 flask. Wear nitril gloves at all times.

3.2. Passage Conditions

1. LTK Kv1.5 cells should be passaged at a 70–90% confluence level (maximum density, 18 million cells/T-75 flask).
2. Remove the medium from the flask and wash cells with 10 ml DPBS without Ca^{2+} and Mg^{2+}.
3. Add 1.5 ml of prewarmed (37 °C) 0.25% Trypsin/1 m*M* EDTA solution, tilt the flask gently to uniformly coat the cells.
4. Incubate at 37 °C for 2 min until the cells detach from the bottom of the flask.
5. Hit the side of the flask to release remaining cells.
6. Add 10 ml of medium and dissociate cells by pipeting up and down with a 10-ml serological pipette.
7. Centrifuge cells for 5 min at 100 *g*.

8. Remove the supernatant and resuspend the cells in 10 ml fresh medium.
9. Prepare a new flask with 12 ml fresh growth medium and add the cell suspension.

Split ratios: 0.75–1 million cells/T-75 flask, 4 days of culture; 1.5–2 million cells/T-75 flask, 3 days of culture; and 3–4 million cells/T-75 flask, 2 days of culture (final cell density, 12–18 million cells/T-75 flask).

3.3. Working Culture

Induction of Kv1.5 channel expression:

1. Dilute the stock solution of dexamethasone (2 m*M* in 100% ethanol; Carl Roth GmbH & Co., Germany, cat. no. 5054) with growth medium to a final concentration of 0.1 μ*M* dexamethasone.
2. Replace the growth medium with 12 ml growth medium containing 0.1 μ*M* dexamethasone for 8–20 h before electrophysiological recordings.

3.4. Cell Preparation for the Flyscreen 8500 (Robot Cultures)

1. Prepare the cells for the robot right before use and transfer the cell suspension into the cell mixer immediately when finished.
2. Make sure the robot is turned on.
3. Defrost an aliquot of 0.05% Trypsin/0.53 m*M* EDTA solution (stored as 2-ml aliquots at −20 °C to guarantee a constant Trypsin activity) in a water bath for 3 min at 37 °C. After this prewarming step, vortex the Trypsin sample until remaining clumps of ice are completely dissolved.
4. Incubate the Trypsin solution sample for additional 3 min at 37 °C to equilibrate the solution to 37 °C.
5. Remove the medium from the flask of your working culture, and wash the cells twice with 10 ml HBSS without Ca^{2+} and Mg^{2+}. After the second washing step, remove HBSS thoroughly.
6. Add 1.6 ml of the prewarmed 0.05% Trypsin/0.53 m*M* EDTA solution, and tilt the flask to uniformly coat the cells.
7. Incubate cells for exactly 2 min at 37 °C in your cell incubator.
8. After the Trypsin incubation, hit the flask on the side sharply to detach the remaining cells.
9. Rapidly add 10 ml L-15 medium (with serum) to inactivate Trypsin.
10. Dissociate the cells consequently by pipeting the cell suspension four times up and down with a 10-ml serological pipette. To determine the total amount of harvested cells, count the cells with a standard cell counter.
11. Centrifuge the cell suspension for 5 min at 100 *g* at room temperature and carefully remove the supernatant.
12. Add an appropriate volume of L-15 Leibowitz medium (without serum) to adjust the cell suspension to 1 million cells/ml. Redissolve the cell pellet by gently pipetting the added L-15 medium four times up and down with a 10-ml serological pipette.
13. Transfer 2.5 ml of the cell suspension into the cell hotel.

3.5. Flip-the-Tip Technology

The key invention of flyion's technology was to invert the general principle of the patch-clamp technique (Lepple-Wienhues, et al., 2003). Rather than approaching a cell with a patch pipette, a few hundred to a thousand cells are automatically dispensed into the back of a FlipTip®, a standard borosilicate patch pipette glued into a plastic jacket (*see* **Fig. 1**).

The FlipTips are 100% quality-controlled and pulled to a pipette resistance of 0.9–1.2 MΩ. An in-house robot produces up to 1000 FlipTips per day. A single package of a FlipTip box holds 48 pieces and is shipped tight-sealed, resulting in a shelf-life of several weeks (*see* **Fig. 2**).

3.6. The Flyscreen® 8500

The main functional components are shown in **Fig. 3**: The tool head (1) consists of the gripper and two pipettes connected to the liquid handling system LHS0 and LHS1 of the robot (2). LHS0 drives washing solution (distilled water) through pipette P0, and LHS1 is connected to a bottle with ES and to pipette 1. The sole function of the gripper is to handle FlipTips in three-dimensional space. Pipette P0 prefills the inner chamber of the recording tip sockets (5) with IS and takes up individual cells from the cell hotel (7) to dispense them into the FlipTips. Pipette P1 prefills FlipTips in the FlipTip rack (6) with ES and applies compounds to the cell taken out from the compound rack (8). The Flyscreen Suite software communicates with the HEKA PatchMaster to

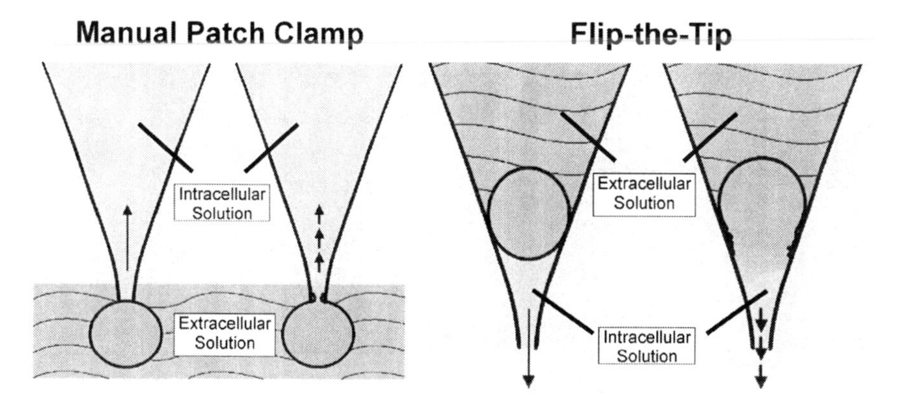

Fig. 1. The basic operation of the "Flip-the-Tip" technology. A cell suspension is dispensed into the open back of a standard patch-clamp pipette. Negative pressure of a few mbars forms the classical GigaOhm Seal and subsequent open whole-cell configuration, similar to the manual patch-clamp technique.

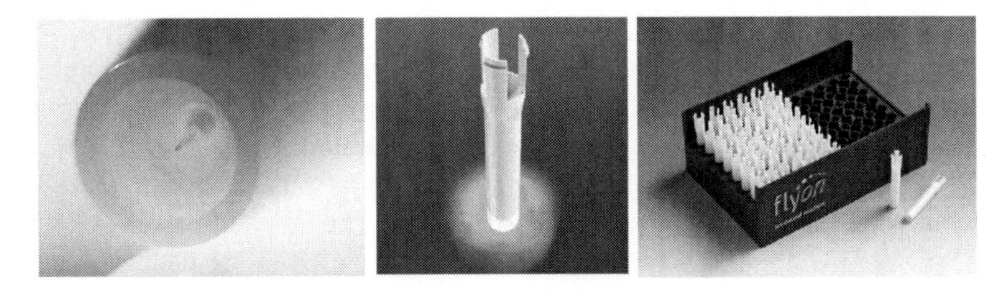

Fig. 2. A standard borosilicate patch-clamp pipette is glued into a plastic jacket, coined the FlipTip®. FlipTips are the recording consumables of the Flyscreen robot, and 48 pieces are packed and sealed tightly in a recording rack. Shelf life of a closed package is several weeks, and FlipTips should be used up the same day of opening.

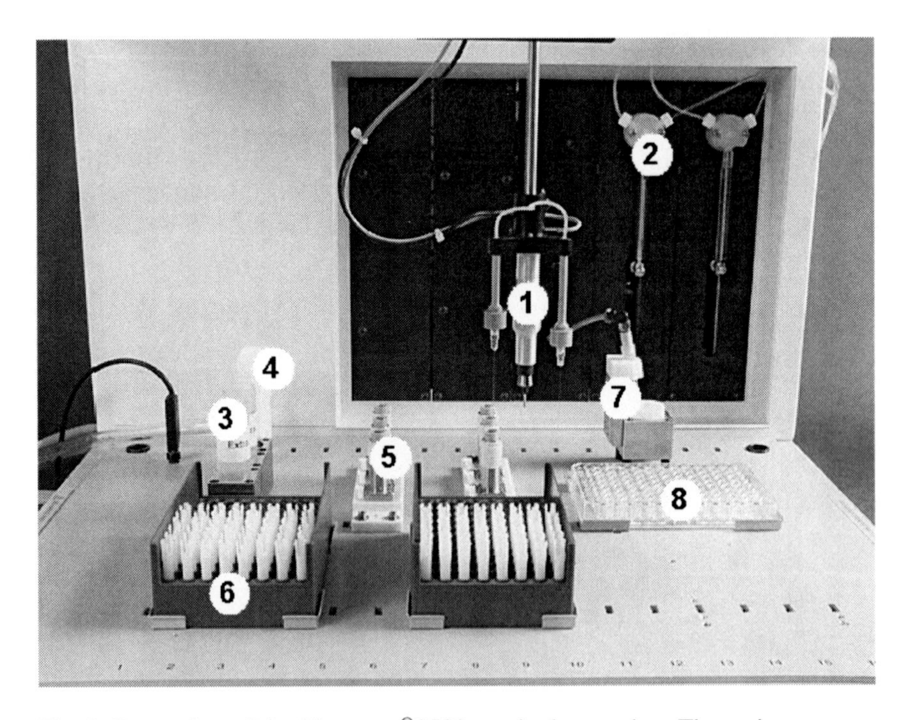

Fig. 3. Front view of the Flyscreen®8500 patch-clamp robot. The main components are numbered and described in detail in the text body. Shown is a six-channel model with 2 × 3 tip socket modules.

control an Electronic Patch Clamp (EPC 10) triple patch-clamp amplifier. In case of a six-channel system, two EPC 10/3 amplifiers are connected to the Flyscreen®8500 (Fejtl, et al., 2005).

As a first step, the inner chambers of the tip sockets are primed with IS by pipette P0. Next, FlipTips are prefilled with ES by pipette P1, taken up by the gripper and placed into the recording tip sockets. P0 fills the FlipTips with ES at the top until electrical contact is established. ES and IS for the recording sockets are provided in dedicated cups (3). After validating the resistance of the pipettes (R_{pip}), the Flyscreen robot dispenses 10 μl of a cell suspension into the open back of a FlipTip. Simply by gravity, the cells move toward the tip. Supported by continuous gentle suction of 8–10 mbars, a single cell finds its way into the very end of the FlipTip to form a 1–5 GΩ seal, similar to manual patch clamping. Brief and strong suction pulses of 100 mbar open the cell to establish the whole-cell configuration. Alternatively, a pore-forming agent such as amphoterecine B can be used to establish the classical perforated patch-clamp condition. In voltage-clamp mode, the response of an ion channel to a voltage step protocol and compound action is recorded (*see* **Fig. 4**). In between tasks, the two pipettes are individually rinsed and washed in an active washing station (4). The recording tip sockets are handled completely independent by

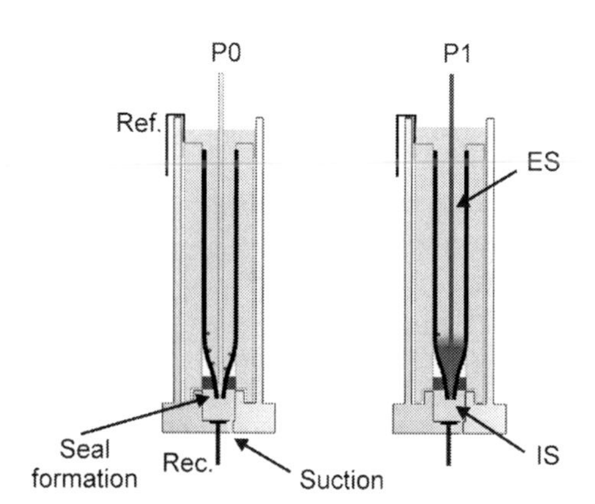

Fig. 4. Shown is a cross-section through a recording tip socket. The inner chamber is filled with intracellular solution (IS), and the FlipTip is filled with extracellular solution (ES). The pipette P0 dispenses cells into the FlipTip, and suction is applied through the suction line connected to the drain hole of the inner chamber. Current is measured between the recording electrode (Rec.) and the reference electrode (Ref.). After whole-cell formation, pipette P1 applies compounds directly onto the cell.

the system. Thus, a true parallel but asynchronous operation is achieved to optimize the throughput.

4. Operation

Assuming a permanent cell culture of the LTK Kv1.5 cells has been established and a working culture has been treated with dexamethasone to induce the Kv1.5 expression, the following steps lead to a successful automated screen with the Flyscreen robot. The described experiment here conducts a screen of the Kv1.5 channel and the application of the potassium channel blocker 4-aminopyridine (Sigma-Aldrich, cat. no. A-0152).

1. Put 1 l of distilled water and 250 ml of ES into a water bath at 45 °C for 45 min to completely degas them.
2. Turn on the robot, PC, monitors, and amplifier. Remove the FlipTips from the tip sockets and clean the tip sockets carefully with a cotton swap tipped in 70% ethanol. Apply a small amount of silicon grease around the very end and at the upper rim of a FlipTip, put it in every tip socket firmly and twist two to three times. This procedure lubricates the rubber O-rings of the tip sockets to completely isolate the inner chamber from the outer part of the FlipTip, separating the intracellular and extracellular compartments, respectively. Clear the hole at the bottom of the tip socket leading to the suction system with the provided metal tip for proper pneumatic access.
3. Start the Flyscreen software and calibrate the three instruments P0, P1, and the gripper.
4. Thaw 10 ml IS and put 5 ml into the dedicated cup. Place the cup into the metal socket labeled IC. Fill 15 ml ES into the dedicated cup. Place it into the metal socket labeled EC. Open a FlipTip box and put it into the FlipTip rack holder.
5. Connect the 1-l water bottle to the LHS0 system at the back of the robot. Fill 50 ml ES in a glass bottle and connect it to the LHS1 system at the back of the robot.
6. Use the "Start-Up" procedure in the Flyscreen software to prime the liquid handling system and the recording tip sockets. Further tasks of the Start-Up procedure are then performed automatically.
7. Prepare a 5 m*M* solution of 4-aminopyridine and dispense 250 μl in each well of a 96-well microtiter plate. Place it into the compound rack holder.
8. Prepare the cells for the robot as described and transfer 2.5 ml of the cell suspension into the cell hotel immediately after preparation. The blue pipette tip will then triturate the cells to avoid aggregation. Replace the cells every 2–3 h.
9. The main user interface of the Flyscreen software allows entry of basic information and a detailed description of the experiment to be performed. Enter description for ES and IS, cell suspension, number of experiments you want to accomplish, number of compound applications per experiment, number of available FlipTips, FlipTip box number and index, a project description, and most importantly, the

sequence file. Edit the compound dialog entry and enter correct volume and concentration of the solutions in use. Save this dialog as a batch file (*see* **Fig. 5**).

10. Press the start button.
11. Put 50 ml distilled water in a separate bottle. Place it in the 45 °C water bath for at least 45 min. The water is needed for the shut-down procedure.
12. After completion of the batch, start a new batch run or continue with the shut-down procedure described next.
13. Remove the cells from the cell hotel and clean the latter with distilled water and 70% ethanol. Dry it then with lint-free tissue and put it back upside down into the designated holder. Remove the IC and EC cups and put 5 ml of FACS Clean Solution (contains 1% free chloride; BD Bioscience, Germany, cat. no. 340345) into a new cup. Place this cup into the holder labeled "Clean." Connect the 50-ml water bottle to the LHS1 system at the back of the robot.
14. Start the "shut-down" procedure in the Flyscreen suite. The shut-down task rinses water through the liquid handling systems and cleans out the tip sockets. The FACS Clean solution chlorides the Ag wires in the inner compartment and at the top of the tip sockets.
15. Connect a 1-l water bottle to the washing station. Insert unused FlipTips into the tip sockets to prevent dust from entering. Empty the waste bottle. Close the

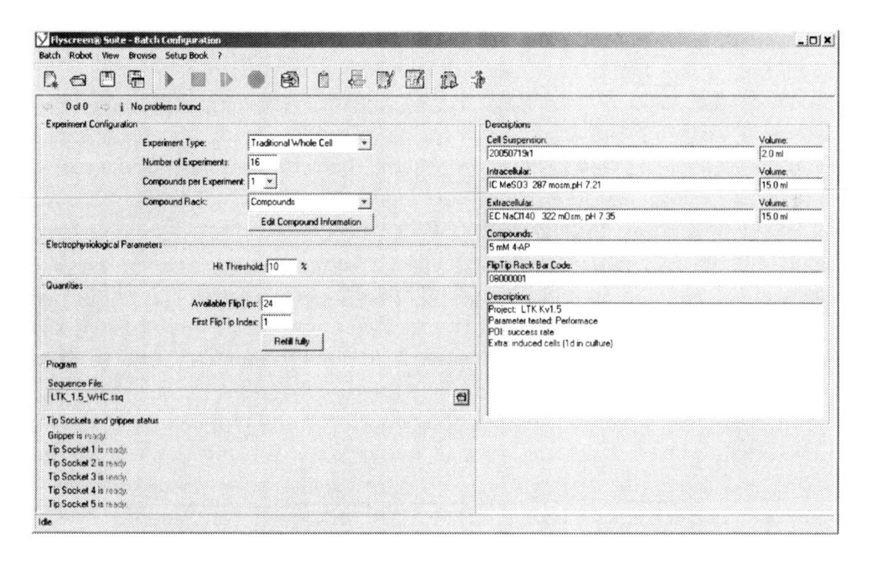

Fig. 5. Flyscreen screenshot of the main dialog window. Main entries are depicted and are self-explanatory. The hit threshold is used to label a successful experiment as a hit or not. A block of an ion channel response by a compound of at least 90% leads to a hit definition in this case.

Flyscreen suite software first before shutting down the robot, the amplifier, the PC, and monitors. Put the dust cover over the robot.

4.1. Data Analysis

In general, data are stored in a MySQL database on an FTP server to be installed either locally on the PC performing the operation or on a network server. Furthermore, several Flyscreen robots can be connected to the same database. This allows the user to analyze the results basically from any PC where the Flyscreen analysis software is installed.

1. Open the Flyscreen suite. Access the batch browser view. Select a batch you want to analyze from the database (*see* **Fig. 6**). The information you see contains all the relevant entries you did while editing the Flyscreen main user interface. Simply click on the batch and the data will be imported from the FTP server into the Flyscreen suite.
2. The tip socket view is a graphical representation of the segment templates executed on each tip socket at a particular time point, similar to a flow chart reader. Double click on any segment you would like to inspect and analyze, that is, the indicated segment template for "multiple compound measurement." This action replays the raw data in the PatchMaster window and shows stored online analysis results in the online analysis window. Inspection of individual traces is done by clicking on the respected trace number. The associated biophysical properties of the cell during this particular measurement are then displayed (*see* **Fig. 7**).
3. Screen all the segment templates named "control measurement" and "multiple compound measurement" of the recorded batch to include all relevant results into a single online analysis view. This is, in general, the peak amplitude of the current but could be any other analysis function provided by the HEKA PatchMaster software, that is, slope or minimum/maximum values (*see* **Fig. 8**).
4. Export all the data points contained in an online analysis window as character separated values (CSV) simply by a single mouse click. Hence with a single action, the data of many recorded cells can be exported at once. Use a third party software (or provided Microsoft® EXCEL macros) to import the CSV and to construct dose–response curves.
5. FlyStat, a Microsoft Access-based toolbox developed and provided by flyion, allows to immediately extract success rates of one or several batches. Open the FlyStat toolbox and select a date range or a specific batch number. Valid FlipTips with regard to a correct pipette resistance, seal and whole-cell rate, and duration of the experiment are given in percentage when you press the "Statistics Table" button. Press the "Export Runs" button for a detailed inspection of the complete experiment with respect to seal resistance, CFast and CSlow values, reason why an experiment was terminated, and to address the most important parameters. Press the "Export Traces" button for an ASCII export of the parameter of interest, starting from the control segment. The time 0 flag indicates that the 0 time base is calculated

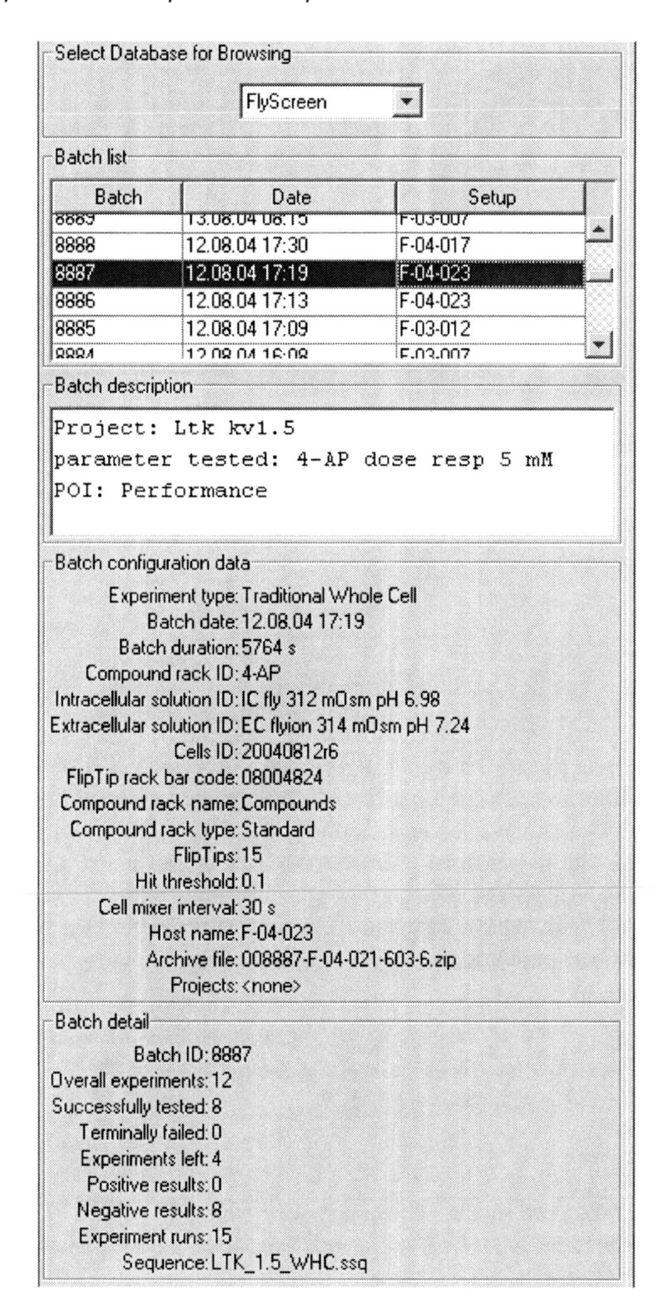

Fig. 6. Database browsing of saved recording batches. Clicking on a batch number (8887) unzips the data from the MySQL database and loads it into the Flyscreen suite. Information that has been edited in the main dialog window linked to this batch is displayed.

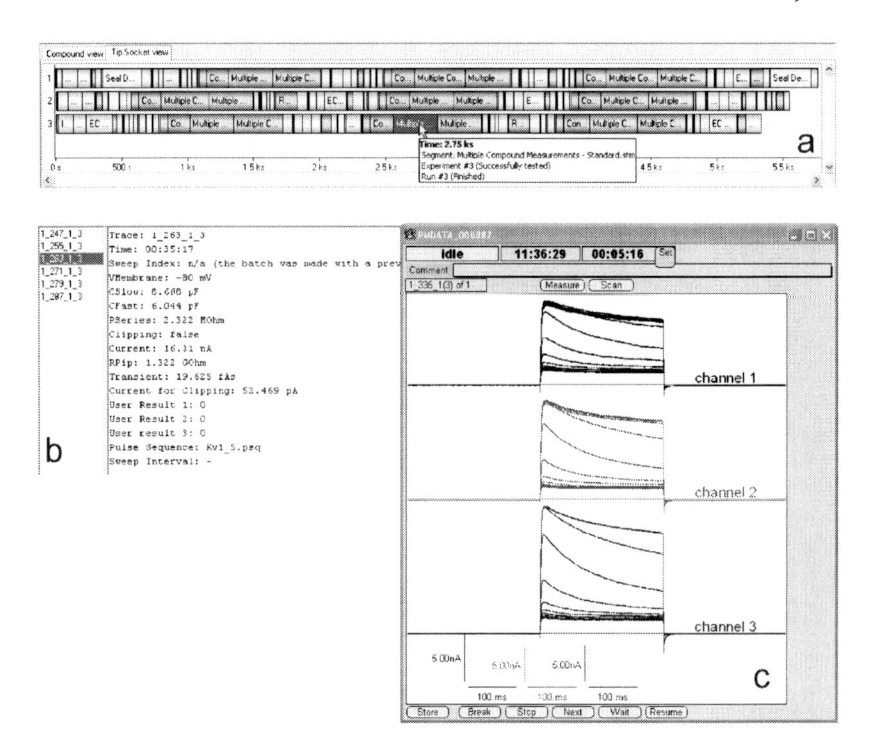

Fig. 7. Tip socket view of the Flyscreen suite. A graphical representation of the complete batch is shown, depicting eight successful experiments out of 12. This represents a success rate of 67% (**A**). A click on an individual trace number reveals the biophysical properties of this cell. In particular, a low SeriesResistance of 2.3 MΩ and a SealResistance of 1.3 GΩ were achieved, proving the high data quality of the Flyscreen system (**B**). The PatchMaster window shows data traces of a successive block of the Kv1.5 potassium channel by 5 m*M* 4-aminopyridine simultaneously on three recording sockets (**C**).

according to the time of the compound application. This allows an overlay of all successful experimental runs to check the time course of the current of interest during compound application (*see* **Fig. 9**).

5. Notes

1. Thawing of the cells should be accomplished within 40–60 s.
2. Permanent cultures of the LTK Kv1.5 cell line should be established approximately 2 weeks prior to a working culture.
3. Do not allow the cells to settle down in the flask, because they will eventually clump.
4. Do not pipet solutions directly onto the cell layer.

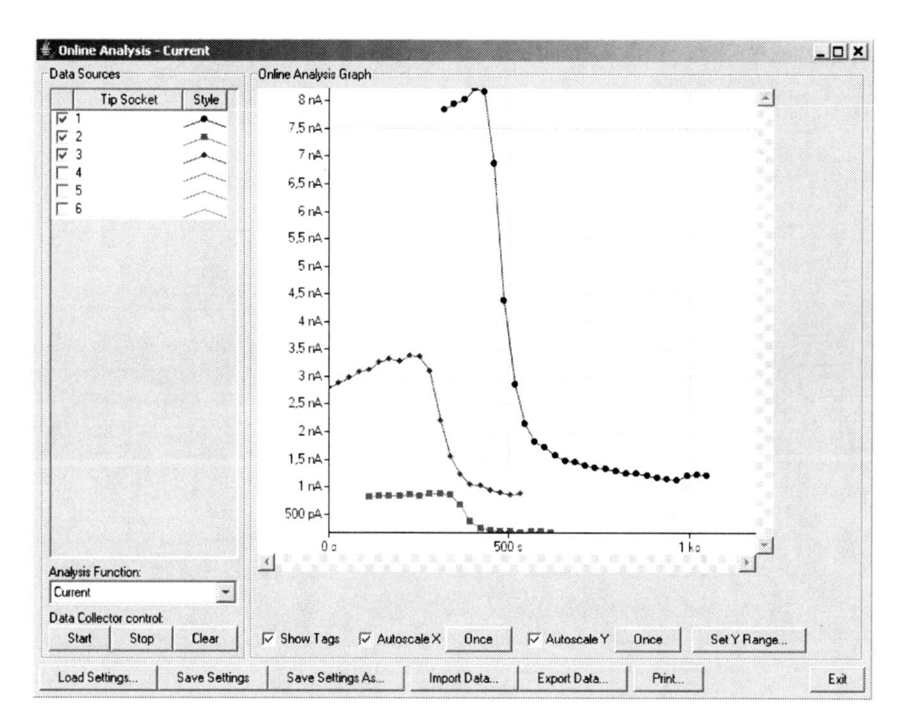

Fig. 8. An analysis window is shown and depicts the current amplitude of a Kv1.5 channel recorded on three tip sockets at the same time. The data of a control segment and a compound application segment are shown, revealing the time course of the blockade by 5 m*M* 4-aminopyridine. Clicking on the "Export Data" button stores the values in character separated value (CSV) format. Several analysis windows can be used simultaneously in online and offline mode to show more than one distinct parameter.

5. Avoid air bubbles while pipeting the cells in the culture flask.
6. Culture medium should be renewed every 2–3 days and cells kept in a humidified incubator at 37 ± 2°C and 8.5% CO_2.
7. Make sure that all solutions are filtered and completely degassed. Check the operation of the dilutor syringes during the "Start-Up" procedure. Remove bubbles either by opening the outgoing tubing valves during the operation or by flushing through a few milliliters of 70% ethanol. Use the washing procedures for the respective pipettes P0 or P1 in case ethanol was used to remove air bubbles.
8. Open a FlipTip box only at the day of usage. Similar to conventional glass pipettes, the FlipTips should be used the same day when the box is opened. The shelf life of a closed FlipTip box is several weeks.
9. Make sure that the FlipTip box and the compound rack are inserted properly into the respective holders of the robot. Otherwise, the gripper may crash into a flip tip or a needle may break because of nonalignment.

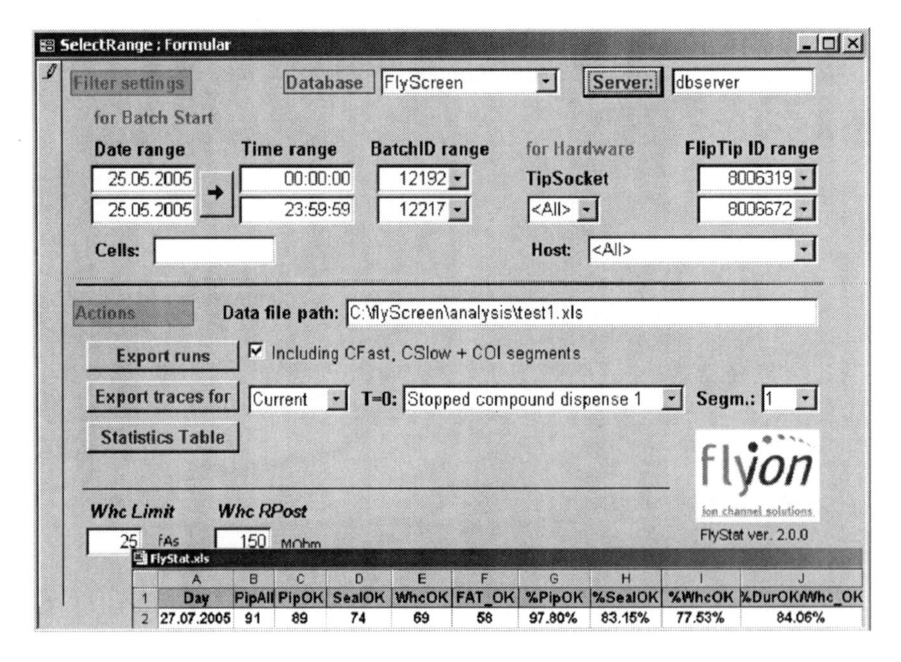

Fig. 9. The Microsoft Access®-based FlyStat toolbox allows for an extensive analysis of several batches at once. An example of a single day summarizing two batches is shown. Ninety-one FlipTips were used, and 89 of them were recognized as OK (97.8%). In this case, a total recording time of 10 min wasss set for a factory acceptance test (FAT), and 58 tries were successful.

10. Replace pipette P0 and P1 on a regular basis. We recommend to change the pipettes at least twice a week. This will ensure best performance.
11. The Sequence file holds the entire outline of a patch-clamp experiment and information about the ion channel to be screened by calling downstream files such as Segment Templates, Pulse Files, and Online Analysis Files (*see* **Fig. 10**). Predefined Segment Templates control operations of the robot, such as pump movements, suction, washing cycles, volume of compound application, and are used to compile a customized Sequence file. The Pulse File defines the actual voltage pulse pattern to evoke the ionic current. The Online Analysis File extracts the defined parameter(s) of interest. The user can modify the content and hence can customize the predefined library of sequences, specific for every ion channel to be studied.
12. Flexible and fixed timing markers are implemented in a sequence for the internal scheduler of the Flyscreen suite. Because a single robot arm is serving three to six tip sockets, an intelligent scheduler optimizes the overall performance. For example, given that a total of 5 min has been set by the user to achieve the

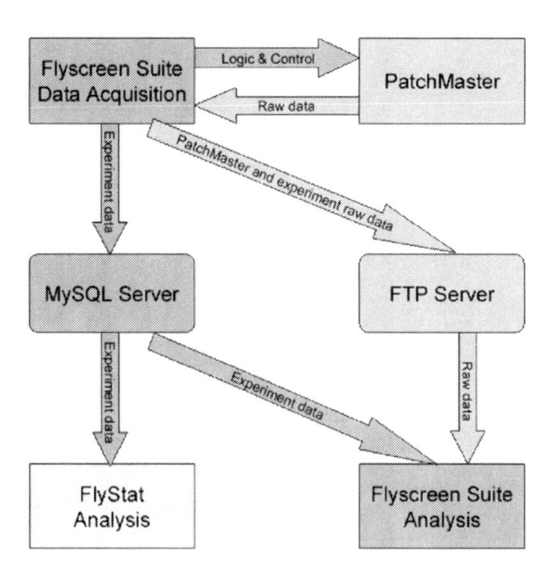

Fig. 10. Schematic representation of the software structure, showing the data handling among the Flyscreen suite, HEKA PatchMaster, the MySQL database, and FlyStat.

GigaSeal state, a cell obviously may reach this state earlier or not. Thus, has a user-set threshold been met before the user limit, the scheduler realizes this event and shifts all following segments forward in time to avoid deadtime. Hence, by building a sequence, the user should consider flexible markers to efficiently use the internal scheduler of the system.

References

1. Lepple-Wienhues A, Ferlinz K, Seeger A, and Schafer A. (2003) Flip the tip: an automated, high quality, cost-effective patch clamp screen. *Recept. Channels* 9(1):13–17.
2. Fejtl, M, Czubayko U, Hümmer A, Krauter T, and Lepple-Wienhues A (2005) Automating true manual patch-clamping. *Genet. Eng. News* 25(14): 37–43.

6

The Roboocyte

Automated Electrophysiology Based on Xenopus *Oocytes*

Christine Leisgen, Mike Kuester, and Christoph Methfessel

Summary

Automated electrophysiological assays are of great importance for modern drug discovery, and various approaches have been developed into practical devices. Here, we describe the automation of two-electrode voltage-clamp (TEVC) recording from *Xenopus* oocytes using the Roboocyte automated workstation, jointly developed by Multi Channel Systems and Bayer Technology Services. We briefly discuss the technology, including its advantages and limitations relative to patch clamp and other TEVC systems. We provide a step-by-step description of typical operating procedures and show that the Roboocyte represents a practical and highly effective way to perform automated electrophysiology in an industrial setting.

Key Words: *Xenopus* oocytes; mRNA expression cDNA expression; two-electrode voltage clamp; electrophysiology; membrane receptors; ion channels; screening; ligand gated; voltage gated; transporters.

1. Introduction

1.1. Automated Electrophysiological Assays

Membrane-based receptors and ion channels are among the most important drug targets, and direct electrophysiological measurement of transmembrane ion currents is the method of choice for analyzing the effects of potential drugs on transmembrane receptors. Therefore, a number of automated techniques have been envisioned, using either pipettes or planar substrates, to automate

From: *Methods in Molecular Biology, vol. 288: Patch-Clamp Methods and Protocols*
Edited by: P. Molnar and J. J. Hickman © Humana Press Inc., Totowa, NJ

patch-clamp recording from cultured cells that heterologously express a target ion channel.

Alternatively, cloned membrane receptors and ion channels can also be studied in *Xenopus* oocytes, a widely used heterologous expression system that is very convenient and well suited for automated procedures.

1.2. The Xenopus Oocyte Expression System

Xenopus oocytes are the immature egg cells of the tropical clawed frog *Xenopus laevis*, which have a long history in studies of oogenesis, fertilization, morphogenesis, and embryonic development. In the 1970s, it was found that oocytes faithfully translate practically any messenger RNA (mRNA) that is injected into their cytoplasm ***(1–3)***. Later, it was discovered that expression of a target protein could also be achieved by nuclear microinjection of complementary DNA (cDNA) with an appropriate promoter ***(4,5)***.

The *Xenopus* oocyte is a heterologous system for highly efficient transcription, translation, and processing of proteins. As true eukaryotic cells, oocytes not only produce the correct peptide sequence, but also perform all the post-translational modifications required to obtain the mature and functional protein, including folding, glycosylation, assembly of subunits, and—in the case of membrane proteins—correct membrane insertion. *Xenopus* oocytes are very large cells (more than 1 mm in diameter), which allow easy handling, injection with cDNA/mRNA, and recording with the two-electrode voltage-clamp (TEVC) method.

TEVC recording dates back more than 50 years when it was introduced by Hodgkin, Huxley, and Katz ***(6)*** in their pioneering studies of signal propagation in axons and the synapse. Compared with patch clamping, the method may seem crude: Instead of pressing the recording pipette against the cell membrane to form a gigaseal, the microelectrodes are simply driven into the cell. For this reason, TEVC works well only for large, robust cells such as the squid axon, skeletal muscle fibers, or, in this case, the *Xenopus* oocyte. Yet, it is a very effective and reliable method, providing high-quality data with comparatively simple equipment.

TEVC recording from oocytes is straightforward to automate, mainly because the cells and their current responses are so large. The cell position can be precisely defined in well plates with conical bottoms. Decisions that need to be made by the experimenter are usually straightforward and easily cast into an algorithmic form.

1.3. Description of the Roboocyte System

The Roboocyte system has been developed by Multi Channel Systems MCS GmbH, Germany in collaboration with Bayer AG, Germany. A prototype was used very successfully since the 1990s in the laboratories of Bayer AG and at the University of Geneva *(7)*. The first commercial version was brought to the market in 2002 and further improved in the following years.

The Roboocyte is a fully automated platform for electrophysiological recording from ion channels or other membrane receptors based on the standard *Xenopus* expression system *(8,9)*. It is highly suitable for functional secondary screening of drug targets. Oocytes are placed into standard 96-well plates with V-shaped wells, where the conical shape causes each oocyte to roll to the center of the well bottom, ensuring that its location is precisely defined *(7)*. The Roboocyte performs an automated injection of cDNA/mRNA into each *Xenopus* oocyte and an automated voltage-clamp recording after the successful expression of the target receptor, sequentially on each well of the 96-well plate. A ready-made measuring head comprises two glass microelectrodes, two silver/silver chloride (Ag/AgCl) reference electrodes, and a perfusion inlet and outlet. It can be reused several times (approximately 1 week of daily use) and allows effortless electrode replacement (*see* **Fig. 1**).

The integrated gravity flow perfusion system is ideal for drug receptor characterizations and quick expression tests. For larger numbers of test solutions, the Roboocyte has optionally integrated an industry-standard liquid handling station from Gilson Inc., USA, which can hold up to 384 different test solutions in deep well plates or glass vials.

Various injection or recording protocols for different receptors and applications can be defined in scripts—small text files written in the dedicated Roboocyte Scripting Language (RSL). This allows flexible recording protocols, including voltage jumps, drug applications, and wash cycles. Based on preselected parameters, the program decides whether a given cell is (still) suitable for further recordings, and, if not, the Roboocyte simply proceeds to the next oocyte in the well plate to continue the measurements. In this way, the system can operate continuously without any human intervention until either all test solutions are processed or all cells contained within a 96-well plate are used up.

It is worthwhile to point out some features of the proprietary voltage-clamp amplifier developed for the Roboocyte. It is very compact and completely integrated into the small-sized robot, and it operates completely under computer control. The integrated ClampAmp is the only digital TEVC amplifier in the

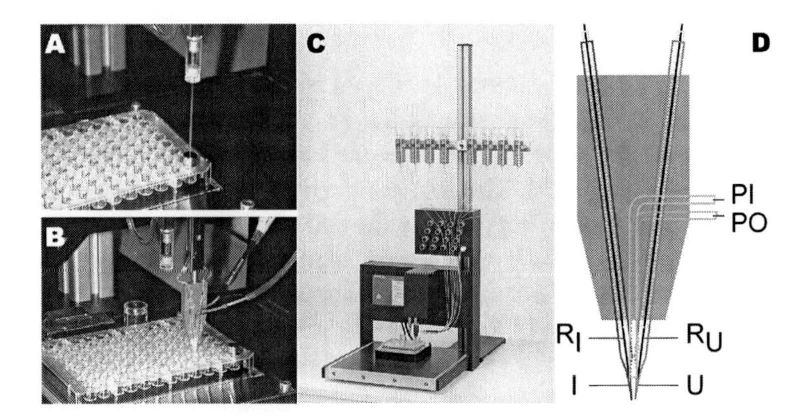

Fig. 1. The Roboocyte. The well plate with 96 oocytes is mounted on the robot's carrier plate. The Roboocyte fully automatically injects the nucleic acid coding for the target ion channel (**A**) and performs the two-electrode voltage-clamp (TEVC) experiment (**B**) as specified by the user-written script. **C** shows the compact main robot (without aspiration pump and optional liquid handler). The preassembled measuring head (**D**) holds two intracellular glass microelectrodes (I, current electrode and U, voltage electrode), two Ag/AgCl wires used as current drain and voltage reference electrodes (R_I and R_U), and two PTFE tubes for the perfusion inlet and outlet (PI and PO, respectively). The PI is connected either to the pinch valve system by a manifold or to the external liquid handler (not shown). A suction pump aspirates the fluid through the outlet and leads it to the waste receptacle (not shown).

market today. It operates either in current-clamp or voltage-clamp mode. It uses a proportional integral (PI)-controlled feedback loop for controlling the current injection and thus establishing the voltage clamp. As with other TEVC amplifiers, one of the two intracellular Ag/AgCl glass micropipettes filled with electrolyte measures the voltage, and the other electrode injects a current proportional to the difference between the measured and the setpoint potentials. Two separate bath electrodes are employed to ensure that, even at large currents, the membrane potential measurement is not affected by an offset.

The dynamic range is $\pm 32\,\mu A$ at 1 nA resolution. The Clamp Amp version 2 is based on the latest microcontroller technology. With a sampling rate of 10 kHz and an overall rise time (10–90%) of $500\,\mu s$, this amplifier supports recording from all kinds of ion channels that can be heterologously expressed in *Xenopus* oocytes, including fast sodium channels, within the limits set by the oocytes membrane capacitance and access resistances.

1.4. Comparison With Other Electrophysiological Techniques

There are two major electrophysiological methods that are standard for analyzing cloned ion channels in vitro: TEVC on *Xenopus* oocytes and patch clamp.

Manual patch clamp is considered the gold standard of electrophysiological assays but has the obvious disadvantages that it requires a high expertise and has a very low throughput. Automated patch-clamp systems can achieve approximately the same throughput as the Roboocyte but do not reach the high technical standard of manual patch clamp. For example, presently available patch-clamp robots do not support all variants of the method and cannot record from native cells and from single channels. Not all machines support the formation of gigaseals required for a good experimental and voltage-clamp control and a biophysical profiling of the receptor, or a fast and continuous laminar buffer flow that is required for the recording from ligand-gated ion channels. Another disadvantage is that automated patch-clamp recording is invariably performed on small, cultured cells and therefore depends on the existence of a suitable cell line that can be easily kept in suspension.

The *Xenopus* oocyte expression system presents a very effective and flexible alternative for recording from all classes of ion channels. The mRNA or cDNA to be expressed is injected directly into the cells a few days before the planned experiment. Oocyte recording is especially favorable whenever a large number of different targets must be investigated, for example, when the selectivity of compounds must be optimized for a number of related receptors from a large family. The culture conditions for oocytes do not require special equipment or knowledge and are always the same, regardless of the expressed receptor. Please see also Note 8.

Despite these advantages of *Xenopus* oocytes, recording with conventional manual TEVC setups, similar to manual patch-clamp recording, is very laborious and time consuming. For example, each oocyte must be placed into the recording chamber; the microelectrodes must be driven into the cell under microscopic control; and the complete experiment, voltage step, and perfusion protocols must be executed under the control of a skilled experimenter.

The OpusXpress from Molecular Devices, USA, using a different approach than the Roboocyte, attempts to compensate for this drawback by multiplying the number of experiments and automating the exchange of test solutions, so that one experimenter can record from eight oocytes in parallel. This parallelization increases the throughput, but in our opinion, this is not yet a fully automated system. The experimenter still needs to supervise the experiments and must manually replace the oocytes in the recording chambers. Moreover, nonviable or nonexpressing oocytes directly affect the throughput. In a sequentially operating machine, like the Roboocyte, unsuitable oocytes can be skipped with virtually no time loss.

The Roboocyte is the only fully automated system in the market that can be operated with only a minimum of setup and hands-on time required and without user intervention and supervision, over day and night. The Roboocyte increases the throughput at least 10-fold compared with a manual TEVC setup, but its main advantage is that the robot completely relieves the user of the time-consuming routine work. This makes the Roboocyte a method suitable for secondary functional screening of compound libraries on a great variety of membrane-based receptors and ion channels.

In summary, the main benefits of the Roboocyte are as follows: the short setup time of the system in the lab, the very low expenses for consumables and operation, and the very short hands-on time because of the full automation. It allows researchers to focus on the experimental design and evaluation of data. The versatility of the system and the bandwidth of applications are unmatched—almost all channel and receptor types can be analyzed with the *Xenopus* expression system. The expression strength and thus the current response amplitude can be modulated to a certain extent by the amount of mRNA that is injected. Stable current and voltage clamps and high current responses (in the range of μA, i.e., three orders of magnitude higher than with patch clamp) lead to very reliable and reproducible results that can be easily analyzed with standard algorithms.

1.5. Typical Applications of the Roboocyte System

Here, we briefly discuss some major classes of membrane-based proteins that can be investigated with the Roboocyte. This discussion cannot be in any way complete, simply because, from 1982 until today, many hundreds of different membrane-based receptors have been studied in oocytes including membrane proteins from invertebrates ***(7)*** and even plants ***(10)***. Today, it cannot be overstated that the expression of membrane receptors in *Xenopus* oocytes with subsequent voltage-clamp recording is one of the best-established "workhorse" methods in receptor research and drug discovery.

1.6. Voltage-Gated Ion Channels

Numerous kinds of voltage-gated ion channels have been expressed and assayed in the oocyte-expression system, including Na^+ channels, many K^+ channels, various Ca^{++} channels, chloride channels, connexins, and many others. The Roboocyte can apply both voltage steps and ramps as specified in appropriate script files and is therefore easily adapted in a completely flexible way to the requirements of different targets. There are very few endogenous voltage-gated channels in the oocyte, and most of these are activated only slowly and after very long stimulation; so, they usually do not interfere with the measurements on expressed ion channels.

One issue with very rapidly acting, voltage-gated channels, like axonal Na^+ channels, in oocytes is the large membrane capacitance that limits the time resolution for large voltage steps to approximately 1 ms. The ClampAmp version 2 has overcome this limitation.

1.7. Ligand-Gated Ion Channels

The measurement of ligand-gated ion channels in *Xenopus* oocytes has a long tradition ***(2,3)***, and the first electrical recordings from cloned ion channels were performed in oocytes ***(11)***. Subsequently, many different kinds of receptors have been studied in this system. It was only later that the transfection of cell lines with ion channels was successful ***(12)*** and became the standard method that it is today.

An advantage of oocyte recording with the Roboocyte is the very large dynamic range, from a resolution limit of 1 nA to the maximum recording capability of 32 μA. This means that the dose–response behavior of a ligand-gated ion channel can be followed over four orders of magnitude and reliably measured at very small ligand concentrations that may not even give a measurable signal in patch-clamp experiments. **Figure 2** shows representative data for an alpha-7-type nicotinic acetylcholine receptor (α7 nAChR), a target that is challenging to measure because of the receptor's very fast desensitization at high agonist concentrations.

1.8. Second-Messenger Coupled Receptors

G protein-coupled receptors (GPCRs), also called 7-transmembrane segment (7TM) or metabotropic receptors, present an extremely important and varied class of drug targets. Therefore, it is very useful that these, too, can easily be expressed in *Xenopus* oocytes.

The activation of an expressed GPCR of the Gq class leads to the activation of an endogenous signal pathway, culminating in the intracellular release of inositol triphosphate (IP3) and of calcium ions. Because *Xenopus* oocytes already have an endogenous calcium-dependent chloride channel, this increase in free calcium activates a strong chloride current, often in an oscillatory manner ***(13)***. Therefore, the expression and activation of Gq-coupled GPCRs can be directly monitored in a TEVC experiment with the Roboocyte.

Other GPCRs of the Gs or Gi classes act through adenylate cyclase that does not lead to any electrical signal in a normal oocyte. Therefore, it is necessary to

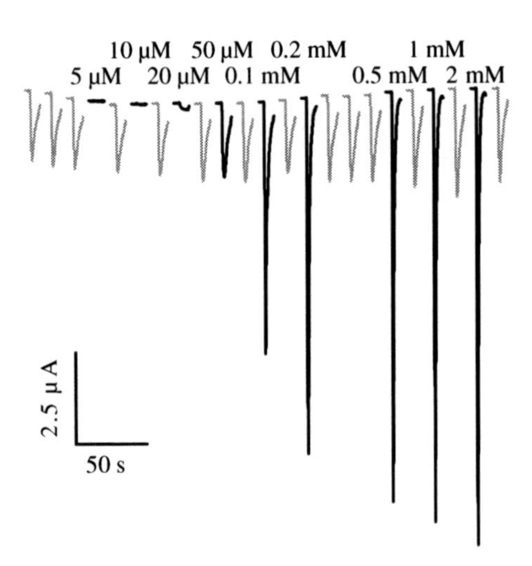

Fig. 2. Ligand-gated ion channel. Typical ion currents recorded from an oocyte expressing an alpha-7 nicotinic receptor at −80 mV. The test compound was acetylcholine (ACh) at the concentrations indicated, and each application was preceded by at least one reference application of 50 µ*M* ACh. The test solutions were applied for 5 s and washed out for 5 min between applications.

coinject another gene that leads to the expression of a suitable effector protein together with the target GPCR. For example, G protein-activated, inward-rectifying (GIRK)-type potassium channels interact with the adenylate cyclase pathway and are modulated by the activation or inhibition of a co-expressed GPCR. In this way, this class of GPCRs can also be studied very effectively in oocytes ***(14)*** with the TEVC method.

It is important to note that, in contrast to ligand-gated or voltage-gated ion channels where the measured current is directly proportional to the opening probability of the channels, such a relationship does not necessarily hold for the activation of currents through a second-messenger pathway. Instead, the dose–response relationships can be nonlinear, as a threshold concentration of the internal messenger is reached and exceeded. Also, various starting levels of intracellular components can also affect the results and the sensitivity of the assay in an unpredictable manner. Therefore, although activation and inhibition of a GPCR is easy to document in oocytes, obtaining a reliable, quantitative dose–response curve can be quite laborious and require many control experiments. Here, it is extremely helpful to employ a robotic system, like the Roboocyte.

1.9. Transporters

Active transporters are another very important class of pharmaceutically relevant membrane proteins, and these also can be expressed in *Xenopus* oocytes. It is clear that a direct electrical signal can be expected only for electrogenic transporters, whose activity results in a net transfer of electrical charge across the plasma membrane. Where a typical ion channel passes millions of ions per second, the exchange rates of transporters are usually much lower. Therefore, from an oocyte expressing a transporter, current signals of only a few nA are to be expected. However, because of the large dynamic range of the oocyte recording system, even these very small currents can also be resolved satisfactorily. **Figure 3** shows an example.

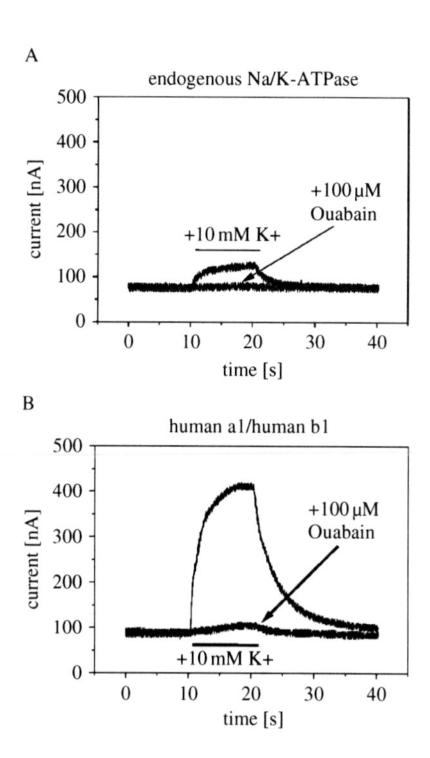

Fig. 3. Recordings from a Na/K ATPase. (**A**) Recording from a noninjected oocyte, showing an endogenous ATPase, which is blocked by Ouabain. (**B**) Recording from an oocyte expressing a human $\alpha 1/\beta 1$ Na/K ATPase, showing substantially larger currents. Note that even the small endogenous currents of only 50 nA can be well resolved with the Roboocyte's digital TEVC amplifier. We thank Dr. Maarten Ruitenberg of IonGate Biosciences GmbH, Frankfurt, Germany, for providing this illustration.

1.10. Other Targets

This brief summary has focused on potential drug targets that, directly or indirectly, produce an electrical response that can be studied with the TEVC method, so that the Roboocyte technology with its TEVC recording mode is directly applicable to all aspects of the experimental procedures. Of course, other proteins that are not electrogenic can also be expressed in oocytes following the injection of mRNA or cDNA, but these must subsequently be studied using biochemical, tracer-flux, or other appropriate methods. Even then, however, the automated oocyte injection and handling procedures made possible by the Roboocyte can significantly improve the throughput and efficiency of such studies.

2. Materials

2.1. Preparation of Xenopus Oocytes

1. Female frogs of *X. laevis*.
2. Shaker holding 50-ml Falcon tubes for enzymatic treatment (Vari-Mix Aliquot Mixer, Type 48700 from Barnstead International, USA).
3. Stereo microscope for quality control of oocytes (Olympus SZH Zoom Microscope from Olympus, Japan).
4. 8-/12-channel pipette or Columbus Microplate Washer (Tecan, Tecan Trading AG, Switzerland) for filling the well plates and washing the oocytes.
5. 96-well plates (nontreated, nonsterile polystyrene, conical bottom, from Nunc, Nalge Nunc International, USA, cat. no. 249570 or from Greiner, Greiner-Bio-One GmbH, Germany, cat. no. 651101).
6. Sieve for grading oocytes by size (a cut-open 50-ml Falcon tube with 800-μm polyamide mesh glued over one end).
7. Oocyte transfer pipette (a cut and fire-polished glass Pasteur pipette with an opening of 1.5 mm).
8. Glass tool for moving oocytes (a glass Pasteur pipette, sealed and pulled over a Bunsen burner, and melted into a golf club-like shape).
9. Petri or cell-culture dishes (100 mm) for sorting and storing oocytes.
10. Petri dishes (60 mm).
11. Beaker (100 ml).
12. Razor blade.
13. Forceps.
14. Parafilm.
15. Collagenase NB8 Broad Range lyophilized (from Cl. histolyticum, >0.9 U/mg PZ activity (Wünsch), <0.2 U/mg Neutral Protease (DMC), product no. 17456 from SERVA Electrophoresis GmbH Germany).
16. Fresh 1.5–2 mg/ml collagenase in Barth's solution without calcium.

17. Gentamicin sulfate salt (potency approximately 600 μg Gentamicin per mg) cat. no. G3632 from Sigma-Aldrich, USA.
18. Gentamicin stock solution (50 mg/ml gentamicin in Barth's solution), stored in 1-ml aliquots at −20 °C.
19. Barth's solution: 88 m*M* NaCl, 2.4 m*M* $NaHCO_3$, 1 m*M* KCl, 0.33 m*M* $Ca(NO_3)_2$, 0.41 m*M* $CaCl_2$, 0.82 m*M* $MgSO_4$, and 5 m*M* Tris/HCl, pH 7.4 with NaOH.
20. Freshly prepared gentamicin working solution: 1 ml of gentamicin stock diluted in 1 l Barth's solution. Alternatively to gentamicin, a mixture of penicillin and streptomycin can also be used to suppress microorganisms.
21. Barth's solution without Ca^{2+}: 88 m*M* NaCl, 2.4 m*M* $NaHCO_3$, 1 m*M* KCl, 0.82 m*M* $MgSO_4$, and 5 m*M* Tris/HCl, pH 7.4 with NaOH.

2.2. cDNA/mRNA Injection

1. Freshly prepared *X. laevis* oocytes as previously described in section 2.1. Note that oocytes already prepared in standard 96-well plates suitable for injection with the Roboocyte are also available from commercial suppliers, for example, from EcoCyte Bioscience, Castrop-Rauxel, Germany, *see* **Note 5**. For cDNA injection, it is important that all oocytes are oriented with the animal pole (brown side) up.
2. Stereo microscope.
3. Vortex mixer.
4. Eppendorf bench top centrifuge.
5. Roboocyte alignment device (with crosshairs).
6. Injection pipettes with a tip size around 5–8 μm (from Multi Channel Systems MCS GmbH).
7. Eppendorf Microloader tips (cat. no. 5242 956.003) for backfilling the injection pipettes.
8. 30–150 ng of cDNA with a typical concentration of 10–50 ng/μl, or 100–750 ng of mRNA with a typical concentration of 50–300 ng/μl, dissolved in distilled water and stored in sterile 0.5- or 1.5- ml Eppendorf tubes at −20 °C until use. For mRNA solutions, special RNase-free water must be used.

2.3. TEVC Recording and Test Compound Application

1. *Xenopus* oocytes plated in standard 96-well plates expressing the target receptor.
2. Stereo microscope.
3. Roboocyte alignment device (with crosshairs).
4. Measuring head (product no. MH) from Multi Channel Systems MCS GmbH.
5. Eppendorf Microloader tips (cat. no. 5242 956.003) for backfilling the micropipettes.
6. Electrolyte solution (e.g., 1.5 *M* K-acetate with 1 *M* KCl, pH 7.2 or 3 *M* KCl).
7. Frog Ringer's solution or equivalent bath solution (115 m*M* NaCl, 2.5 m*M* KCl, 1.8 m*M* $CaCl_2$, and 10 m*M* HEPES, pH 7.2/osmolarity: 240 mOsm/kg).
8. Test compounds dissolved in bath solution.

3. Methods

3.1. *Preparation of* Xenopus *Oocytes*

3.1.1. Oocyte Removal and Defolliculation

1. Remove the appropriate amount of ovarian tissue surgically from one side of an anesthetized frog. Please refer to standard protocols on this subject ***(15)*** (*see* **Note 1**).
2. Transfer the ovarian lobes into a new large Petri or cell-culture dish (e.g., 100-mm Falcon) filled with Barth's solution without Ca^{2+}.
3. Divide the tissue into smaller pieces (approximately 0.5 cm) with a razor blade and forceps.
4. Transfer the clumps into 50-ml Falcon tubes containing 40 ml of 1.5–2.0 mg/ml collagenase in Barth's without Ca^{2+}. A volume of up to 7.5 ml of tissue can be put into a single tube (*see* **Notes 2** and **3**). For more tissue, use additional tubes. Otherwise, the collagenase digest would take too much time.
5. Place the tubes onto the shaker and let them shake gently at 20 °C. After 90 min (and then every 15 min), check the progress and briefly shake the tube vigorously by hand to separate loose oocytes from the tissue.
6. Typically, 120 min after the beginning of the treatment, practically all oocytes should be isolated and the majority of them should have already lost their follicles. If not, continue the collagenase treatment for another 30 min.
7. When the majority of oocytes have been released from their follicles, wash the oocytes extensively with Barth's solution (minimum of five times with 30 ml) to wash out the collagenase completely. Then fill up the tube (approximately to 40 ml) with Barth's solution and put it back onto the shaker for 10 min.
8. Change the solution to Barth's without Ca^{2+} and put it onto the shaker again for another 10 min.
9. Practically all oocytes should be defolliculated now. Shake the tube vigorously to remove the follicle cells completely, if necessary.
10. Wash the oocytes two times with 30 ml of Barth's solution.

3.1.2. Oocyte Selection and Plating

1. Fill a 100-ml beaker with 80 ml Barth's solution. Place the oocyte mesh sieve into the beaker for grading the oocytes by size. The mesh of the sieve should be completely immersed in the fluid.
2. Pipette an amount of oocytes onto the sieve. Approximately half the sieve should be covered with oocytes. Too many oocytes on the sieve will lead to an inefficient filtration (*see* **Note 4**).
3. Gently move the sieve about 2 cm up and down (in the liquid) to separate the oocytes by size.
4. Transfer the oocytes of appropriate size (that did not go through the sieve) into a 60-mm Petri dish filled with Barth's + gentamicin.

5. The filtered oocytes are incubated at 18 °C for 1 h.
6. Use a stereo microscope and the golf club-shaped glass tool to visually check each single oocyte for the following criteria: no visible damage of the cell, well-separated colors (dark and light brown), no residues of follicular tissue, and size of about 1.2 mm.
7. Prepare the plates by filling each well with 200 μl Barth's + gentamicin.
8. Aspirate a number of oocytes with a transfer pipette.
9. Carefully drop one oocyte into each of the wells of a well plate. Owing to the higher density of the vegetal pole, the oocytes will usually settle with the animal pole facing up if the (sticky) follicle cells were completely removed. If occasionally more than one cell drops into a well, this can be corrected later.
10. Immediately check the position of each oocyte under a stereo microscope and correct it carefully with the golf club-shaped glass tool if necessary. Generally, a manual correction will not be needed for more than 5% of the oocytes.
11. Cover the well plate with its lid, seal the rim with Parafilm to reduce evaporation, and store it at 18 °C until injection.
12. Although it is possible to inject the cells right after plating, they should not be washed or otherwise treated until they have been allowed to adhere to the well bottoms (about 2–3 h). It is not a problem if oocytes are injected on the next day, after being stored overnight at 18 °C. For best results, wash the oocytes approximately every second day with an 8-channel or 12-channel pipette or an automated plate washer (e.g., Tecan Columbus).

3.2. cDNA/mRNA Injection

The cDNA or mRNA to be injected (for one or several well plates) is manually filled into a ready-to-use or a custom-made injection pipette. According to the user-defined injection protocol, the cDNA/mRNA is then pushed automatically into the nucleus or the cytoplasm by compressed air *see* **Note 6**. The injection volume is determined by the product of injection time and injection pressure. With the Roboocyte, one can use cDNA or mRNA concentrations and injection volumes as described in the standard literature for the *Xenopus* expression system.

1. Take the cDNA/mRNA sample out of the freezer at least 20 min before the planned injection.
2. Let the sample stand at room temperature for at least 15 min. It needs to be completely thawed before use.
3. Mix the sample vigorously for a short moment, with a vortex mixer or manually by striking the side of the tube with a finger several times.
4. Spin the sample in an Eppendorf centrifuge (or similar) at 2000g for 5 min immediately before the injection to precipitate any remaining particles that might clog the

injection pipette. If this gives insufficient results, it may be advisable to try out spinning at full speed (>10,000g) for 5 min (*see* **Note 7**).

5. Start the Roboocyte program, create a new file, and select either cDNA or mRNA for injection.
6. In the virtual well plate view, select either the complete well plate (96 oocytes) or mark an individual selection of oocytes for injection by clicking on the icons representing the wells.
7. Mount the well plate onto the carrier of the robot.
8. Backfill the injection pipette with approximately 5 μl cDNA/mRNA solution per plate (40–50 nl per oocyte) by using a Microloader pipette tip.
9. Insert the back end of the pipette into the injection needle holder on the injection axis and fasten the screw.
10. Using the stereo microscope, align the needle tip to the crosshairs of the alignment device on the well plate.
11. Check the injection volume by applying a test shot under microscopic control and adjust the injection time and pressure if necessary. A small, spherical droplet of solution will appear at the pipette tip if all settings are correct. The volume can be estimated from the diameter of the droplet.
12. Load the injection protocol and start the injection sequence by clicking the Start button.
13. After the injection, the well plates are covered, sealed, and incubated at 18 °C for 2–7 days until a sufficient expression of the target receptor has been achieved. The expression should be checked every second day by running a suitable expression test recording protocol.

3.3. TEVC Recording and Test Compound Application

1. Start the Roboocyte program. Select and load the well plate file corresponding to the plate that will be used for the experiment.
2. Prepare the test solutions required for the planned application. If the pinch valves are to be used, fill the required test solutions into the appropriate reservoirs, prime the lines with the liquid, and check that the liquid flows when the valves are activated, that is, that there are no air bubbles in the tubing. The final step should be a flush with the perfusion liquid without test compound (usually valve no. 1). When using the optional Gilson liquid handler, fill the test solutions into the appropriate reservoirs, wells, or tubes on the rack. Using the manual software controls, briefly flush the tubing with the bath solution to ensure that the line contains only plain Ringer solution at the start of the experiment.
3. Enter the test compound names and concentrations into the liquid handling worksheet of the Roboocyte program. Assign each compound to the appropriate reservoir.
4. In the virtual well plate view, select either the complete well plate (96 oocytes) or mark an individual selection of oocytes for recording by clicking on the icons

representing the wells. At this point, it is a good idea to exclude wells containing damaged oocytes or empty wells.
5. Mount the well plate containing the oocytes onto the carrier of the robot.
6. Either reuse the already installed measuring head or if a replacement is necessary, backfill the micropipettes with the electrolyte solution, insert the Ag/AgCl wires, and plug the measuring head connectors into their sockets on the recording axis of the robot.
7. Align the micropipette tips to the crosshairs on the well plate under microscopic control.
8. Load the recording and perfusion protocol (RSL script) and start the experiment by clicking the Start button.

This will cause the robot to begin performing the protocol defined by the RSL script in the first of the selected wells. The experimental protocol, as defined by the user in the script, is then run automatically without any further user input required until all compounds have been tested, all oocytes have been processed, or until a user-defined break-point stops the run, for example, because the electrode impedance is out of range. This flexibility makes the Roboocyte such a powerful instrument.

As an example for a typical recording protocol, we describe briefly the measurement of a ligand-gated ion channel. The detailed parameters are those chosen for an α7 nAChR and can of course be adapted for other types of channels (*see* **Fig. 4**).

The first steps are to reset the liquid-junction potential to zero and to check the electrode resistance. It should be <1 MΩ for the current pipette but is less critical for the voltage recording pipette. Then, the oocyte is impaled with both intracellular electrodes in current clamp mode, with the holding current set to zero. The pipettes are lowered into the oocyte step-by-step until a negative membrane potential (e.g., less than −7 mV) is detected with both pipettes. Obviously, if a negative potential cannot be detected after a predetermined number of steps, the procedure is stopped automatically and the measuring head is automatically moved into the next well of the plate.

When a cell has been successfully impaled, the amplifier is switched to the voltage-clamp mode, and the membrane potential is clamped at the desired value (typically −80 mV for nAChR). The condition of the oocyte is monitored by a leak current check that is repeated between individual recordings.

An expression test with a reference compound (e.g., 50 μ*M* ACh) follows, where a minimum response (e.g., −200 nA) is required. Generally, we perform up to three initial reference measurements to ensure that the oocyte gives a stable response.

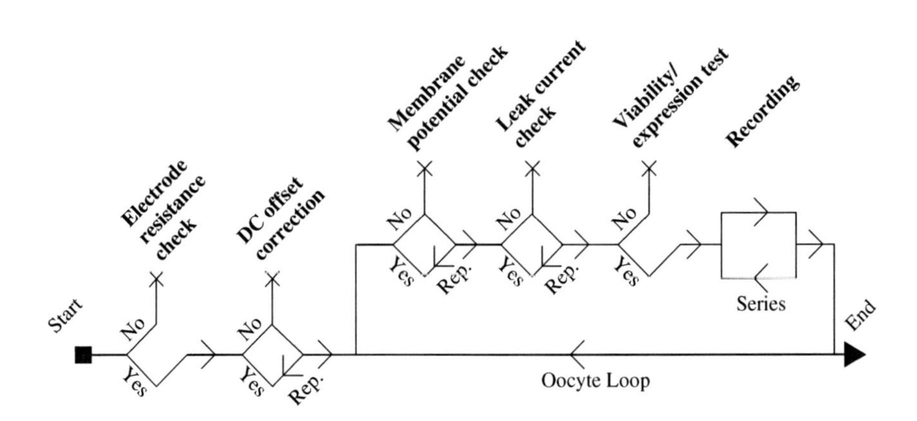

Fig. 4. Nested structure of Roboocyte Scripting Language (RSL) recording protocols. The flow chart illustrates how the elements of an RSL script form a nested structure, allowing a complex recording protocol with various tests and response-dependent actions. In this way, the robot can respond appropriately to the electrical signals that it records in the course of an experiment. The complexity of a protocol is up to the user's choice.

If the oocyte meets all these user-defined requirements, the test compound cycle begins. Compounds are applied consecutively to the same oocyte after a complete washout of the previous test solution. Before each test compound application, the condition of the oocyte is monitored, and the presence of a correct reference response is verified. If a reference response deviates more than 15% from the previous one, then up to three reference responses are elicited until the response is stable to make sure that desensitization or run-down effects of the target do not falsify the results. When the data are evaluated, the amplitudes are usually normalized against the preceding reference application.

The duration of the test compound application clearly depends on the target under investigation. For example, here the compound was applied for 5 s and washed out for 5 min after the recording.

3.4. Data Analysis and Results

The recorded current traces can be viewed online while a script is still in progress. They are permanently stored in a local Roboocyte plate file after each recorded sweep. The parameters extracted from the recordings, such as amplitude or area under the curve, together with the relevant information about compounds, dosage, and applied potential, are filed into a Microsoft Access database whenever a "save" command is executed in the script or whenever the script is stopped. Dose/response and voltage/current relationships are automatically plotted in the offline analysis window of the Roboocyte program. For more extensive analysis, data can be exported to a separate, project-defined database where the compound library is managed as well or exported to an analysis and graphing software like Origin or IGOR. By defining suitable scripts and queries, the analysis of extensive results can be made quite efficient. This is highly recommended to handle the large amount of data generated by the Roboocyte.

To compare data from different experiments and to eliminate the effects of variability from cell to cell, or of a run-down in the course of an experiment, it is important to normalize the current signals to a standard reference response that is obtained with a defined concentration of a known agonist.

3.5. Data Analysis for Ligand-Gated Ion Channels

To plot the dose–response curve of a ligand-gated ion channel, the data are usually plotted in semi-logarithmic coordinates and fitted to the expression:

$$I(C) = I_{\max} \times \frac{1}{1 + \left(\frac{EC_{50}}{C}\right)^n} \quad (1)$$

The parameters of this fit are as follows: EC_{50} is the concentration giving a half-maximal response; n is the Hill coefficient expressing the steepness of the curve; and I_{max} is the maximum current achieved at saturating concentrations of the agonist.

In practice, a number of effects and artifacts can make reliable measurements at higher ligand concentrations difficult. Although the Roboocyte amplifier was designed to handle large current signals, the maximum current response obtained from oocytes at high agonist concentrations can exceed the upper limit of 32 μA. More critically, it is practically impossible to apply a test compound simultaneously over the entire surface of the oocyte. Therefore, especially for channels that inactivate or desensitize very rapidly, the recorded peak current may be much less than the theoretical maximum, because the response is "smeared out" in time. Finally, agonists of ion channels, such as nicotinic receptors, occasionally act as noncompetitive blockers of the ion channel pore at very high concentrations. All of these effects can lead to poorly reproducible current amplitudes at the higher end of the dose–response range.

On the other hand, things are much better at low agonist concentrations where the large dynamic range large dynamic range from 1 nA to more than 30 μA is an advantage of oocyte recording with the Roboocyte. This makes it possible to measure reliably current responses that are only 1% of the maximal current or even less. This is practically impossible with patch-clamp recording from small cells. It suggests a different mode of evaluation of agonistic dose–response curves, which is described next.

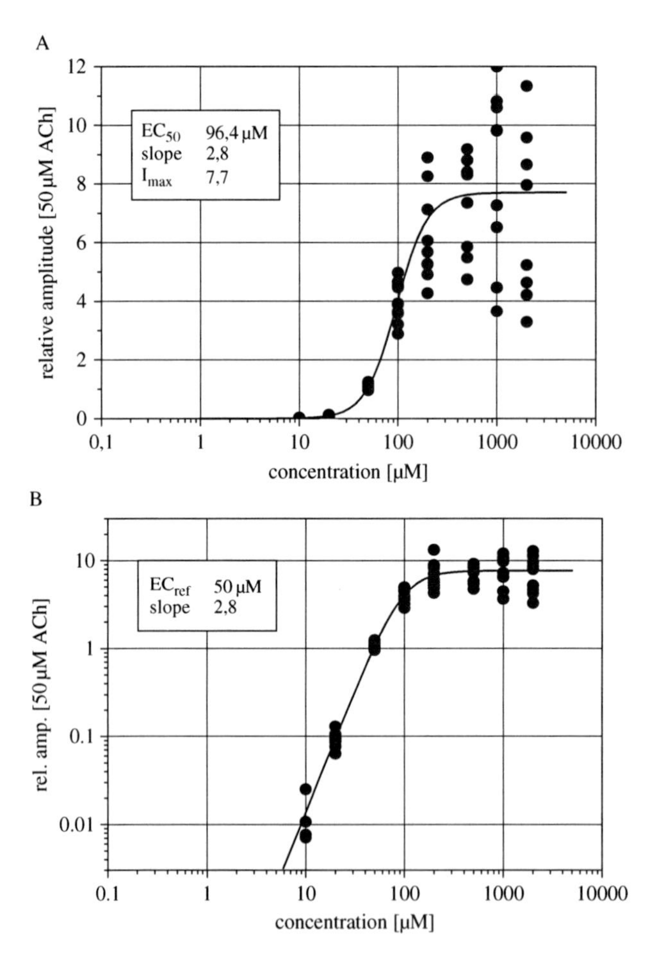

Fig. 5. Analysis of dose–response data for a ligand-gated channel. (**A**) Dose–response curve plotted in customary semi-logarithmic coordinates. Note that data points scatter widely at higher agonist concentrations, mostly because the α7-type receptors desensitize very rapidly. (**B**) Same data plotted in double log representation. The data from lower concentrations fit very well to the straight-line approximation. Note that EC_{ref}, in this case, is by definition 50 μM, as ACh is the ligand under investigation and also the reference standard. Experimental conditions are as in **Fig. 2**.

We find it convenient to present dose–response data for agonists of ligand-activated channels in a double logarithmic format (*see* **Fig. 5**).

All current responses are normalized to the preceding reference response elicited by a defined concentration of a standard agonist, arbitrarily chosen to give a robust but nonsaturating signal. In our example, for an α7 nAChR, the reference is 50 μ*M* ACh.

For very small current responses, $I << I_{max}$, the classical Hill plot becomes equivalent to the expression

$$\log \frac{I(C)}{I_{max}} \approx n \times \log \frac{C}{EC_{50}} \tag{2}$$

which, in double logarithmic coordinates, becomes a straight line passing through EC_{50} at I_{max}, with its slope given by the Hill coefficient. Because, for the reasons previously mentioned, we usually cannot determine I_{max} directly, we also cannot determine an EC_{50} value and prefer to express the potency of a ligand as EC_{ref}. This is defined as the concentration of ligand where the straight line passes through the 100% value of the response, elicited by the reference dose of ACh. In other words, the EC_{ref} value gives us an estimate of the concentration of test compound that is equipotent with 50 μ*M* ACh. Like the classical EC_{50} values, these EC_{ref} numbers can easily be used for ranking compounds according to their potency. However, to determine EC_{ref}, we emphasize data collected at low concentrations of test compounds, where oocyte data are especially reproducible, and we do not need any measurements at saturating levels of agonist. This makes it much more convenient to collect large amounts of data, requires less test compound, and allows us to obtain many more measurements from one cell. Note again that this straight-line relationship is only valid at low concentrations, which might not even give a measurable signal in patch-clamp experiments.

4. Notes

1. Pieces of ovary are obtained from anesthetized female *X. laevis* frogs through a small incision in the abdomen that is sutured immediately. The wound heals quickly, and after a few months, the same frog can be used again.
2. The surgically removed tissue is placed in collagenase solution for 1–2 h until individual oocytes separate and lose their follicle cell layer. The digestion should be completed after about 2–2.5 h. The final concentration of the collagenase solution has to be optimized according to the collagenase batch and experimental conditions. Do not prepare these solutions in advance as collagenase activity may decrease rapidly even if the solution is stored at −20 °C. The digestion with collagenase is a critical step in the oocyte preparation. If the activity is

too high, or the incubation time too long, the oocyte quality may suffer. If the collagenase treatment is insufficient, remaining pieces of follicular tissue will make the oocytes "sticky." In turn, this will lead to a bad centering in the well (i.e., the oocyte may attach to the wall rather to the well bottom) and may result in a decreased service life of the measuring heads.

3. One prominent problem of all electrophysiological recording techniques is the quality of the biological material. One can therefore not stress enough that only the best material can generate best results. It is advisable to spend some time on the optimization of the oocyte preparation, as oocytes of higher quality will last longer and therefore save costs. For example, it was possible in our laboratory to test up to 60 compounds consecutively on a single oocyte. Our experience showed that the oocyte quality also influences the lifetime of the intracellular recording electrodes. When oocytes are in bad shape, they tend to clog the micropipette tips, leading to an increase of the electrode impedance.
4. The oocyte size is also an important issue. The bigger the cells, the higher is the impalement success rate and the longer the cell will last during an experiment. Therefore, we grade the cells for size through a nylon mesh, followed by a manual selection of healthy-looking round cells with a minimum size of 1.2 mm and a well-defined pigmentation. Usual procedure in our laboratory is to inject the oocytes on a Friday and allow them to incubate over the weekend, so that measurements can begin on Monday. Healthy oocytes can be used for at least 1 week and sometimes well into the following week. For best results, *Xenopus* oocytes should be recorded in a temperature-controlled room at a temperature of not more than 20 °C. Also, it is recommended for longer experiments to include an automated wash routine that replaces the bath solution in all unused wells roughly every 3 h. If enough well plates with target-positive cells are available, it may be advisable to store some of them in the refrigerator at the end of the week to have them available in the subsequent work week.
5. An important consideration especially for an automated system is to ensure that there is always enough biological material provided for the experiments, as a system standing still produces unnecessary costs. Here, using commercial providers for *Xenopus* oocytes can save time and expenses. For example, oocytes supplied by EcoCyte Bioscience provided the same high quality as oocytes that were freshly prepared in our own laboratory.
6. Traditionally, mRNA is used for heterologous expression of proteins in *Xenopus* oocytes. It has the advantage that the current response can be modulated by the amount of mRNA that is injected, as the expression strength normally increases with the amount of transcript. However, injection of cDNA presents a useful alternative and is rapidly becoming more popular. Plasmid cDNA is cheaper to produce and easily obtained in large quantities sufficient for many experiments. It is also easier to handle, as it is not susceptible to the degradation by RNases that are ubiquitously present in the environment. On the other hand, cDNA must be injected into the

nucleus of the oocyte. This is quite difficult to achieve when manually injecting cDNA, resulting in decreased success rates. One technique described in the literature is to bring the nucleus to the surface of the cells by gentle centrifugation *(7)*, but this treatment is not helpful for the viability of the cells. We have checked the typical position of the germinal vesicle in the oocytes—usually at about one-third of the diameter under the brown animal pole of the cell—and optimized the penetration depth for maximized expression. The penetration depth for cDNA injection is preset to 300 μm (measured from the bottom of the well plate) and to 800 μm for mRNA. These settings have been shown to be the optimum values for the Roboocyte. Interestingly, even though the nucleus is known to be located near the animal pole, which is positioned face up, cDNA has to be injected more deeply for optimum results. It may be that the nucleus is pressed down by the injection needle until the built-up pressure is sufficient for the needle to penetrate the membrane of the nucleus. The overall expression rate achieved by this method of cDNA injection is typically better than 80% of the oocytes, depending of course on the properties of the target receptor. Trials with colored injection solution have shown that the nucleus was stained in 95% of the cells. mRNA is injected into the cytoplasm. Generally, it is not important where in the cytoplasm the RNA is released, although some targets may be inserted preferentially in the animal (brown) or vegetal (white) side of the cell. The default injection distance of 800 μm (from the well bottom) is a safe value even for smaller oocytes.

7. Although the chemical purity of the cDNA or mRNA sample is not critical, it has to be very clean and completely free of particles. Protein contaminations, cellular debris, residues of the extraction kit or from gels, or a high salt content can lead to clogging of the injection pipettes. Samples should be diluted with distilled water, not with saline solutions, to avoid the formation of salt crystals. Centrifugation of the sample immediately before use is recommended to spin down any remaining particles. Make sure you take up the liquid from the top of the sample and do not touch the bottom of the tube with the pipette tip to avoid aspirating any residues.
8. One question that arises occasionally is whether the ion channels or receptors expressed in *Xenopus* oocytes are actually identical to those found in a native tissue of interest or whether there may be biochemical or functional differences arising, for example, from different post-translational processing or a different lipid environment. First of all, this question can be asked about every heterologous expression system, including the cell lines generally used for High Throughput Screening (HTS) or automated patch-clamp assays. In the case of oocytes, a very detailed comparison of bovine nicotinic acetylcholine receptors expressed in oocytes with their native counterparts revealed no differences in pharmacology and gating behavior ***(15)***. Later, some subtle differences were described for a neuronal nicotinic receptor, but these may well have been due to the involvement of different receptor subunits. In general, the pharmacological properties, which are of interest in industrial drug discovery, are sufficiently congruent in the native and in the heterologous

expression system to allow satisfactory decisions to be made on the basis of the experimental results.

References

1. Gurdon J.B., Lane C.D., Woodland H.R., and Marbaix G. (1971) Use of frog eggs and oocytes for the study of messenger RNA and its translation in living cells. *Nature* **233**, 177–182.
2. Barnard E.A., Miledi R., and Sumikawa K. (1982) Translation of exogenous messenger RNA coding for nicotinic acetylcholine receptors produces functional receptors in Xenopus oocytes. *Proc. R. Soc. Lond. B* **215**, 241–246.
3. Gundersen C.B., Miledi R., and Parker I. (1984) Messenger RNA from human brain induces drug- and voltage-operated channels in *Xenopus* oocytes. *Nature* **308**, 421–424.
4. Gurdon J.B. and Melton D.A. (1981) Gene transfer in amphibian eggs and oocytes. *Ann. Rev. Genet.* **15**, 189–218.
5. Bertrand D., Cooper E., Valera S., Rungger D., and Ballivet M. (1991) Electrophysiology of neuronal nicotinic acetylcholine receptors expressed in *Xenopus* oocytes following nuclear injection of genes or cDNAs. in: Conn, M. (ed.) *Methods in Neurosciences, Vol. 4.* Academic Press, New York 174–193.
6. Hodgkin, A.L., Huxley A.F., and Katz, B. (1952) Measurement of current-voltage relations in the membrane of the giant axon of *Loligo*. *J. Physiol.* **116**, 424–448.
7. Schulz R., Bertrand S., Chamaon K., Smalla K.H., Gundelfinger E.D., and Bertrand D. (2000) Neuronal nicotinic acetylcholine receptors from *Drosophila*: two different types of α-subunits coassemble within the same receptor complex. *J. Neurochem.* **74**, 2537–2546.
8. Schnizler K., Kuester M., Methfessel C., and Fejtl M. (2003) The Roboocyte: automated cDNA/mRNA injection and subsequent TEVC recording on *Xenopus* oocytes in 96-well microtiter plates. *Receptors Channels* **9**, 41–48.
9. Pehl U., Leisgen C., Gampe K., and Guenther E. (2004) Automated higher-throughput compound screening on ion channel targets based on the *Xenopus laevis* oocyte expression system. *Assay Drug Dev. Technol.* **2**, 515–524.
10. Schroeder, J.I. (1994) Heterologous expression and functional analysis of higher plant transport proteins in *Xenopus* oocytes. (M. Montal, ed.) *Methods: A Companion to Methods in Enzymology* **6**, 70–81.
11. Sakmann B., Methfessel C., Mishina M., Takahashi T., Takai T., Kurasaki M., Fukuda K., and Numa S. (1985) Role of acetylcholine receptor subunits in gating of the channel. *Nature* **318**, 538–543.
12. Sine S.M. and Claudio T. (1991). Stable expression of the mouse nicotinic acetylcholine receptor in mouse fibroblasts. *J. Biol. Chem.* **266**, 13679–13689.
13. Takahashi T., Neher E., and Sakmann B. (1987) Rat brain serotonin receptors in *Xenopus* oocytes are coupled by intracellular calcium to endogenous channels. *Proc. Natl. Acad. Sci. U.S.A.* **84**, 5063–5067.

14. Sharon D., Vorobiov D., and Dascal N. (1997). Positive and negative coupling of the metabotropic glutamate receptors to a G protein activated K+ channel, GIRK, in *Xenopus* oocytes. *J. Gen. Physiol.* **109**, 477–490.
15. Mishina M., Takai T., Imoto K., Noda M., Takahashi T., Numa S., Methfessel C., and Sakmann B. (1986) Molecular distinction between fetal and adult forms of the muscle acetylcholine receptor. *Nature* **321**, 406–411.

II

Physiology

7

Infrared-Guided Laser Stimulation as a Tool for Elucidating the Synaptic Site of Expression of Long-Term Synaptic Plasticity

Gerhard Rammes, Matthias Eder, Walter Zieglgänsberger, and Hans-Ulrich Dodt

Summary

Long-term potentiation is a synaptic mechanism thought to be involved in learning and memory. Long-term depression (LTD), an activity-dependent decrease in synaptic efficacy, may be an equally important mechanism that permits neural networks to store information more effectively. Two forms of LTD have been identified in the mammalian central nervous system, which are induced by the synaptic activation of *N*-methyl-D-aspartate (NMDA) and metabotropic glutamate (mGlu) receptors, respectively. Whereas the expression mechanisms of NMDA receptor-dependent LTD have been demonstrated to be postsynaptic, those of mGlu receptor-dependent LTD have not been clearly identified. In order to address this issue, a variety of different electrophysiological methods have been used. A very elegant way to realize this experimental approach is provided by the development of photolytic application of glutamate, which allows the temporally and spatially highly specific activation of any neuron or any part of the neuron. By means of simultaneous application of electrical and photolytic stimulation techniques, it has been demonstrated that mGlu receptor-induced LTD is compatible with a presynaptic mechanism of expression.

Key Words: Brain slice; caged compounds; infrared-guided laser stimulation; photostimulation; synaptic plasticity; LTD; DHPG; hippocampus; dendrite.

From: *Methods in Molecular Biology, vol. 288: Patch-Clamp Methods and Protocols*
Edited by: P. Molnar and J. J. Hickman © Humana Press Inc., Totowa, NJ

1. Introduction

Two main forms of long-term depression (LTD) have been identified in the mammalian central nervous system. One form, known as homosynaptic LTD, requires the activation of *N*-methyl-D-aspartate (NMDA) receptors, Ca^{2+} entry into the postsynaptic neuron, and the activation of ser/thr protein phosphatases and is known to be expressed postsynaptically ***(1)***. The second form involves activation of metabotropic glutamate (mGlu) receptors, but the enzymes and expression mechanisms involved have not yet been clearly identified. To study these processes, the transient activation of mGlu receptors with a group I mGlu receptor agonist, 3,5-dihydroxyphenylglycine (DHPG), can be used ***(2–4)***. This so-called DHPG-induced LTD occludes synaptically induced mGlu receptor-dependent LTD ***(5)*** and is being used to investigate signaling and expression mechanisms of this type of synaptic plasticity ***(5–11)***. Unlike NMDA receptor-dependent LTD, DHPG-induced LTD does not require either Ca^{2+} ***(12)*** or ser/thr protein phosphatases ***(8)*** but involves activation of $G\alpha_q$ ***(7)*** and protein synthesis ***(5)***. The locus of expression of DHPG-induced LTD is highly controversial with evidence for both presynaptic ***(12)*** and postsynaptic changes ***(9,10)***. For elucidating the synaptic site of expression of mGlu receptor-dependent LTD, the precise spatial and temporal stimulation by photolytic uncaging of glutamate on neuronal membranes is well suited. Thus, receptor currents were evoked in CA1 hippocampal neurons by alternating photostimulation and electrical activation of synaptic afferents before and after the induction of DHPG- induced LTD.

2. Materials

1. Extracellular: 125 m*M* NaCl; 2.5 m*M* KCl; 25 m*M* $NaHCO_3$; 1.25 m*M* NaH_2PO_4; 1 m*M* $MgCl_2$; 2 m*M* $CaCl_2$; 25 m*M* D-glucose, equilibrated with 95% O_2/5% CO_2. Made up fresh for every experiment.
2. The patch solution comprised the following: 130 m*M* K-gluconate, 5 m*M* KCl, 0.5 m*M* EGTA, 2 m*M* Mg-ATP, 10 m*M* HEPES, 5 m*M* D-glucose. Made up in 50 ml and frozen at −80°C in 1-ml aliquots.
3. The beam of the laser was directed through a multimode optical fiber into the epifluorescence port of the microscope (Zeiss, Germany) and focused through a ×60/0.9 NA objective lens (Olympus, Hamburg, Germany) to an optical spot of 1 μm diameter (*see* **Fig. 1A, B**).
4. The output power of the argon-ion laser (maximum = 90 mW) could be adjusted with a remote control in increments of 1 mW to evoke a depolarizing response of about 3–5 mV at the soma in response to 3-ms shuttered light pulses (Uniblitz shutter; Vincent Associates, Rochester, NY).

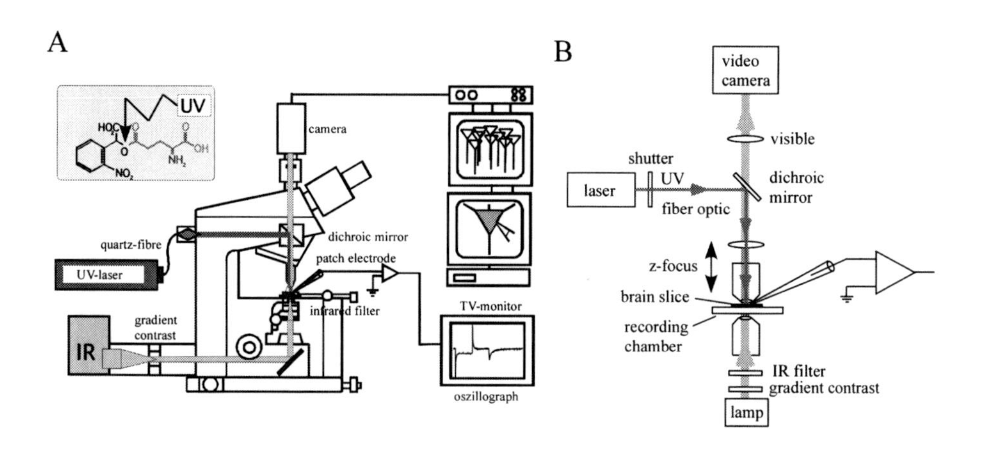
A
UV
camera
quartz-fibre
dichroic mirror
patch electrode
UV-laser
infrared filter
gradient
contrast
IR
TV-monitor
oszillograph
B
video
camera
visible
shutter
UV
laser
dichroic
mirror
fiber optic
z-focus
brain slice
recording
chamber
IR filter
gradient contrast
lamp

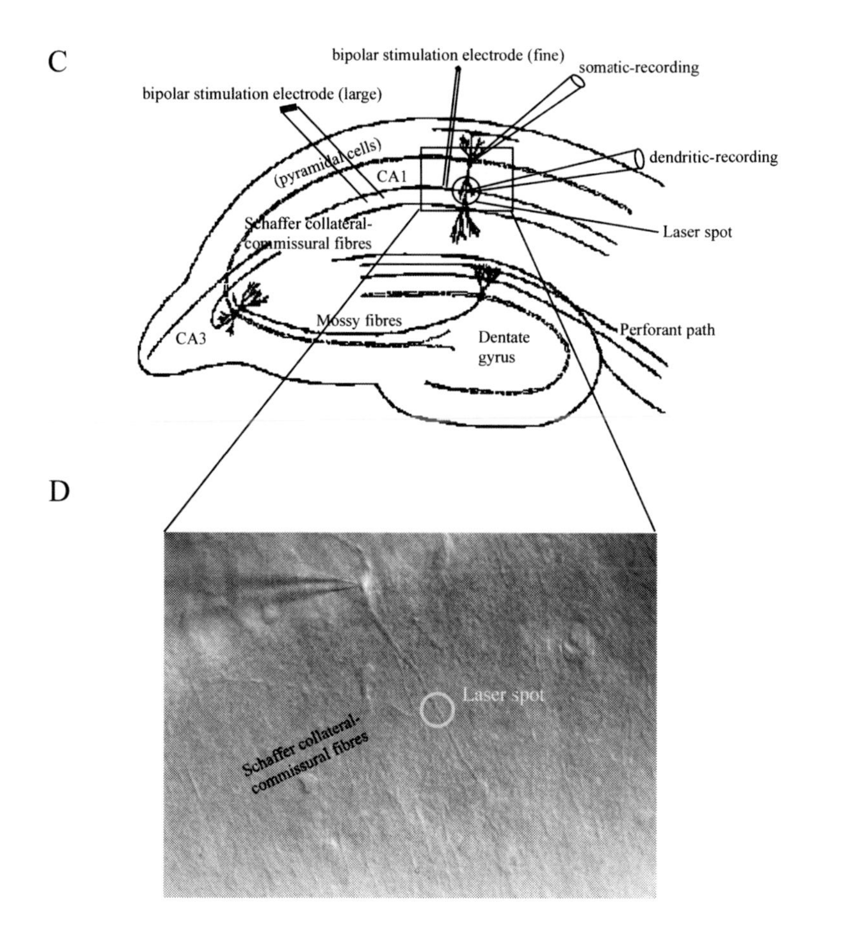
C
bipolar stimulation electrode (fine)
somatic-recording
bipolar stimulation electrode (large)
(pyramidal cells)
CA1
dendritic-recording
Schaffer collateral-
commissural fibres
Laser spot
Mossy fibres
Perforant path
CA3
Dentate
gyrus
D
Laser spot
Schaffer collateral-
commissural fibres

5. When the laser spot was positioned at the dendrite, approximately 70 μ*M* from the soma, uncaging of glutamate elicited a response of about 200 pA. The caged glutamate used in this study has a quantum yield of 0.14, and the half-life of the major component of the photolytic reaction is in the microsecond range ***(13)***. A further important property of γ-carboxy-2-nitrobenzyl (γ-CNB)-caged glutamate is its high stability; thus, at a concentration of 0.5 m*M*, it does not desensitize glutamate receptors by spontaneous hydrolysis and/or contamination with free glutamate ***(14)***. Furthermore, it also has been shown that 1 m*M* γ-CNB-caged glutamate neither desensitizes glutamate receptors in cultured hippocampal neurons ***(13)*** nor influences the properties of layer V pyramidal neurons of 2- to 3-week-old rats ***(15)***.
6. To make sure that this is also valid for our experiments in which 250 μ*M* caged glutamate was used, we performed additional control experiments. We tested whether 1 m*M* caged glutamate affects the amplitude of glutamate receptor-mediated excitatory postsynaptic currents (EPSCs). Consistent with the results from **refs.** ***13*** and ***15***, the amplitude does not change significantly after the addition of caged glutamate (98 ± 4% as compared with control; $p = 0.72$ [15 EPSCs in the absence and presence of caged glutamate were averaged for four neurons/three animals]). Furthermore, exposure of the cells to the brief periods of uncaging light used in our experiments has no effect on neuronal activity ***(16)***.
7. The microscope was mounted on a motorized, three-dimensional stage (Luigs & Neumann, Germany). Neurons were visualized by means of the infrared (IR) videomicroscopy technique ***(16,17)*** with the gradient contrast system (*see* **Fig. 1A, B**).

Fig. 1. (**A** and **B**) Experimental set-up used for infrared (IR)-guided laser stimulation. Neurons in the brain slice were visualized by illumination with IR light and the gradient contrast system. At the same time, light pulses from a ultraviolet (UV) laser were fed by a quartz fiber into the microscope and directed by a dichroic mirror onto the recorded neuron. The laser spot of an optical diameter of 1 μm formed by the objective in the specimen plane was made visible before the experiment by a fluorescent paper, and its position was marked on the television monitor. By positioning the site of the neuron to be stimulated at this point, the laser stimulation could be precisely guided by visual control. Figures slightly changed according to **ref.** ***16***. (**C**) Schematic diagram of the transverse hippocampal slice showing the major excitatory projections and the placements of the stimulation and recording electrodes. The stimulation electrodes were placed at a fixed position to either stimulate many afferents of the Schaffer collaterals or only those fibers that are on the level of the dendritic patch. Currents were recorded from single pyramidal cells by using patch-clamp electrodes placed in the somatic or dendritic region. (**D**) CA1 region as visualized by IR microscopy shows the recording electrode and the position of the laser spot.

8. Stimulation electrodes: Bipolar tungsten electrode insulated to the tip with 2 μm × 75 μm tip diameter or fine insulated platinum bipolar electrodes (2 $\mu M \times 25\,\mu M$, FHC, USA).
9. Currents were recorded using a switched voltage-clamp amplifier (SEC-10L, npi Electronics, Tamm, Germany).
10. A Power Macintosh G3 computer collected data by a data acquisition and evaluation program (Pulse v. 8.5, Heka electronic GmbH, Germany).
11. Analysis of amplitudes, charges, and kinetics was performed using macros written for Igor Pro 4.01 (WaveMetrics, USA).
12. Drugs: The following drugs were added also to the Artificial cerebrospinal fluid (ACSF): 2,3-dihydroxy-6-nitro-7-sulfonyl-benzo(F)quinoxaline (NBQX, Sigma, Germany), S-DHPG (Tocris, UK), and γ-CNB-caged glutamate (Molecular Probes, the Netherlands). All drugs were solved in bidest H_2O and aliquoted.

3. Methods

3.1. Procedure

1. All experiments were performed at room temperature.
2. Brain slices were placed in a submerged recording chamber of an "infrapatch" set-up (Luigs & Neumann) (*see* **Fig. 1A, B**) and perfused with carbogenated ACSF at a flow rate of 2–3 ml/min.
3. Individual CA1 neurons from stratum pyramidale were visually identified using IR videomicroscopy and the "gradient contrast" system ***(16)***.
4. Whole-cell recordings (seal resistance greater than 1 GΩ) from somata were performed in voltage-clamp mode (−60 mV holding potential) using a discontinuous single-electrode voltage-clamp amplifier (SEC-10LX, npi Electronics).
5. Recordings were accepted only if the holding current was less than 100 pA. Current was low-pass filtered (1 kHz).
6. The patch-clamp electrodes with open-tip resistances of 4–6 MΩ (6–8 MΩ for dendritic recordings) were pulled from borosilicate glass capillaries (1.5 mm outer diameter and 0.86 mm inner diameter; Harvard Apparatus, UK) on a DMZ-Universal puller (Zeitz-Instruments, Germany).
7. The neurons (dendrites) included ($n = 20$) ***(10)*** had an average resting potential of -61 ± 0.7 mV (-62 ± 0.3 mV) and an input resistance of 98 ± 4 MΩ (454.9 ± 76.2 MΩ).
8. The series resistance 18 ± 1 MΩ for somatic and 38 ± 1 MΩ for dendritic whole-cell recordings was monitored continuously during the recordings, which was terminated if changes of more than 10% were observed.
9. Although dendritic recordings were not confirmed by filling the neuron with a fluorescent dye, a number of other observations suggested that all recordings were made from dendrites; upon breaking the patch membrane, increases in capacitance were observed, which were much larger than those expected for small structures, such as presynaptic terminals or glial cells that might lie over the dendrite.

10. Schaffer collateral-commissural fibers were electrically stimulated at 0.033–1 Hz using either a tungsten or a fine insulated platinum bipolar electrode.
11. The stimulation electrode was placed near the apical dendrite (approximately 30–50 μm lateral distance) at the level of the laser stimulation (*see* **Fig. 1C, D**). This way, synapses, located in a defined region of the dendrite, could be reliably activated. The width of the electrical stimuli was 0.05 ms.
12. Complete wash-in of the drugs was achieved within 5–10 min depending on the substance. Data were only collected after complete wash-in was confirmed by a stable baseline.
13. For photostimulation, glutamate was applied to visually identified neuronal structures by flash photolysis (wavelength = 351–364 nm) of caged glutamate (0.25 m*M*).
14. The duration of the shuttered light pulses was 3 ms.
15. Exposure of cells to the brief periods of uncaging light alone used in our experiments has no effect on neuronal activity ***(16,18)***.
16. During the application of caged glutamate, a total amount of 5.5 ml ACSF containing the caged compound was continuously recirculated and oxygenated.

4. Results

4.1. Focal Glutamate Responses

Initially, photolysis to rapidly uncage L-glutamate over a small region of dendrite of a CA1 neuron was performed to establish how closely one could mimic the synaptic activation of alpha-Amino-3-hydroxy-5-metyl-4 iroxazole proprionic acid (AMPA) receptors in hippocampal slices. Using a laser intensity of 30% maximum, a 3-ms pulse focused on a 1-μm spot elicited a current that was similar, in both time-course and amplitude, to synaptic currents induced in the same neurons by stimulation of Schaffer collateral-commissural fibers ***(14)***. Both responses were blocked in parallel by 5 μ*M* NBQX ($n = 6$) (*see* **Fig. 2A**).

4.2. Homosynaptic LTD

Homosynaptic LTD has been induced through low-frequency stimulation (LFS) either by pairing 60 shocks delivered at 0.5 Hz at −40 mV (through the fine platinum electrode) or by delivering 900 shocks at 1 Hz (through the tungsten electrode) in bridge mode with no current injection. Whole-cell voltage-clamp recordings (−70 mV) were obtained locally from the dendritic region, where the photolysis was performed and neighboring fibers stimulated (*see* **Fig. 1C, D**). EPSCs and responses to glutamate photolytically released by laser light were obtained alternately throughout an experiment. LFS led to a reduction in amplitude of both the EPSCs and responses to focal L-glutamate.

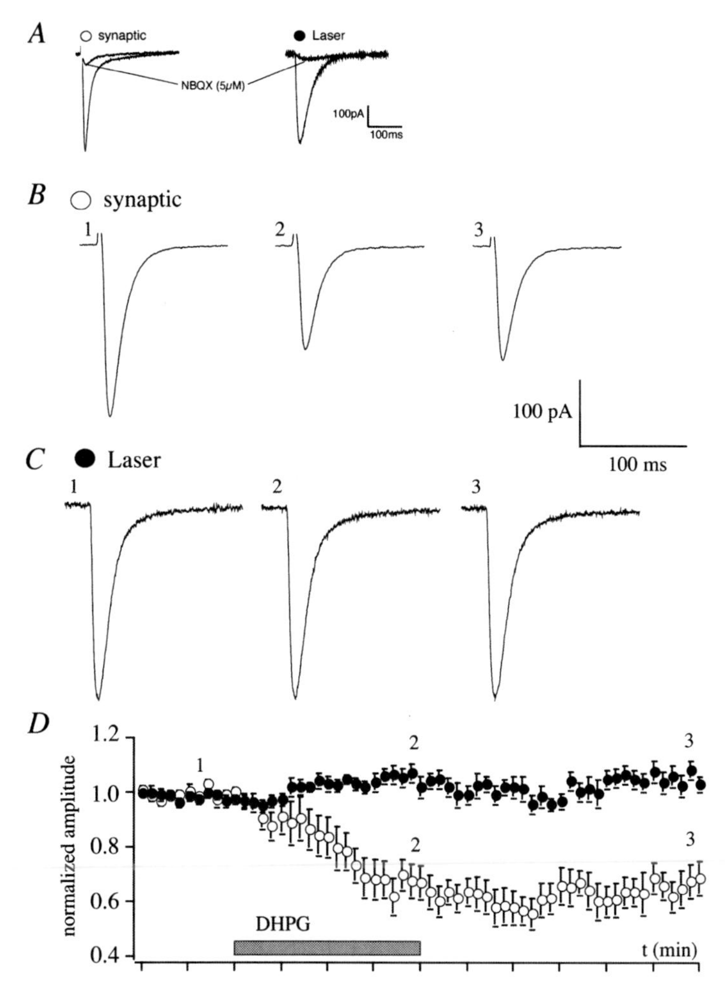

Fig. 2. (**A**) EPSCs (left) and current responses to uncaged glutamate (right) are blocked by 2,3-dihydroxy-6-nitro-7-sulfonyl-benzo(F)quinoxaline (NBQX) (5 μ*M*). (**B**) EPSCs (averages of 10 successive traces) obtained before (1), during (2), and 30 min following (3) the application of 3,5-dihydroxyphenylglycine (DHPG) (30 μ*M*; applied for time indicated by bar). (**C**) Inward currents induced by the uncaging of glutamate at the corresponding time points for the same neuron. (**D**) Pooled data for seven experiments plotting EPSC amplitude (open symbols) and laser-induced responses (filled symbols) versus time. The numbers (1–3) refer to the times at which the traces (in **B** and **C**) were obtained. All figures according to **ref. *20***.

These data show that focally uncaged L-glutamate can be used to accurately monitor the behavior of synaptic AMPA receptors during synaptic plasticity. Homosynaptic LTD was not associated with any change in the kinetics of EPSCs or in input resistance of the neurons. Access resistances were stable throughout.

4.3. DHPG-Induced LTD

In the last set of experiments, LTD has been induced by applying S-DHPG (30 μ*M*; 20 min) to slices obtained from juvenile rats. Again, EPSCs and responses to focal glutamate were obtained alternately. DHPG induced a robust LTD in which responses were depressed to $65 \pm 6\%$ of control, 30 min following wash-out of the compound. In contrast, responses to focally uncaged L-glutamate were never depressed by this treatment; 30 min following wash-out of DHPG, responses were $108 \pm 6\%$ of control, $n = 7$ (*see* **Fig. 2B–D**). DHPG-induced LTD was not associated with any change in the kinetics of EPSCs or in input resistance of the neurons. Access resistances were stable throughout.

The combination of IR gradient-contrast videomicroscopy, the photolytic application of neurotransmitter, and electrophysiological recording provides a powerful tool to examine the subcellular distribution, densities, and properties of functional neurotransmitters in the living brain slice *(19)*. Owing to its non-invasive application and high spatial and temporal resolution, photostimulation offers an elegant way to probe postsynaptic sensitivity during LTD. As such, the experiments presented here demonstrate that homosynaptic LTD can be explained by a postsynaptic alteration in AMPA receptor function but DHPG-induced LTD cannot. Because photostimulation mimics synaptic transmission without activation of presynaptic terminals, this result is consistent with a postsynaptic expression of homosynaptic LTD. Conversely, DHPG-induced LTD is not associated with a decrease of the postsynaptic glutamate sensitivity of CA1 pyramidal neurons. These data are most simply explained on the basis of postsynaptic and presynaptic expression mechanisms for NMDA and mGlu receptor-dependent LTD, respectively *(20)*.

5. Notes

1. Because caged compounds are quite expensive, caged glutamate was applied at a concentration of 250 μ*M* in a continuously recirculated volume of 5.5 ml ACSF.
2. A crucial prerequisite for investigating the expression site of synaptic plasticity through the presented technique requires identical receptors activated by both photostimulation and the presynaptic release site.
3. The probability that the number and location of receptors activated by presynaptic release were identical with those activated by photostimulation was increased

with the following measures: (1) the laser spot was positioned on the dendrite approximately 100 μm from the soma at the level of the fine bipolar electrode (*see* **Fig. 1C, D**)—this was based on the assumption that the afferent fibers are anatomically ordered in parallel to the CA1 somata layer and (2) a bipolar electrode was used, which stimulates almost all afferents innervating the recorded neurone (tungsten electrode).

References

1. Bear, M.F. and W.C. Abraham. (1996) Long-term depression in hippocampus. *Ann Rev Neurosci.* **19**, 437–462.
2. Palmer, M.J., A.J. Irving, G.R. Seabrook, D.E. Jane, and G.L. Collingridge. (1997) The group I mGlu receptor agonist DHPG induces a novel form of LTD in the CA1 region of the hippocampus. *Neuropharmacology* **36**, 1517–1532.
3. Fitzjohn, S.M., A.E. Kingston, D. Lodge, and G.L. Collingridge. (1999) DHPG-induced LTD in area CA1 of juvenile rat hippocampus; characterisation and sensitivity to novel mGlu receptor antagonists. *Neuropharmacology* **38**, 1577–1583.
4. Camodeca, N., N.A. Breakwell, M.J. Rowan, and R. Anwyl. (1999) Induction of LTD by activation of group I mGluR in the dentate gyrus in vitro. *Neuropharmacology* **38**, 1597–1606.
5. Huber, K.M., J.C. Roder, and M.F. Bear. (2001) Chemical induction of mGluR5- and protein synthesis-dependent long-term depression in hippocampal area CA1. *J Neurophysiol.* **86**, 321–325.
6. Fitzjohn, S.M., R.A. Morton, F. Kuenzi, T.W. Rosahl, M. Shearman, H. Lewis, D. Smith, D.S. Reynolds, C.H. Davies, G.L. Collingridge, and G.R. Seabrook. (2001) Age-related impairment of synaptic transmission but normal long-term potentiation in transgenic mice that overexpress the human APP695SWE mutant form of amyloid precursor protein. *J Neurosci.* **21**, 4691–4698.
7. Kleppisch, T., V. Voigt, R. Allmann, and S. Offermanns. (2001) G(alpha) q-deficient mice lack metabotropic glutamate receptor-dependent long-term depression but show normal long-term potentiation in the hippocampal CA1 region. *J Neurosci.* **21**, 4943–4948.
8. Schnabel, R., I.C. Kilpatrick, and G.L. Collingridge. (2001) Protein phosphatase inhibitors facilitate DHPG-induced LTD in the CA1 region of the hippocampus. *Br J Pharmacol.* **132**, 1095–1101.
9. Snyder, E.M., B.D. Philpot, K.M. Huber, X. Dong, J.R. Fallon, and M.F. Bear. (2001) Internalization of ionotropic glutamate receptors in response to mGluR activation. *Nat Neurosci.* **4**, 1079–1085.
10. Xiao, M.Y., Q. Zhou, and R.A. Nicoll. (2001) Metabotropic glutamate receptor activation causes a rapid redistribution of AMPA receptors. *Neuropharmacology.* **41**, 664–671.

11. Watabe, A.M., H.J. Carlisle, and T.J. O'Dell. (2002) Postsynaptic induction and presynaptic expression of group 1 mGluR-dependent LTD in the hippocampal CA1 region. *J Neurophysiol.* **87**(3):1395–1403.
12. Fitzjohn, S.M., M.J. Palmer, J.E.R. May, A. Neeson, S.A.C. Morris, and G.L. Collingridge. (2001) A characterisation of long-term depression induced by metabotropic glutamate receptor activation in the rat hippocampus in vitro. *J Physiol (Lond).* **537**(2):421–430.
13. Wieboldt, R., K.R. Gee, L. Niu, D. Ramesh, B.K. Carpenter, and G.P. Hess. (1994) Photolabile precursors of glutamate: synthesis, photochemical properties, and activation of glutamate receptors on a microsecond time scale. *Proc Natl Acad Sci USA.* **91**, 8752–8756.
14. Eder, M., W. Zieglgansberger, and H.U. Dodt. (2002) Neocortical long-term potentiation and long-term depression: site of expression investigated by infrared-guided laser stimulation. *J Neurosci.* **22**, 7558–7568.
15. Schubert, D., J.F. Staiger, N. Cho, R. Kotter, K. Zilles, and H.J. Luhmann. (2001) Layer-specific intracolumnar and transcolumnar functional connectivity of layer V pyramidal cells in rat barrel cortex. *J Neurosci.* **21**, 3580–3592.
16. Dodt, H.U., M. Eder, A. Frick, and W. Zieglgansberger. (1999) Precisely localized LTD in the neocortex revealed by infrared-guided laser stimulation. *Science.* **286**(5437):110–113.
17. Dodt, H.U. and W. Zieglgansberger. (1994) Infrared videomicroscopy: a new look at neuronal structure and function [Review] [43 refs]. *Trends Neurosci.* **17**, 453–458.
18. Dodt, H.U., M. Eder, A. Schierloh, and W. Zieglgansberger. (2002) Infrared-guided laser stimulation of neurons in brain slices. *Sci STKE.* **2002**, PL2.
19. Eder, M., W. Zieglgansberger, and H.U. Dodt. (2004) Shining light on neurons–elucidation of neuronal functions by photostimulation. *Rev Neurosci.* **15**, 167–183.
20. Rammes, G., M. Palmer, M. Eder, H.U. Dodt, W. Zieglgansberger, and G.L. Collingridge. (2003) Activation of mGlu receptors induces LTD without affecting postsynaptic sensitivity of CA1 neurons in rat hippocampal slices. *J Physiol.* **546**, 455–460.

8

Single-Cell RT–PCR, a Technique to Decipher the Electrical, Anatomical, and Genetic Determinants of Neuronal Diversity

Maria Toledo-Rodriguez and Henry Markram

Summary

The patch-clamp technique has allowed detailed studies on the electrical properties of neurons. Dye loading through patch pipettes has allowed characterizing the morphological properties of the neurons. In addition, the patch-clamp technique also allows harvesting mRNA from single cells to study gene expression at the single-cell level (known as single-cell reverse transcription–polymerase chain reaction [RT–PCR] *(1–3)*). The combination of these three approaches allows determination of the Gene expression, Electrophysiology and Morphology (GEM) profile of neurons (gene expression, electrophysiology, and morphology) using a single patch pipette and patch-clamp recording. This combination provides a powerful technique to study and correlate the neuron's gene expression with its phenotype (electrical behavior and morphology) *(4–7)*. The harvesting and amplification of single-cell mRNA for gene expression studies is a challenging task, especially for researchers with sparse or no training in molecular biology (*see* **Notes 1** and **2**). Here, we describe in detail the GEM profiling approach with special attention to the gene expression profiling.

Key Words: Single-cell RT–PCR; multiplex PCR; ion channel; calcium-binding protein; neuropeptide; morphology; electrophysiology.

1. Introduction

Neurons, the building blocks of the nervous system, largely differ in terms of their anatomical, electrophysiological, and gene expression properties, indicating unique functional roles for any given cell type. In last 15 years, there has been an explosive increase in the amount of electrical, morphological, and

From: *Methods in Molecular Biology, vol. 288: Patch-Clamp Methods and Protocols*
Edited by: P. Molnar and J. J. Hickman © Humana Press Inc., Totowa, NJ

gene expression data available from neurons. Nevertheless, despite this increase, the principles of diversification linking different data sets have not been clearly defined. Two key obstacles have impeded the derivation of these principles of diversification—(1) technical: it was and still is technically very challenging to study single neurons simultaneously at the electrical, morphological, and gene expression levels and (2) data set size: owing to the technical difficulties too few neurons have been studied, making statistical analysis difficult (*see* **Note 3**). To explore the principles that guide neuronal diversity, other groups and we developed a multidisciplinary approach that allowed studying of single neurons simultaneously at the morphological, electrophysiological, and gene expression level *(5–7)*. For this purpose, whole-cell patch-clamp recordings are combined with single-cell reverse transcription (RT), multiplex polymerase chain reaction (PCR), and morphological identification (*see* **Figs. 1** and **3**).

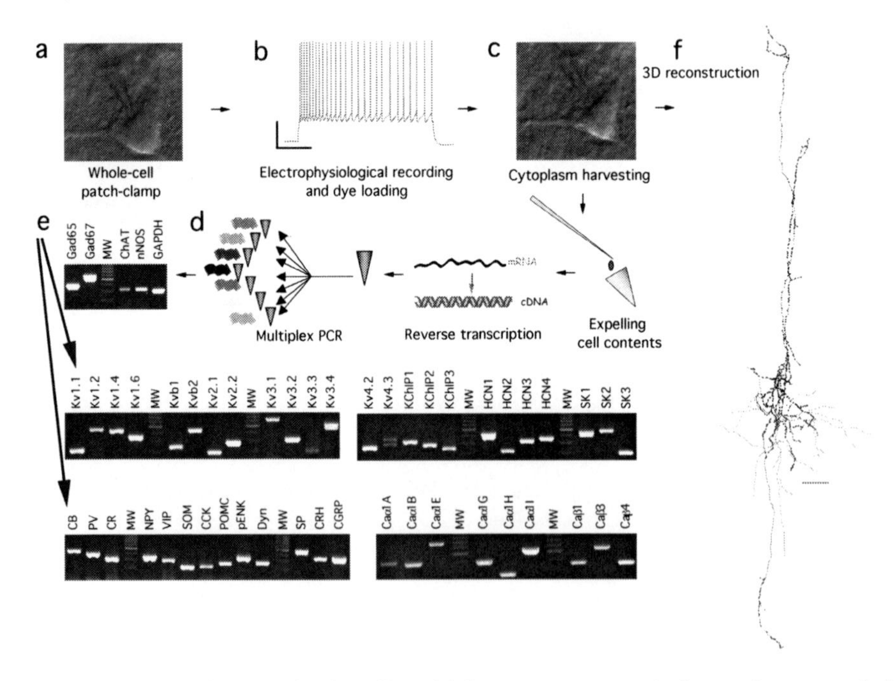

Fig. 1. The step-by-step single-cell multiplex reverse transcription–polymerase chain reaction (RT–PCR) technique. (**A**) Whole-cell patch-clamp recordings from different classes of neocortical neurons are obtained. (**B**) A complex set of stimulation protocols are employed. Scale bars: current 20 mV, time 500 ms. (**C**) At the end of each recording, the neuron's cytoplasm is extracted for subsequent RT. (**D**) Multiplex RT–PCR is performed to simultaneously detect the expression of 51 mRNAs from each neuron.

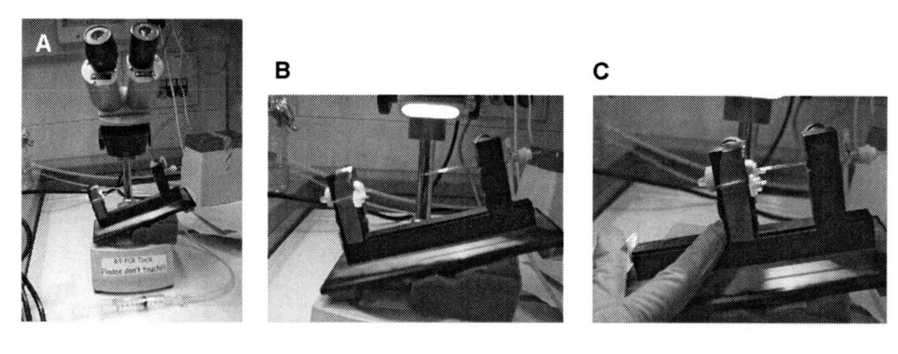

Fig. 2. Expelling the single-cell contains in the patch pipette, device, and procedure. **(A)** Device to expel the single-cell contains after harvesting into the patch pipette. **(B)** Patch pipette secured to the pipette holder. **(C)** Braking the pipette tip against the microcentrifuge tube bottom.

2. Materials

2.1. Solutions

1. Extracellular solution: 125 m*M* NaCl, 2.5 m*M* KCl, 25 m*M* glucose, 25 m*M* $NaHCO_3$, 1.25 m*M* NaH_2PO_4, 2 m*M* $CaCl_2$, and 1 m*M* $MgCl_2$ (Sigma, Israel).
2. RNase-free intracellular solution containing biocytin: 100 m*M* potassium gluconate, 20 m*M* KCl, 4 m*M* ATP-Mg, 10 m*M* phosphocreatine, 0.3 m*M* GTP, 10 m*M*

Fig. 1. **(E)** Agarose gels showing the 51 mRNA species simultaneously amplified from 250 pg total brain mRNA. These include the voltage-activated K^+ channels (Kv1.1/2/4/6, Kvβ1/2, Kv2.1/2, Kv3.1/2/3/4, Kv4.2/3, KchIP1/2/3); the K^+/Na^+ permeable hyperpolarization-activated channels (HCN1/2/3/4); the calcium-activated K^+ channels (SK1/2/3); the voltage-activated Ca^{2+} channels (Caα1A/B/E/G/E/I, Caβ1/3/4); the calcium-binding proteins (calbindin [CB], parvalbumin [PV], calretinin [CR]); the neuropeptides (neuropeptide Y [NPY], vasoactive intestinal peptide [VIP], somatostatin [SOM], cholecystokinin [CCK], proenkephalin [pENK], proopiomelanocortin [POMC], dynorphin [Dyn], substance P [SP], corticotropin-releasing hormone [CRH], and caltitonin gene-related peptide [CGRP]); the enzymes (glutamic acid decarboxylase [Gad65 and Gad67], choline acetyltransferase [ChAT], nitric oxide synthase [nNOS], and glyceraldehyde-3-phosphate dehydrogenase [GAPDH]). *See* **Table 1** for the size of the PCR product as predicted by its mRNA sequence. MW: 100 bp ladder molecular weight marker. **(F)** During recording, neurons are also loaded with a dye allowing subsequent morphological classification and 3D computer anatomical reconstruction. Soma and dendrites in light grey and axon in black. Scale bar: 100 μm.

HEPES (pH 7.3, 310 mOsmol, adjusted with sucrose), and 0.5% biocytin (Sigma). The intracellular solution is prepared under RNase-free conditions; water is autoclaved; glassware and pH meter are cleaned with NaOH (10*N*); and chemicals are opened using gloves and RNase-free tools from the first time. After preparation, the intracellular solution is filtered, tested for RNase contamination by incubating the solution overnight at 37°C with RNA. The next morning, the RNA is run in an agarose gel. The intracellular solution is confirmed RNase free if the two sharp bands of the ribosomal RNA appear in the gel. Finally, the solution is aliquoted in single-use vials and stored at −20°C.

3. Fixative: 0.1 *M* phosphate buffer (2.65 g/l $NaH_2PO_4 \cdot H_2O$, 14 g/l K_2HPO_4; pH 7.4) containing 2% paraformaldehyde, 1% glutaraldehyde, and 0.3% saturated picric acid.
4. ABC–Triton X-100 solution: 20 μl A + 20 μl B + 10 μl 10% Triton X-100 in 1 ml Phorphate Buffer (PB) (ABC-Elite; Vector Labs, UK). Prepare at least 30 min ahead of use.
5. Diaminobencidine (DAB) solution: mix 17.5 mg DAB (1 ml), 200 μl 1% cobalt chloride, 150 μl 1% ammonium nickel sulfate, and 0.1 *M* PB to 25 ml. Then add 0.03% H_2O_2 (10 μl 30% H_2O_2 in 1 ml PB) for 1 ~ 2 min. (Note: DAB reaction needs to be done under covering.)
6. RT mix: 4 μl 5× Superscript III buffer (Invitrogen, Israel), 1 μl 0.1 *M* Dithiothreitol (DTT) (Invitrogen, Israel), 2 μl 5 m*M* dNTP (Promega), 0.4 μl RNAsin (40 U/μl; Promega), and 0.5 μl Superscript III (200 U/μl; Invitrogen, Israel).

2.2. RNase-Free Tools

1. RNase-free pipettes: bake capillars before pulling at 200°C overnight. Clean puller with RNase-Zap (Ambion) according to the manufacturer's instructions. Use gloves and autoclaved tweezers to handle the capillars.
2. RNase-free electrode wire: rechlorinate the electrode at the beginning of each recording session or when suspected RNase contamination.

3. Methods

3.1. Sample Preparation and Patch-Clamp Recordings

1. Rapidly decapitate the rat (use anesthesia when working with adult animals), extract the brain, and cut the two hemispheres apart.
2. Section neocortical slices (sagittal; 300 mm thick) on a vibratome (DSK; Microslicer, Japan) filled with iced extracellular solution.
3. After sectioning, leave the slices in extracellular solution with oxygenation at 34°C for 30 min (incubate the chamber in a water bath). Afterward, remove the chamber to room temperature.
4. Identify neurons using Infra-Red Differential Interference Contrast (IR-DIC) microscopy *(8)*.

5. Perform somatic whole-cell recordings (using RNase-free pipettes–resistance 3 MΩ–and intracellular solution [containing biocytin]). For a good intracellular filling, stay in the neuron at least 20 min.

3.2. Histological Procedures

1. After recording, fix the slices for 24 h in cold fixative.
2. Rinse slices several times (10 min each) in PB.
3. Transfer slices into PB containing 3% H_2O_2 for 30 min.
4. Rinse five to six times in PB (10 min each).
5. Incubate slices overnight at 4°C in ABC–Triton X-100 solution.
6. Wash sections several times in PB (10 min each).
7. Develop with DAB solution under visual control using a bright-field microscope (Zeiss Axioskop) until all processes of the cells appeared clearly visible (usually 2–4 min).
8. Stop the reaction by transferring the sections into PB.
9. After washing in the same buffer, mount the slices in aqueous mounting medium (IMMCO Diagnostics, Israel).

3.3. Single-Cell RT–Multiplex PCR

1. At the end of the recording, aspirate the cell's cytoplasm into the recording pipette under visual control by applying gentle negative pressure. Only process cells whose seal was intact throughout the recording and whose nucleus was not harvested.
2. Withdraw the electrode from the cell to form an outside-out patch to prevent contamination as the pipette is removed.
3. Secure the pipette in a pipette holder (*see* **Fig. 2A**, **B**) and apply positive pressure. Under visual control, bring the pipette tip close to the bottom of an RNase-free microcentrifuge tube and break the tip against the tube bottom (*see* **Fig. 2C**).
4. To the single-cell contents, add 0.4 μl RNAsin (40 U/μl; Promega), 4 μl oligo-dT primer (25 ng/μl; Invitrogen), and RNase-free water up to 12.5 μl. Incubate at 70°C for 5 min.
5. Cool on ice for 1 min and briefly centrifuge to bring the contents to the tube bottom.
6. Add 16 μl RT mix and incubate for 50 min at 50°C.
7. Heat inactivate the reverse transcriptase by incubating at 70°C for 10 min. At this step, the cDNA can be frozen and stored at −20°C before further processing.
8. Optimize each multiplex PCR conditions using total RNA purified from rat brain, so that a PCR product could be detected from 250 pg to 1 ng total RNA without contamination caused by unspecific amplification. **Table 1** summarizes the primer pairs we included in the different multiplexes, the name and accession number of the genes amplified, and the length of the PCR product. In this case, four different multiplex PCRs were performed to test the expression of 51 mRNA species from each cell. Each multiplex PCR is performed using a different amount of cDNA depending on the multiplex complexity and the number of mRNA molecules

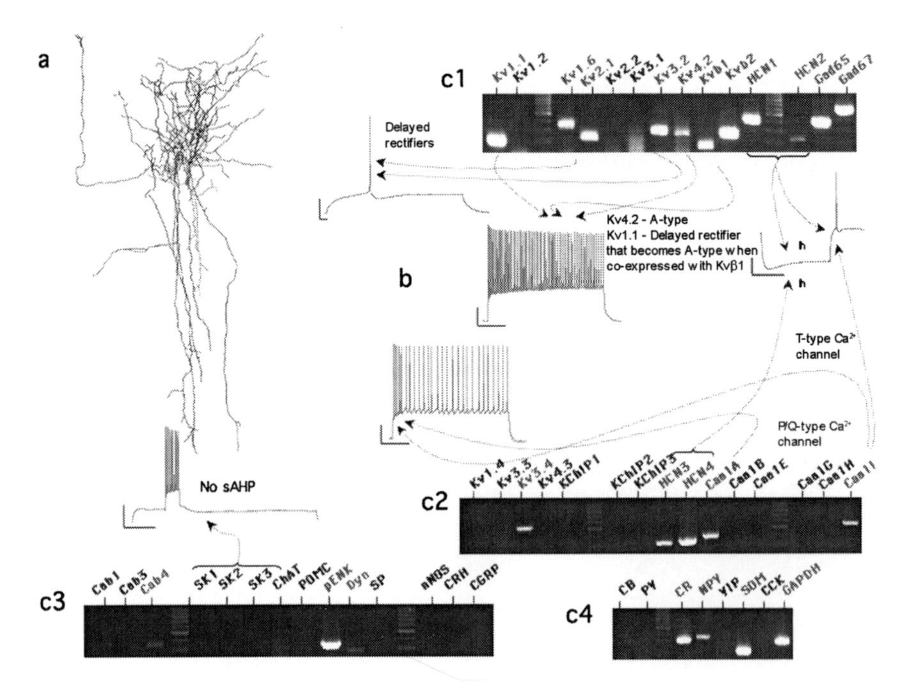

Fig. 3. A representative example of the Gene expression, Electrophysiology and Morphology (GEM) profiling of a layer three bitufted cell. **(A)** 3D computer reconstruction. Scale bar: 100 μm. **(B)** Representative responses to step current injections. Scale bars: current 20 mV, time 500 ms. **(C1–4)** Agarose gels showing the mRNA expression profile of 51 K^+ and Ca^{2+} channels, Ca^{2+} binding proteins, neuropeptides, and enzymes. For abbreviations, *see* **Fig. 1** legend. Soma and dendrites in light gray and axon in black.

suspected to be expressed by the cell. During calibration, different combinations of genes are distributed between the pools, and different primer pairs are tested until an even amplification of all genes in the pool was obtained (*see* **Notes 1** and **2**). The genes co-amplified in each multiplex PCR were (for abbreviations, *see* **Fig. 1** legend; for primers and amplicon length, *see* **Table 1**)

- Pool I–CB, PV, CR, NPY, VIP, SOM, CCK, and GAPDH.
- Pool II–SP, ChAT, POMC, pENK, Dyn, nNOS, CRH, CGRP, and Caα1E.
- Pool III–Kv1.1, Kv1.2, Kv1.6, Kv2.1, Kv2.2, Kv3.1, Kv3.2, Kv4.2, Kvβ1, Kvβ2, HCN1, HCN2, Gad65, and Gad67.
- Pool IV–Kv1.4, Kv3.3, Kv3.4, Kv4.3, KChIP1, KChIP2, KChIP3, HCN3, HCN4, Caα1A, Caα1B, Caα1G, Caα1H, Caα1I, Caβ1, Caβ3, Caβ4, SK1, SK2, and SK3.

Table 1
List of All Primers Used in This Study

mRNA (reference)	GeneBank number		Primes from 5′ to 3′	Amplicon size (bp)
GAPDH *(17)*	M17701	Sense	GCCATCAACGACCCCTTCAT	315
		Antisense	TTCACACCCATCACAAACAT	
Gad65 *(16)*	M72422	Sense	TCTTTTCTCCTGGTGGTGCC	391
		Antisense	CCCCAAGCAGCATCCACAT (Gad down)	
Gad67 *(16)*	M76177	Sense	TACGGGGTTCGCACAGGTC	600
		Antisense	Gad down	
ChAT *(22)*	M88488	Sense	GGCCATTGACAACCATCTTCTG	321
		Antisense	CTTGAACTGCAGAGGTCTCTCAT	
nNOS	U67309	Sense	CTTCCGAAGCTTCTGGCAAC	328
		Antisense	GCTGGATGGCTTTGAGGACAT	
CB *(16)*	M27839	Sense	AGGCACGAAAGAAGGCTGGAT	432
		Antisense	TCCCACACATTTTGATTCCCTG	
PV *(16)*	M12725	Sense	AAGAGTGCGGATGATGTGAAGA	389
		Antisense	ATTGTTTCTCCAGCATTTTCCAG	
CR *(16)*	X66974	Sense	CTGGAGAAGGCAAGGAAAGGT	311
		Antisense	AGGTTCATCATAGGGACGGTTG	
NPY *(16)*	M15880	Sense	GCCCAGAGCAGAGCACCC	362
		Antisense	CAAGTTTCATTTCCCATCACCA	
VIP *(16)*	X02341	Sense	TGCCTTAGCGGAGAATGACA	290
		Antisense	CCTCACTGCTCCTCTTCCCA	

(Continued)

Table 1
(Continued)

mRNA (reference)	GeneBank number		Primes from 5′ to 3′	Amplicon size (bp)
Som ***(16)***	K02248	Sense	ATCGTCCTGGCTTTGGGC	208
		Antisense	GCCTCATCTCGTCCTGCTCA	
CCK ***(16)***	K01259	Sense	CGCACTGCTAGCCCGATACA	216
		Antisense	TTTCTCATTCCGCCTCCTCC	
SP ***(21)***	M15191	Sense	GAGCATCTTCTTCAGAGAATCGC	468/513
		Antisense	TCGCTGGCAAACTTGTACAACTC	
POMC	K01878	Sense	ATAGACGTGTGGAGCTGGTG (in exon 2)	253
	J00759	Antisense	TTCCTCCGCACGCCTCTGC (in exon 3)	
PENK ***(21)***	M28263	Sense	CAAACAGGATGAGAGCCACTTGC	370
		Antisense	GCTTCTGCAGCTCCTTTGCTTC	
Dyn ***(18)***	M32781	Sense	GCCATAGGGGGATTTGGTAGC	266
	M32783	Antisense	AACCTCAGAGGGGATCACAAG	
CRH	M54987	Sense	AACTCAGAGCCCAAGTACGTTGA	335
		Antisense	TCACCCATGCGGATCAGAATC	
CGRP	L29188	Sense	GGTCGGGAGGTGTGGTGAAG (in exon 5)	420
	L00111	Antisense	TCGACAGGGTGGTTTATGGGG (in exon 6)	
Kv1.1 ***(19)***	M26161	Sense	CCGCCGCAGCTCCTCTACT	209
		Antisense	CAAGGGTTTTGTTTGGGGGCTTTT	
Kv1.2 ***(19)***	X16003	Sense	GAAAAGTAGAAGTGCCTCTACCATAA	458
		Antisense	TTGATATGGTGTGGGGGCTATGA	
Kv1.4 ***(19)***	X16002	Sense	CTGGGGGACAAGTCAGAGTATCTA	434
		Antisense	ACTCTCCTCGGGACCACCT	

Kv1.6	X17621	Sense	GGGAACGGCGGTCCAGCTA	351
		Antisense	GTGCATCTCATTCACGTGACTGAT	
Kv2.1 ***(20)***	X16476	Sense	CAACTTCGAGGCGGGAGTC	229
		Antisense	TCCAGTCAACCCTTCTGAGGAGTA	
Kv2.2 ***(20)***	M77482	Sense	ACCAGGAGGTTAGCCAAAAAGACT	446
		Antisense	AGGCCCCTTATCTCTGCTTAGTGT	
Kv3.1 ***(20)***	X62840	Sense	CCAACAAGGTGGAGTTCATCAAG	640
		Antisense	TGGTGTGGAGAGTTTACGACAGATT	
Kv3.2 ***(20)***	X62839	Sense	ACCTAATGATCCCTCAGCGAGTGA	302
		Antisense	CAAAATGTAGGTGAGCTTGCCAGAG	
Kv3.3	M84211	Sense	GAGACCCCCGTCCCAATG	179
		Antisense	CGGGGGAAGGGGCATAGTC	
Kv3.4 ***(19)***	X62841	Sense	TCAGGCACACGGGACAGAAAC	418/522
		Antisense	GGGCAGAGGACTTGGGAGACATA	
Kv4.2 ***(19)***	S64320	Sense	CCGAATCCCAAATGCCAATGTG	265
		Antisense	CCTGACGATGTTTCCTCCCGAATA	
Kv4.3	U42975	Sense	GGGCAAGACCACGTCACTCA	296/386
		Antisense	CTGCCCTGGATGTGGATGGT	
Kvb1 ***(19)***	X70662	Sense	AAGGGAGAAAACAGCAAAACAAGC	170
		Antisense	TGGCACCAAGGTTTTCAATGAGTT	
Kvb2	X76724	Sense	ACAGTGGCATCCCACCCTACT	283
		Antisense	GTGGACGATGGAGGACGACAAT	
KChIP1	AB046443	Sense	AAAGGCGACCCTCCAAAGATAAG	330
		Antisense	GGACAGTTCCTCTCAGCAAAATCG (KChIP1 *long*)	

(Continued)

Table 1
(Continued)

mRNA (reference)	GeneBank number		Primes from 5′ to 3′	Amplicon size (bp)
KChIP2	AF269283	Sense	RCARKYCCTKTACCGAGGCTTCA (KChIP up)	289
		Antisense	CAAGCATTTCCTCCTTTGTGATAC	
KChIP3	AB043892	Sense	KChIP up	249
		Antisense	GTCGTAGAGATTGAAGGCCCACT	
HCN1	AF247450	Sense	CCTCAAATGACAGCCCTGAATTG	405
		Antisense	TCGGTGTGGAACTACCAGGTGT	
HCN2	AF247451	Sense	CTCTCCGGCAACGCGTGTG	211
		Antisense	AGTCCCTGCGGTCCGGACT	
HCN3	AF247452	Sense	TGCCCCTCTCCCCTGATTC	335
		Antisense	TTCCAGAGCCTTTGCGCCTA	
HCN4	AF247453	Sense	AACCTGGGGGCTGGACAGA	462
		Antisense	CTGGGCAGCCTGTGGAGAG	
SK1	U69885	Sense	GCATGTGCACAACTTCATGATGGA (SK up)	449
		Antisense	TGGGCGGCTGTGGTCAGGTG	
SK2	U69882	Sense	SK up	461
		Antisense	CGCTCAGCATTGTAGGTGACATG	
SK3	U69884	Sense	SK up	192
		Antisense	CATCTTGACACCCCTCAGTTGG	
Caa1A *(18)*	M64373	Sense	GAGCGGCTGGATGACACAGAAC	420
		Antisense	CTGGCGACTCACCCTGGATGTC	

Caa1B	M92905	Sense	TTGGCTCCTTCTTCATGCTCAAC	409
		Antisense	GATAAGGAACCGGAACATCTTCTC	
Caa1E ***(18)***	L15453	Sense	AGACTGTGGTGACTTTTGAGGACC	693
		Antisense	GAGCTATGGGGCACCATGGCTT	
Caa1G	AF027984	Sense	TGGGCTCCTTCTTCATGATCAAC (CaT up)	407
		Antisense	GGAACTCTGAGCGTCCCATTAC	
Caa1H	AF290213	Sense	CaT up	271
		Antisense	CTGCTGGGATCCACCTTCTTAC	
Caa1I	AF086827	Sense	CaT up	556
		Antisense	AGGTCCGAGGAGACCCCATC	
Cab1	X61394	Sense	CCCTAAACTGCTGTGGGTGGA	359
		Antisense	CCCAGCTCTGCTCCCCAAAG	
Cab3	M88751	Sense	ACTGACCACCTCCTGCCCTAC	555
		Antisense	GTCCTGCCTCACCTGCACTG	
Cab4	L02315	Sense	GCTATGGTATTTGTTTGCTGGAAG	351
		Antisense	GACTGCAGAAGGAACAACACCTC	

For abbreviations, *see* **Fig. 1** legend. R=A or G; K=G or T; Y=C or T.

9. The first amplification round consists of 10 min hot start at 95°C followed by 25 cycles (94°C for 40 s, 56°C [pools I, II, and IV] or 58°C [pool III] for 40 s, and 72°C for 1 min) performed with a programmable thermocycler (Eppendorf, Germany). For each pool, all genes are simultaneously amplified in a single tube containing 1/10 (pools I and IV) or 2/5 (pools II and III) of the RT product, 100 n*M* of each of the primers, 200 μ*M* of each dNTP (Promega), 20 μl solution Q (Qiagen, Germany), and 5 U HotStarTaq DNA polymerase (Qiagen) in a final volume of 100 μl.
10. The second round of PCR consists of 40 cycles (94°C for 40 s, 56°C [pools I, II, and IV] or 58°C [pool III] for 40 s, and 72°C for 1 min). In this case, each gene is individually amplified in a separate test tube containing: 1 μ*M* of its specific primers, 2 μl of first PCR product (template), 200 μ*M* of each dNTP, 1 *M* betaine (Sigma), and 1 U TaqZol DNA polymerase (Tal-Ron, Israel), in a final volume of 20 μl.
11. The products of the second PCR are analyzed in 1.5% agarose gels using ethidium bromide.
12. Amplification specificity can be randomly verified by restriction analysis.

3.4. Controls for the Single-Cell RT–Multiplex PCR

1. For each PCR amplification, controls for contaminating artifacts are performed using sterile water instead of cDNA.
2. A control for nonspecific harvesting of surrounding tissue components is randomly employed by advancing pipettes into the slice and retrieving without seal formation and suction.
3. Amplification of genomic DNA can also be avoided by designing primers overspanning introns and/or never harvesting the cell nucleus.
4. Controls in which the RT is omitted should give negative results (no bands).

4. Notes

1. The detection of specific mRNA transcripts from a single cell is a technical challenge because of the very low amount of mRNA contained in a single neuron (approximately 10^6 molecules transcribed from about 15,000 different genes [*9*]). Besides, this very small number of mRNA molecules may be (1) lost in one of the numerous steps from the harvesting of the neuron's cytoplasm to the final PCR amplification or (2) destroyed by endogenous or contaminating RNases. Depending on the number of genes to be studied, the investigator can choose between different approaches.

   a. One gene per cell *(2)*.

      - In this case, the entire contents harvested from the cell are reverse transcribed and PCR amplified for one single gene.
      - Advantages: it is the simplest approach to set (requires only to design one primer pair) and has the lowest chance to obtain false-negatives.

- Disadvantages: it is not possible to study co-expression of two or more genes, and if the gene is not detected, it is impossible to know if the gene is not expressed in that specific cell or there was a failure in the harvesting or amplification (false-negative).

b. Multiple genes from a single cell.
Depending on the way that the cytoplasmic contents are processed:

i. Splitting the cell's cytoplasm into as many reactions as genes to be tested and then amplifying each gene independently ***(10)***.

- Advantages: (1) Still is a relative simple and straightforward approach (each primer pair is designed independently). (2) Because an internal control can be included (house keeping gene, a gene always expressed in every cell), it is possible to control for false-negatives.
- Disadvantages: as stated before, the number of mRNA molecules contained by a single neuron is small. If the cell contents are spilt into too many fractions, it may happen that, although the gene was expressed, the aliquot for testing the expression of that gene did not contain any of its mRNA molecules (giving rise to false-negatives).

ii. Multiplex PCR ***(11)***. In this case, the genes whose expression is going to be tested are pre-amplified in a common PCR before being split (into as many reactions as genes to be tested), and subsequently, they are amplified independently (as described in the previous sample).

- Advantages: (1) The pre-amplification step ensures that, regardless of the number of fractions, any aliquot may contain at least a few copies of each mRNA species expressed by the neuron. (2) Because an internal control can be included (house keeping gene, a gene always expressed in every cell), it is possible to control for false-negatives.
- Disadvantages: The primer design becomes complex as it is required to avoid cross-interactions between the different primer pairs and amplification products. The number of genes that can be included in the multiplex PCR is limited as the probability to encounter false-negatives because of amplification failures rises as the number of primer pairs included in the multiplex pool because of (1) competition for reagents, (2) interference between the multiple primers (unspecific annealing of one primer with others or with the PCR products from other genes), and (3) difficulty in designing primers that will optimally melt and anneal in the same conditions (temperature, salt concentrations, times, etc.) ***(12)***.

iii. Degenerated primers ***(13)***. In the particular case where the genes to amplify show a high degree of similarity in their sequence (as in ion channel or receptor families), a single pair of degenerated primers can be designed for the first PCR (located in regions identical or nearly identical for all the genes). Subsequently, the identity of each individual gene can be determined by (1) differences in the size of the PCR products (if the regions amplified have a different size for different members of the family), (2) using restriction enzyme analysis where each selected enzyme specifically cuts the cDNA of only one gene, and (3) performing second PCR using nested primers (primers located inside the previously amplified region) specific for each gene.

- Advantages: (1) Relative simple and straightforward primer design (only one primer pair has to be designed for the first PCR, whereas for the second PCR, each primer pair is designed independently).
- Disadvantages: depending on the sequence similarity of the different genes to be amplified, many times, the degenerated primers will not amplify all the genes with the same efficiency (*see* **Fig. 4** comparison between the multiplex and degenerated protocols to amplify the members for the Kv1 family).

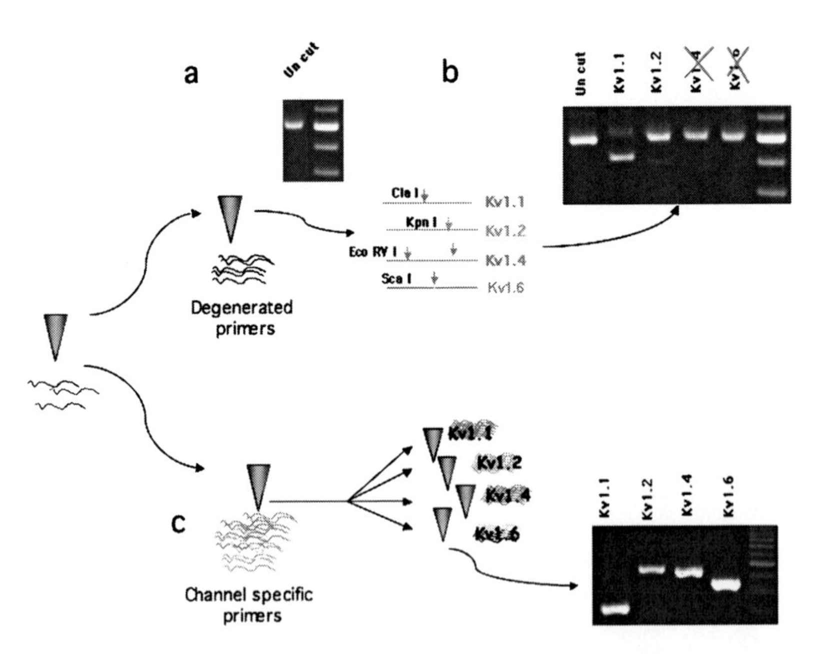

Fig. 4. Kv1 family members amplified using degenerated primers channel-specific primers. **(A)** Simultaneous amplification of four members of the Kv1 family using degenerated primers. **(B)** Determination of the identity of each Kv1 family member by restriction enzyme analysis. **(C)** Simultaneous amplification of the Kv1 family members Kv1.1, Kv1.2, Kv1.4, and Kv1.6 using gene-specific primers.

2. As it can be seen, the higher the number of genes to be tested, the higher the (1) challenge of designing the experiment, (2) number of manipulations and experiment duration, and (3) number of false-negatives. For beginners (in particular beginners with low or no experience in molecular biology), we recommend to start by the easiest (one gene per cell), then move to testing multiple genes from a single cell using published primer sequences and only then launch to design their own multiplex mixtures.
3. The analysis of single-cell GEM data presents special opportunities but requires consideration of several sources of systematic and random error, in particular testing for the effects of false-negatives in the gene expression. This can be overcome by the use of statistics. Moreover, the cell-wise correlation of GEM information allows efficient application of modern statistical classification techniques with fewer samples and lower noise than might be required from mixed cellular tissue extracts. The current best set of statistical techniques is distributed for free by The Institute for Genomics Research Multi-Experiment Viewer ***(14)***. Examples of application of these techniques can be found in ***(5,6,15)***.

5. Conclusions

The combination of patch-clamp recordings with morphological identification and gene expression characterization provides a novel system approach to relate gene expression profiles, neuronal morphology, and electrical behavior. Using these techniques, it is possible to derive precise information about (1) the correlation and developmental changes between gene expression and behavior and (2) the alterations leading to pathologies in developmental, degenerative, and psychological brain disorders.

Acknowledgments

The authors thank Prof. Phil Goodman for helpful insights on the statistical approaches to the single-cell RT–PCR data analysis. We thank Shaoling Ma, Claudia Herzberg, Raya Eilam, and Tal Hetzroni for their technical assistance. This work was supported by the National Alliance for Autism Research and a European Union grant.

References

1. Eberwine, J., H. Yeh, and K.C.Y. Miyarisho, Analysis of gene expression in single live neurons. *Proc Natl Acad Sci USA*, 1992; 89(7): p. 3010–14.
2. Lambolez, B., et al., AMPA receptor subunits expressed by single Purkinje cells. *Neuron* 1992; 9(2): p. 247–58.
3. Sucher, N.J. and D.L. Deitcher, PCR and patch-clamp analysis of single neurons. *Neuron* 1995; 14(6): p. 1095–100.

4. Ceranik, K., et al., A novel type of GABAergic interneuron connecting the input and the output regions of the hippocampus. *J. Neurosci.* 1997; 17(14): p. 5380–94.
5. Toledo-Rodriguez, M., et al., Neuropeptide and calcium binding protein gene expression profiles predict neuronal anatomical type in the juvenile rat. *J. Physiol.* 2005; 567: p. 401–13.
6. Toledo-Rodriguez, M., et al., Correlation maps allow neuronal electrical properties to be predicted from single-cell gene expression profiles in rat neocortex. *Cereb. Cortex* 2004; 14(12): p. 1310–27.
7. Wang, Y., et al., Anatomical, physiological, molecular and circuit properties of nest basket cells in the developing somatosensory cortex. *Cereb. Cortex* 2002; 12(4): p. 395–410.
8. Dodt, H.U. and W. Zieglgansberger, Visualizing unstained neurons in living brain slices by infrared DIC-videomicroscopy. *Brain Res.* 1990; 537(1–2): p. 333–6.
9. Sargent, T.D., Isolation of differentially expressed genes. *Methods Enzymol.* 1987; 157: p. 423–32.
10. Foehring, R.C., et al., Unique properties of R-type calcium currents in neocortical and neostriatal neurons. *J. Neurophysiol.* 2000; 84(5): p. 2225–36.
11. Ruano, D., et al., Kainate receptor subunits expressed in single cultured hippocampal neurons: molecular and functional variants by RNA editing. *Neuron* 1995; 14(5): p. 1009–17.
12. Edwards, M.C. and R.A. Gibbs, Multiplex PCR: advantages, development, and applications. *PCR Methods Appl.* 1994; 3(4): p. S65–75.
13. Plant, T.D., et al., Single-cell RT-PCR and functional characterization of Ca2+ channels in motoneurons of the rat facial nucleus. *J. Neurosci.* 1998; 18(23): p. 9573–84.
14. Saeed, A.I., et al., TM4: a free, open-source system for microarray data management and analysis. *Biotechniques* 2003; 34(2): p. 374–8.
15. Cauli, B., et al., Classification of fusiform neocortical interneurons based on unsupervised clustering. *Proc. Natl. Acad. Sci. USA* 2000; 97(11): p. 6144–9.
16. Cauli, B., et al., Molecular and physiological diversity of cortical nonpyramidal cells. *J. Neurosci.* 1997; 17(10): p. 3894–906.
17. Aranda-Abreu, G., et al., Embryonic lethal abnormal vision-like RNA-binding proteins regulate neurite outgrowth and tau expression in PC12 cells. *J. Neurosci.* 1999; 19(16): p. 6907–17.
18. Glasgow, E., et al., Single cell reverse transcription-polymerase chain reaction analysis of rat supraoptic magnocellular neurons: neuropeptide phenotypes and high voltage-gated calcium channel subtypes. *Endocrinology* 1999; 140(11): p. 5391–401.
19. Song, W.J., et al., Somatodendritic depolarization-activated potassium currents in rat neostriatal cholinergic interneurons are predominantly of the A type and attributable to coexpression of Kv4.2 and Kv4.1 subunits. *J. Neurosci.* 1998; 18(9): p. 3124–37.

20. Baranauskas, G., T. Tkatch, and D.J. Surmeier, Delayed rectifier currents in rat globus pallidus neurons are attributable to Kv2.1 and Kv3.1/3.2 K(+) channels. *J. Neurosci.* 1999; 19(15): p. 6394–404.
21. Surmeier, D.J., W.J. Song, and Z. Yan, Coordinated expression of dopamine receptors in neostriatal medium spiny neurons. *J. Neurosci.* 1996; 16(20): p. 6579–91.
22. Yan, Z. and D.J. Surmeier, Muscarinic (m2/m4) receptors reduce N- and P-type Ca2+ currents in rat neostriatal cholinergic interneurons through a fast, membrane-delimited, G-protein pathway. *J. Neurosci.* 1996; 16(8): p. 2592–604.

9

Mechanosensitive Ion Channels Investigated Simultaneously by Scanning Probe Microscopy and Patch Clamp

Matthias G. Langer

Summary

Mechanosensitive ion channels play an important role for the perception of mechanical signals such as touch, balance, or sound. Here, a new experimental strategy is presented providing well-defined access to single mechanosensitive ion channels in living cells. As a representative example, the investigation of mechanosensitive transduction channels in cochlear hair cells is discussed in detail including all essential technical aspects. Three different techniques were combined: atomic force microscopy (AFM) as a device for local mechanical stimulation, patch clamp for recording the current response of mechanosensitive ion channels, and differential interference contrast (DIC) microscopy equipped with an upright water-immersion objective lens. A major challenge was to adapt the mechanical design of the AFM setup to the small working distance of the light microscope and the electrical design of the AFM electronics. Various protocols for the preparation and investigation of the organ of Corti with AFM are presented.

Key Words: Cochlea; hair cells; stereocilia; organ of Corti; mechanoelectrical transduction; transduction channel; AFM; patch clamp.

1. Introduction

Hair cells are mechanosensory cells in the inner ear responsible for one of the major processes of hearing: the transduction of sound into an electrical signal *(1)*. In particular, the hair bundle is the antenna for mechanical stimuli. Incoming sound deflects the hair bundles of cochlear hair cells, thereby opening

From: *Methods in Molecular Biology, vol. 288: Patch-Clamp Methods and Protocols*
Edited by: P. Molnar and J. J. Hickman © Humana Press Inc., Totowa, NJ

mechanosensitive transduction channels located in the stereocilia. Recently, transient receptor potential cation channel, subfamily A, member1 (TRPA1) *(2)* was identified as a major component of the protein complex forming the transduction channel. However, molecular identity of other involved molecules such as the tip link and their interaction is still uncertain. Commonly, for functional transduction-channel recordings, entire hair bundles are displaced by various methods such as the water jet or glass fibers, whereas patch clamp measures the resulting current responses. However, single-channel recordings using these methods are limited to hair bundles with only few functional transduction channels. Therefore, a new instrument overcoming this limit was developed allowing single-channel recordings in intact hair bundles with many functional transducer channels. Atomic force microscopy (AFM), patch clamp, and differential interference contrast (DIC) light microscopy were combined in one setup giving access to both the mechanical and the electrical properties of single transduction channels. A mechanical design was developed scanning the AFM probe, while keeping the sample and attached patch-clamp pipette at rest. AFM very locally measures and applies force to cochlear hair bundles and additionally gives the opportunity to image the investigated specimen with nanometer resolution. This opens the perspective for measurement of single-channel gating from the channel's resting position to the point of maximum open probability. An AFM scan protocol was developed allowing displacement and imaging of individual stereocilia, opening of single transduction channels, and calculation of the local hair bundle stiffness.

2. Materials

2.1. Hair Cells From Neonatal Rats

1. Animals: Pregnant Wistar rats were purchased from Charles River GmbH, Germany.
2. For isolation of the organ of Corti, postnatal rats were killed by cervical dislocation in accordance with the animal care guidelines of the University of Ulm within 2–12 days after birth.

2.2. Mounting Organs of Corti Onto Glass Coverslips

Organs of Corti were cut into three sections (basal, medial, and apical) and either mounted on Cell-Tak®-coated (Becton Dickinson GmbH, Germany) glass coverslips (diameter 10 mm, thickness 0.3 mm) or clamped by a fine glass fiber.

2.3. Preparation Buffer and Cell Culture Medium

1. During dissection, (4-(-2 hydroxyethyl)-1-piperazineethanesulfonic acid (HEPES)–Hanks' solution was used containing 5.4 m*M* KCl, 0.5 m*M* $MgCl_2 \cdot 6H_2O$, 0.4 m*M*

$MgSO_4 \times 7H_2O$, 141.7 m*M* NaCl, 1.6 m*M* $CaCl_2 \times 2\ H_2O$, 10.0 m*M* HEPES, 3.4 m*M* L-glutamine, and 6.3 m*M* D-glucose solved in 1000 ml pure water, pH 7.3, filtrated sterile.
2. Those samples not used directly after preparation are transferred into Ø 35-mm Falcon dishes filled with 3 ml culture medium (MEM D-VAL, Gibco BRL Life Technologies Ltd, UK, with 10% heat-inactivated fetal calf serum and 10 m*M* HEPES buffer, pH 7.2) supplemented at 310 K and 5% CO_2. No antibiotics were added to the medium to prevent ototoxic effects.

2.4. Extracellular Solutions

During the experimental procedure, cochlear cultures are bathed at room temperature (297 K) in a solution containing 144 m*M* NaCl, 0.7 m*M* NaH_2PO_4, 5.8 m*M* KCl, 1.3 m*M* $CaCl_2$, 0.9 m*M* $MgCl_2$, 5.6 m*M* D-glucose, 10.0 m*M* HEPES NaOH, pH 7.4, and osmolarity 304 mOsm.

The solution is exchanged before and after investigation using a conventional perfusion system in order to keep cells in good condition.

2.5. Intracellular Solutions

1. Intracellular solution: Patch-clamp electrodes were filled with a solution containing 135.0 m*M* KCl, 3.5 m*M* $MgCl_2$, 0.1 m*M* $CaCl_2$, 2.5 m*M* Na_2ATP, 5.0 m*M* ethylene glycol tetraacetic acid (EGTA), and 5.0 m*M* HEPES (pH adjusted to 7.4).
2. Microloaders: The patch-clamp pipette is filled from the back side with intracellular solution using a 5-ml syringe (Becton Dickinson GmbH), which is connected through a syringe filter (Minisart RC 4, Sartorius AG, Germany) to a 20-μl microloader tip (Eppendorf, Germany).

2.6. AFM Probes

1. Stiffness measurements: Triangular-shaped Si_3N_4-cantilever beams (type: gold-coated MLCT-AUHW–nitride microlevers, VEECO Instruments GmbH, Germany) with spring constants ranging from 0.01 to 0.02 *N*/m with pyramidal-shaped tips are used for stiffness measurements and displacement of single stereocilia. The typical tip radius is 50 nm (*see* **Note 1**).
2. High-resolution imaging: For higher spatial resolution and imaging applications, sharp nitride microlevers (type: MSCT-AUHW, VEECO Instruments GmbH) with a nominal tip radius of 10 nm are used.

2.7. Patch-Clamp Equipment

1. Patch-clamp electrodes: Patch-clamp electrodes were pulled from 1 mm Q100-70-75 quartz glass without filament, outer diameter 1.0 mm, inner diameter 0.70 mm, length 75 mm (Science Products GmbH, Germany).

2. Pipette fabrication: Patch-clamp pipettes were pulled using a Sutter P-2000 laser puller (Sutter Instrument, USA). The typical resistance is about 4 MΩ.
3. Patch-clamp amplifier: For whole-cell voltage-clamp measurements, a WPC-100 patch-clamp amplifier with separate PA-50 headstage (ESF Electronic, Germany) is used.
4. Micromanipulation: A three-axis translation stage (two-axis translation stage from Spindler & Hoyer, Goettingen, Germany combined with a single-axis translation stage M-461-X-M, Newport GmbH, Germany) provides positioning of the patch-clamp electrodes with submicron resolution.

3. Methods

Local investigation of cochlear hair bundles was technically approached from the metrological and the preparative side. A novel setup combining AFM and patch clamp was developed allowing manipulation and imaging of single stereocilia of cochlear hair bundles and simultaneous measurement of evoked current responses from mechanosensitive transducer channels. One major problem that had to be solved was to combine AFM with an upright high-resolution optical microscope providing optical control of the AFM probe's position during approach to the sample. Another key feature of this setup is the possibility to simultaneously perform patch-clamp and AFM measurements in physiological environment. For proper positioning of the patch-clamp electrode to the hair cell body, an AFM design providing free access with the pipette to the sample is essential. Commercially available AFM microscopes either do not allow free access from outside with a pipette to the AFM probe or do not give free access to the AFM cantilever with an upright water-immersion objective. Important technical details necessary to solve these problems will be emphasized in the following paragraph.

For combined force/patch-clamp measurements at inner hair cells (IHCs) and outer HCs (OHCs), two different preparation methods have been established allowing the AFM stylus to approach the tips of individual stereocilia from the top.

3.1. Preparation and Culturing Organs of Corti

1. According to a method described by Sobkowicz et al. *(3)* and Russell and Richardson *(4)* for mice, organs of Corti are isolated from 2- to 12-day-old Wistar rats.
2. Rats were killed by cervical dislocation.
3. Immediately after decapitation, the frontal part of the skull is separated from the head by sharp scissors in order to reduce bacterial contamination of the cultured specimen.
4. The skull is split into two halves by sharp scissors and transferred to HEPES–Hanks' solution.

5. The brain is extracted from both hemispheres using fine forceps.
6. Further preparation steps are observed under a binocular microscope (magnification ×40, Leica MZ6, Germany).
7. Both temporal bones (from the left and right hemisphere) were extracted by resection of the surrounding bone and muscle.
8. The middle ear is opened and the cochlea identified by its characteristic shape protruding into the tympanum. The cochlea can be opened easily by fine forceps because of the lack of ossification of the temporal bone in neonatal rats.
9. The cartilaginous or in later ages bony shell of the cochlea can be removed by two fine tweezers. In neonatal rats, usually the whole ductus spiralis containing the organ of Corti and the stria vascularis remains as a spiral structure intact.
10. Using two fine tweezers, the organ of Corti is separated from the stria vascularis like opening a zip. During this procedure, the two compartments separated by the membrane of Reissner are freely accessible, whereas the tectorial membrane still covers the hair cell stereocilia.
11. Under optimal conditions, the resulting spiral structure contains the whole length of the organ of Corti and can be split into three pieces (basal, medial, and apical) using very fine tweezers.
12. Obtained C-shaped segments of the organ of Corti are mounted on small glass coverslips using adhesive Cell-Tak or small borosilicate glass fibers (diameter 150–250 μm) keeping hair cells in mechanically stable condition for AFM and patch-clamp recordings.

3.2. Mounting Techniques for the Organ of Corti

Depending on the type of cell being investigated, two different methods for mounting the organ of Corti (*see* **Fig. 1**) are used.

1. Investigating OHCs: For AFM studies of OHCs, the organ is placed with the basilar membrane onto Cell-Tak-coated glass coverslips (diameter 10 mm). This configuration provides stereociliary hair bundles of OHC mostly standing upright in respect to the supporting glass coverslip.
2. Investigating IHCs: Hair bundles of IHC are linearly arranged at an angle of about 45° in respect to the supporting glass coverslip. Moreover, these hair bundles are located very near to the steep edge of the inner sulcus preventing easy access with the AFM probe to the IHC's hair bundle. In order to get access to these sensory hair bundles with the AFM, probe sections of the organ of Corti were turned upside down (with the tectorial membrane facing the support) to the Cell-Tak-coated glass coverslip. The sample was transferred into Ø 35-mm Falcon dishes filled with 3 ml MEM culture medium supplemented at 310 K and 5% CO_2. After about 24 h, the outer border of the organ is bend up by about 90°. After 48 h after preparation, the border is turned down by 180°. Using this configuration, IHCs are standing upright and can easily be approached from the top by the AFM tip.

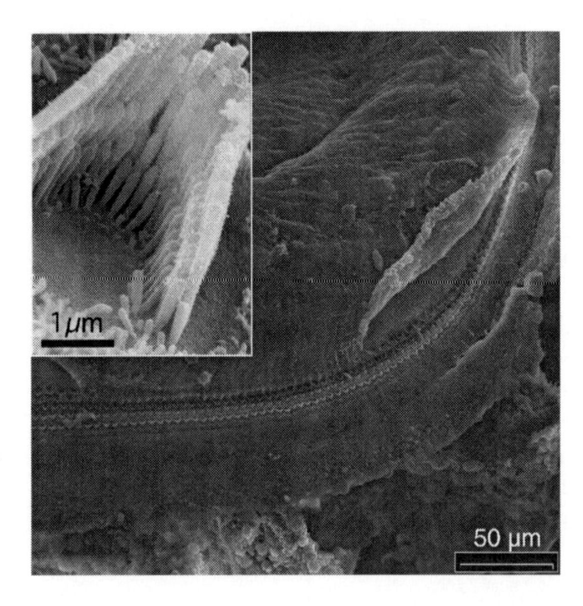

Fig. 1. Scanning electron microscope image of a section of the organ of Corti from a postnatal rat (day 5 after birth). At this early stage of development, the tectorial membrane does still not cover hair bundles of outer hair cells (OHCs). OHCs are arranged in three rows and are free accessible by the atomic force microscopy probe. In the upper left, the structure of a V-shaped hair bundle from an OHC (day 9 after birth) is depicted at higher magnification. Hair bundles of OHCs are arranged in three rows with different length and are connected by lateral links and tip links. Tip links connect the taller stereocilia with adjacent stereocilia of the shorter row and are thought to directly pull at the mechanotransducer channels.

3.3. Light Microscopy

The whole setup can be divided into three major units, namely, the optical microscope, the AFM, and the patch-clamp setup, depicted in **Fig. 2**. All components are mounted on a vibration isolation table (Physik Instrumente GmbH, Germany) reducing mechanical vibrations in the frequency range between 15 and 100 Hz.

Fig. 2. Mechanical design of the combined atomic force microscopy (AFM)/patch-clamp setup. (**A**) Side view from the left side depicting the AFM head (1) including the three-axis piezoelectric tube scanner (2 and 3) and the XY-translation stage positioning the specimen chamber (4). Patch-clamp pipette and AFM cantilever are monitored from

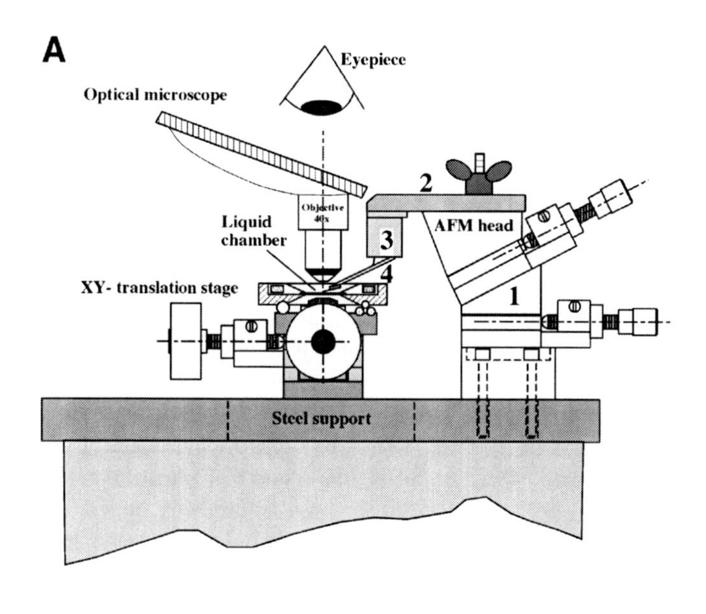

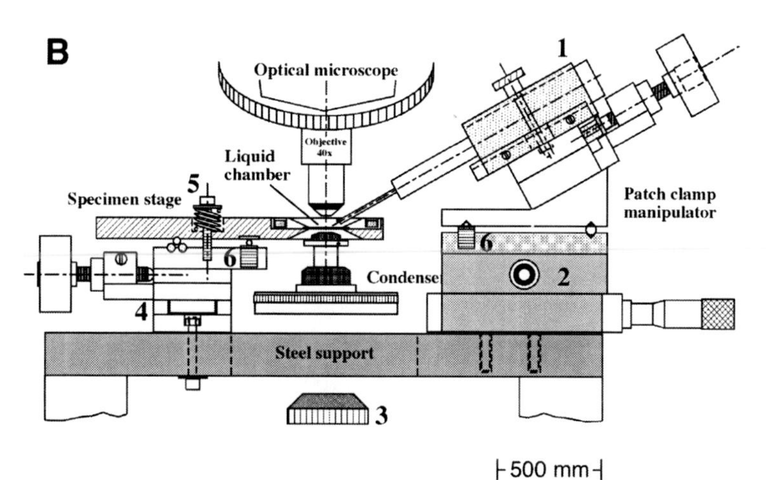

⊢500 mm⊣

Fig. 2. **(B)** the top through the microscope eyepiece. Front view of the AFM/patch-clamp setup. The patch-clamp preamplifier headstage (1) is mounted on the top of a three-axis translation device (2). The specimen is illuminated from below (3) using a 100-W halogen light source. Both patch-clamp headstage (1) and specimen translation stage are equipped with an identical piezostack actuator (6) lifting patch-clamp electrode and specimen synchronously up and down. The specimen chamber can be positioned in two directions by means of a XY-translation stage (4).

1. Positioning of the light microscope: For observation of cochlear hair cells and laser detection of the AFM cantilever displacement, an upright DIC microscope (Axioskop FS, Zeiss, Germany) is used. The entire light microscope is mounted on a custom-built XY-translation stage allowing independent positioning of the objective with respect to the AFM head and the patch-clamp electrode. The translation stage can be disconnected from the micrometer screw and moved back by 200 mm providing free access to the AFM cantilever and the specimen chamber. XY-translation stage and front micrometer screw are connected by strong NdFeB magnets (BramagN-magnets, Peter Welter GmbH, Germany) providing fast connection and disconnection of both components.
2. DIC light microscopy: Samples are observed in a light microscope (Axioskop FS I, Zeiss) featuring infrared DIC and an Achroplan ×40/0.75 water-immersion objective (Zeiss) with 1.92 mm working distance. The outside of the objective is coated with a polymer facilitating undisturbed electrophysiological measurements of ion channel currents. The sample is illuminated in transmission by a 100-W halogen light source. DIC provides information from a thin optical section of the organ of Corti (total thickness 50–250 μm), which helps to identify hair bundles in the apical region and supports proper coarse approach of the AFM tip to the top of hair bundles, being only 1.5–2.0 μm in length.
3. Eyepiece and camera: CCD camera (Model C2400, Hamamatsu Photonics, Japan) and eyepiece are mechanically separated from the optical microscope reducing mechanical vibrations of the microscope (*see* **Fig. 3**). Both components are mounted on two aluminum columns, located on the left and the right of the light microscope, and can be adjusted in height.

3.4. AFM Setup

Patch-clamp setup, specimen chamber, and AFM head are mounted on a U-shaped steel plate supported by two 190 mm high and 20 mm thick aluminum columns depicted in **Fig. 2**.

1. Specimen chamber: For investigation, sensory hair cells are transferred into a liquid chamber consisting of a conical Polychlortrifluorethylene (PCTFE) ring and a glass coverslip with a diameter of 22 mm. The glass coverslip is cemented to the bottom of the ring using low-melting wax. The entire chamber is mounted with three NdFeB magnets on a magnetic steel plate having an aperture of 15 mm. A micrometer screw-driven XY-translation device (M-461-XY-M, Newport GmbH) allows the adjustment of the specimen chamber in both horizontal directions.
2. Vertical positioning of the specimen chamber: Vertical fine approach of the sample toward the AFM tip is done by means of a piezoelectric actuator lifting the specimen chamber with subnanometer resolution. The steel plate supporting the specimen chamber is pivoted on three ball bearings, whereas one ball bearing is supported by a piezoelectric stack (pa/qna/10/8×8, Marco GmbH, Germany). Elongation of

Fig. 3. Photograph depicting the eyepiece and CCD camera, which are separated from the light microscope. This design reduces the mass at the front end of the light microscope and shifts the resonance frequency of the light microscope to higher frequencies. Moreover, the resolution of the optical detection system, which is mounted on the light microscope, is improved. The aluminum support used to mount both components is connected to two aluminum columns allowing vertical adjustment of eyepiece and CCD camera.

the piezoelectric stack lifts the steel support by about 20 μm at a driving voltage of 200 V. As vertical movement of the specimen chamber is small compared with the distance from the axis of rotation, lateral movements can be neglected.

3. AFM head: The AFM head is placed in front of the light microscope. It consists of a three-axis translation device (M-461-XYZ-M, Newport GmbH) with the vertical translation unit tilted by 26° in respect to the horizontal plane. For enhanced adjustment in the submicron range, Polytetrafluorethylen (PTFE) knobs being

25 mm in diameter were added to the standard bolt head. A removable steel plate is mounted on top of the translation stage, acting as a support for a piezoelectric tube scanner *(5)* (material VP-A55, Ø 10.2 mm, length 25.4 mm; Valpey Fisher Corp., USA). The outer electrode of the piezoelectric tube is segmented into eight commensurate electrodes electrically connected in a way that the walls at the end of the piezoelectric tube remain perpendicular to the direction of motion *(6)*. The AFM tip consequently moves in an even plane in contrast to common piezoelectric tube scanners with only four outer electrodes scanning on a curved plane. The maximum scan size is 6.5 μm for maximum driving voltages of ±160 V applied to the outer-segmented electrodes. The maximum axial elongation of the piezoelectric tube is 3.7 μm, which is sufficiently high for vertical fine approach of the AFM tip with nanometer resolution. The AFM tip is positioned below the objective (*see* **Fig. 2**) using a thin titanium beam being connected to the piezoelectric tube scanner. The mechanical resonance frequency of the entire scan head was found to be 1.9 kHz. For combination of AFM and patch clamp, it is very important to consider that after a seal between patch-clamp pipette and hair cell has formed relative movements between these two components have to be prevented. This problem was solved scanning the cantilever rather than the sample.

4. Optical detection: As the AFM cantilever is movable in all three directions, an optical detection has been developed being sensitive only to bending of the cantilever beam rather than lateral scan movements. In our setup, the optical deflection method was combined with an upright light microscope (*see* **Note 2**). Compared with other detection methods, it has the advantage of being exclusively sensitive to angular changes if the laser beam is oriented perpendicular to the reflective surface of the AFM cantilever. An interferometrical detection was not chosen because for a slightly tilted AFM cantilever, lateral scan movements result in changes of the optical path although no force acts on the AFM tip. Another disadvantage of an interferometer is the nonlinear relationship between intensity and displacement. Force interaction between AFM tip and specimen is detected using a modified optical deflection method *(7,8)*. The main components such as the focusing optics, beamsplitter, and detector are attached to the connector for the fluorescent light source at the backside of the light microscope (*see* **Fig. 4**). A 10-mW pigtailed laser diode module (Schäfter&Kirchhoff, Germany) with a wavelength of 675 nm and single-mode fiber optics was used as a laser source providing a round-shaped Gaussian beam profile.

The laser light passes the following optical components:

1. Focusing lens with FC connector with a focal length of 280 mm and an added antireflection coating for 675 nm. This component focuses the laser light in the back focal plane of the ×40 water-immersion objective.
2. 50/50 beamsplitter cube transmits 50% of the laser light to the AFM cantilever while the other 50% gets lost.

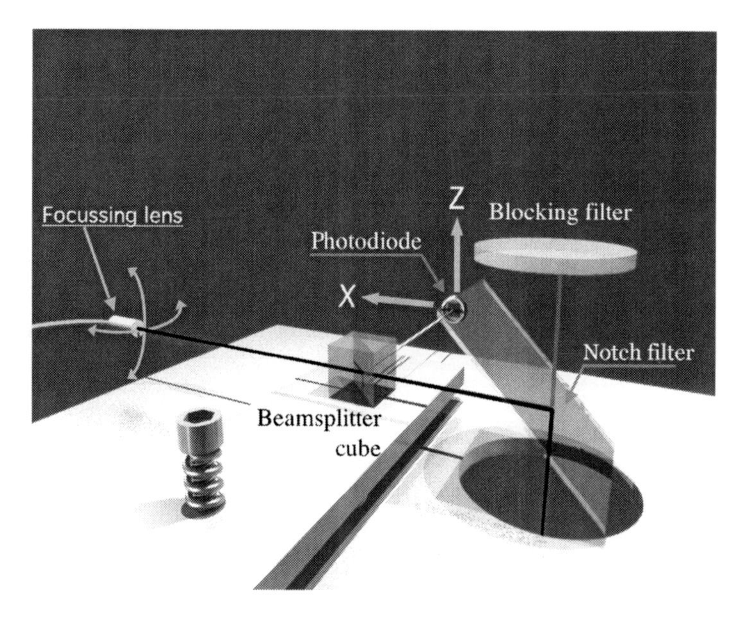

Fig. 4. Optical components of the optical lever detection. The design basically consists of a pigtailed laser diode (1 mW, wavelength 675 nm) connected to a single-mode fiber. The end of the fiber is connected to a focusing lens using a FC connector. The laser light is transmitted through a 50/50 beamsplitter cube and reflected to the objective at an angle of 90° using a laser line notch filter. The focal length of the focusing lens was matched to the back focal plane of the objective such that the laser light is collimated by the objective. Vertical translational movements of the atomic force microscopy (AFM) cantilever therefore do not result in changes of the laser signal on the photodiode. The laser light reflected by the AFM cantilever surface propagates the same way back to the beamsplitter cube and is reflected to a position-sensitive four-quadrant photodiode. A blocking filter protects the experimenter's eyes against high laser radiation.

3. Interference filter (Chroma Technology Corp., USA) selectively reflecting the laser light with 98% to the microscope objective. Visible light from the sample above a wavelength of 700 nm and below 600 nm can directly pass the mirror and is monitored in the eyepiece or by CCD camera.
4. Water-immersion objective (×40/0.75, Zeiss) acts as a collimator for the laser light resulting in a laser spot of 10 μm in diameter on the AFM cantilever (*see* **Note 3**).
5. Reflective gold-coated surface of an AFM cantilever is a Si_3N_4-AFM cantilever coated with 5 nm gold, and a spring constant of 0.01 *N*/m was used for experiments.

6. As the AFM cantilever bends because of the force interaction with cochlear hair bundles, the angle of reflection changes. This change is detected using the following optical components:
7. A ×40/0.75 water-immersion objective as described in step 4. Laser light reflected by the AFM cantilever is detected through the same objective.
8. The identical interference filter as described in step 3 reflects the laser light the same way back.
9. A beamsplitter cube reflects 50% of the laser light onto a four-quadrant photodiode. The other 50% gets lost but does not reenter the laser diode. This is very important for high laser power stability as reentered laser light induces intensity fluctuations.
10. Concave lens with a focal length of −10 mm matches the laser spot diameter with the active surface of the photodiode.
11. On the four-quadrant photodiode (Model S4349, Hamamatsu Photonics), bending of the AFM cantilever is detected as a vertical (because of attractive or repulsive forces) or lateral shift (because of frictional forces) of the laser spot.
12. Changes in the detection signal (*see* **Note 1**) while scanning the AFM cantilever are prevented adjusting the collimated laser beam perpendicular to the AFM cantilever surface (*see* **Fig. 5**). For adjustment, a precision kinematic mirror mount is used tilting the focusing lens of the fiber optics in two orthogonal directions by about ±2°. The reflected laser light is positioned to the center of the four-quadrant photodiode adjusting the photodiode by a XY-translation stage.

3.5. Patch-Clamp Setup

1. Mechanical design: Patch-clamp electrode and headstage of the patch-clamp amplifier are positioned by means of a three-axis translation stage. For long-term stability and low-drift, high-precision micrometer screws with 13 mm travel are used (*see* **Note 4**). The translation stage moving the electrode in vertical direction was tilted by 26° in respect to the horizontal plane. After a cell of interest was identified in the light microscope and a seal with the electrode was formed, no further relative movements or vibrations between patch-clamp electrode and cell membrane may occur. However, after a seal is formed, the hair cell still has to be approached to the AFM tip. This problem was solved by synchronizing the movements of the specimen chamber and the patch-clamp headstage (*see* **Note 5**), each supported by three stainless steel spheres. Both headstage and specimen chamber are equipped with electrically identical piezoelectric stacks moving both components synchronously up and down (*see* **Note 6**).
2. Electrical design: An electrically grounded chopper cylinder shields the piezoelectric tube scanner of the AFM preventing electrical crosstalk to the patch-clamp electrode

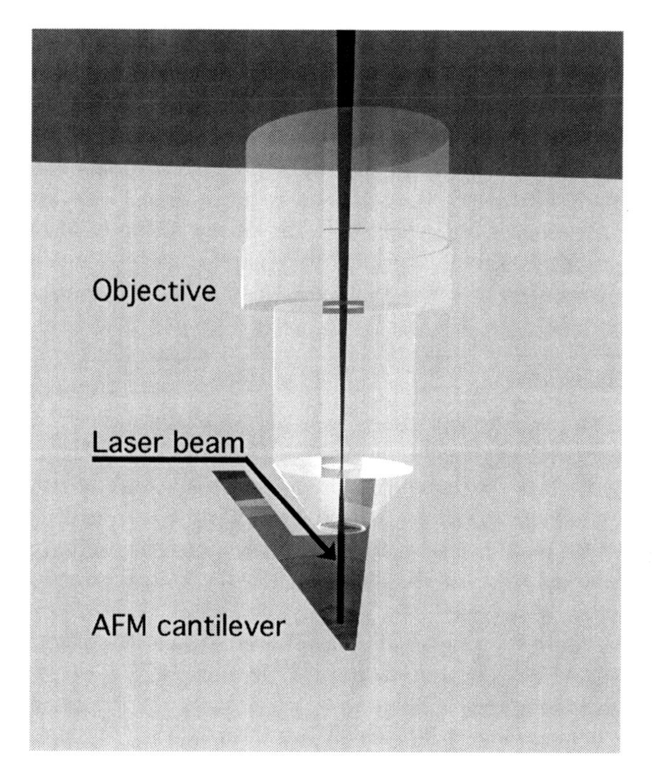

Fig. 5. Principle of optical lever detection. The beam of a laser diode is focused to the back focal plane of the objective. The collimated laser beam is reflected by the gold-coated surface of the atomic force microscopy (AFM) cantilever. For optical detection of the AFM cantilever deflection, the laser light is detected through the objective and positioned on a four-quadrant photodiode. Force exerted to the AFM tip bends the cantilever thus changing the angle of the reflected laser beam. Repulsive force acting vertically on the AFM tip shifts the laser spot on the photodiode in vertical direction, whereas lateral forces shift the laser spot on the photodiode horizontally.

(*see* **Note 7**). Differences in the electrical potential between the titanium beam of the AFM head and the reference electrode in the liquid chamber are prevented connecting the titanium beam with the reference electrode. Cables connecting the high-voltage amplifier with the piezoelectric scanner are located more than 200 mm apart from the patch-clamp electrode. A Faraday cage encompassing the entire setup shields the patch-clamp electrode from external electrical noise (in particular, 50 and 60 Hz noise).

3. Local drug application and perfusion: Our setup is equipped with a hydraulic-driven three-axis translation device (Narashige Scientific Instrument Lab., Japan) allowing to position perfusion pipettes (tip diameter 10–50 μm) with a travel range of 10 × 10 × 10 mm. Perfusion pipettes were fabricated of 2-mm borosilicate glass (Science Products GmbH) using a DMZ-Universal puller (Zeitz-Instrumente Vertriebs GmbH, Germany). The perfusion system is driven by hydrostatic pressure. A simple valve controls the liquid flow. The silicon tubing is attached to the open end of the perfusion pipette. For local drug application, the pipette is attached to the three-axis translation device and approached to the investigated hair cell. The typical distance between hair bundle and pipette tip was 20–50 μm.

3.6. Data Acquisition

1. Electronics: The driving electronics consists of a modular 19″ rig, namely, the feedback electronics keeping the acting force during measurements at constant level, three analog displays for monitoring the photodiode signals, a power supply, five identical high-voltage amplifiers (output ±160 V) each driving a separate electrode of the segmented piezoelectric tube, and a high-voltage amplifier (output 0–200 V) with two potentiometers, one for coarse and one for fine adjustment of sample stage and patch-clamp headstage.
2. Analog digital/digital analog (ADDA) converter: Data are digitized using an eight-channel 16-bit ADDA converter board ITC 16 (Instrutech Corp., USA) connected through a Bus Peripheral Component Interconnect (PCI) card to a G4 Power Macintosh Computer with 1.33 GHz (Apple Computer Inc., USA).
3. Software: Stimuli for AFM and patch clamp are generated using the custom-made software program "Pulse++ 1.7" (developed by Ulrich Rexhausen, Institute of Physiology II, Germany, and Klaus Bauer, Max Planck Institute for Medical Research, Germany). This software supports user-defined voltage patterns for the piezoelectric AFM scanner and voltage protocols for voltage-clamp measurements. A triangular-shaped signal with parabolic-shaped top and bottom is used as a scanning signal for the AFM tip. The parabolic shape reduces higher harmonics in the frequency spectrum, which might induce resonant oscillations of the AFM scanner. A second step-like signal is generated moving the AFM tip from one scan line to the next one. A maximum of four independent time-correlated voltage signals can be generated giving the opportunity to simultaneously drive four different devices during a single experiment. The "Pulse++" software additionally supports simultaneous recording of up to four signals at a maximum bandwidth of 25 kHz (*see* **Note 8**).

3.7. Displacement of Single Stereocilia by AFM

Hair bundles of intact OHCs were investigated by AFM in the medial and basal turn of 2- to 12-day-old rat cochlea.

1. AFM mode: Individual stereocilia are displaced by the AFM tip using the constant-height mode (with the feedback electronics switched off) (*see* **Note 9**).

2. Approaching the AFM tip: A hair bundle of interest was visually selected in the light microscope and directly positioned below the AFM tip. The specimen chamber is approached to the AFM tip by means of a piezoelectric stack, whereas the AFM cantilever is repetitively scanned in the same line (scan range 4.5 μm). The approach is stopped as soon as the force applied by the stereocilia causes significant deflection of the Si_3N_4 cantilever (approximately 20–50 nm). During the approach, the detector signal is monitored on a digital storage oscilloscope (HM 1007, Hameg GmbH, Germany) and the scope window of the data acquisition software.
3. Scanning protocol: The sharp stylus of an AFM cantilever (*see* **Fig. 6**) scans in inhibitory and excitatory direction across the tip of a single stereocilium of the tallest row ***(9)*** (*see* **Fig. 6**). Buckling of the AFM cantilever is prevented using V-shaped cantilever beams scanned at an angle of 90° in respect to the longitudinal axis of the cantilever. For excitatory displacement, the tip link's tension is increased opening the connected transduction channel, whereas in inhibitory direction, those channels being still open at rest are closed. Force interaction between the stylus and tips of single stereocilia is detected with nanometer resolution using the optical lever method ***(8,10)***. As the stylus laterally pushes against a single stereocilium, the resulting repulsive force bends the AFM cantilever upward and displaces the stereocilium horizontally (*see* **Fig. 6A**). Maximum cantilever deflection occurs when the AFM tip is in contact with the uppermost part of the stereocilium (*see* **Fig. 6B**). Further movement of the AFM stylus in the same scan direction reduces the cantilever's pressure on the stereocilium's surface, until the bended stereocilium eventually snaps off from the AFM tip. At this particular point, the restoring force due to the stereocilium's spring constant overcomes the adhesion force between AFM tip and the stereocilium (*see* **Fig. 6C**). At a given position, the AFM tip inverts its scan movement and displaces the identical stereocilium in inhibitory direction. After each line scan, a small voltage offset is added to the piezoelectric tube moving the AFM stylus to the next scan line.
4. AFM imaging: Successively recorded line scans performed at different points of the hair bundle were aligned next to each other and displayed separately for excitatory and inhibitory displacement as grayscale images (*see* **Fig. 7**). The AFM images reveal discrete bright spots representing the tips of individual stereocilia of the investigated hair bundle. The spatial resolution of individual stereocilia in the AFM image supports local force interaction of the AFM tip with individual stereocilia.

3.8. Stiffness Measurements at Single Stereocilia

For quantitative determination of stereocilia stiffness as a function of displacement, the same scanning protocol as described in **Subheading 3.7.** is used. However, several limiting conditions have to be taken into account:

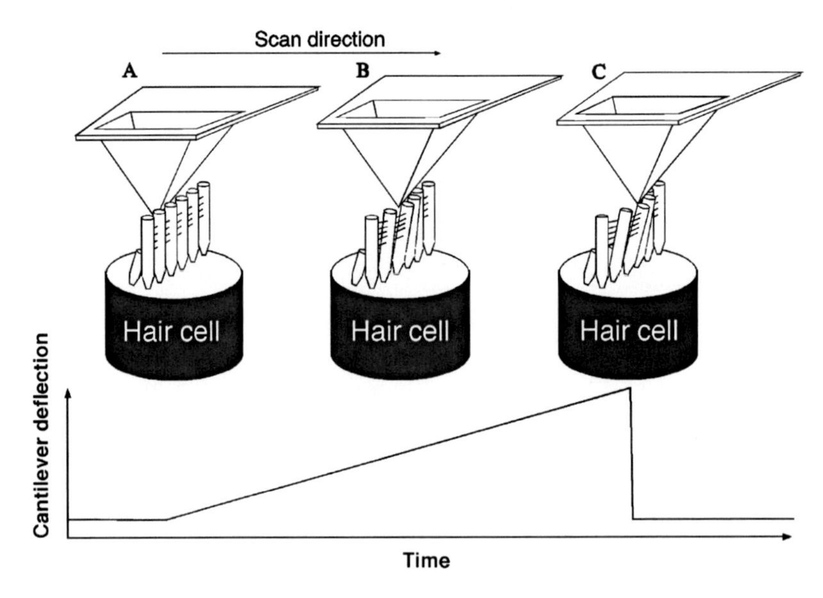

Fig. 6. Displacement of stereocilia by atomic force microscopy (AFM). The AFM cantilever is scanned at constant height by means of the piezoelectric tube scanner. At resting position, the recorded AFM signal corresponds to a flat line superimposed with electrical and thermal noise (**A**). As the lateral surface of the AFM tip makes contact to the tip of a stereocilium (**B**), the AFM cantilever is bended upward (*see* schematic trace below). Maximum repulsive force is detected when the sharp AFM tip and the top of the corresponding stereocilium are in contact (**C**). As the AFM cantilever scans more to the right, the stereocilium jumps back to the resting position, whereas the AFM cantilever jumps down to its initial position.

1. Adsorbates (like pieces of the tectorial membrane) sticking to the tips of stereocilia might disturb the force interaction between AFM tip and sample surface (*see* **Note 10**).
2. The AFM tip must exhibit a clean surface.
3. The AFM tip geometry in particular the angle between sidewall of the tip and scanned surface must be well known.
4. Buckling of the AFM tip, which is due to the exerted force, must be prevented.
5. A mechanically stable preparation during the measurement is required.
6. Before experiments, the direction of AFM tip motion and axis of symmetry of the OHC bundles were aligned parallel rotating the specimen chamber.
7. Slope of force curve: For stiffness measurements, single stereocilia of the tallest row are displaced in excitatory and inhibitory direction ***(11)***. A typical example of an AFM trace is depicted in **Fig. 8**. The ordinate displays the vertical deflection of the

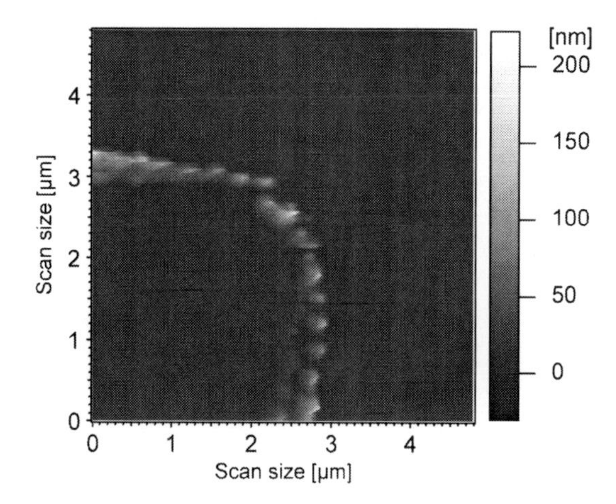

Fig. 7. Atomic force microscopy (AFM) image depicting stereociliary tips of an outer hair cell (postnatal rat, medial section, and 5 days after birth). The white spots correspond to the force interaction between AFM tip and individual stereocilia. The image demonstrates the capability of AFM to address single stereocilia under physiological conditions.

AFM cantilever while scanning an individual stereocilium. When being in contact with the sample, the positive slope of the force curve depends on the elasticity of the investigated sample ***(11)***. As the stiffness increases as for example because of chemical fixation, the positive slope of the force curve also increases ***(11)***. In other words, for the same horizontal force exerted by an AFM tip, a soft stereocilium shows a higher displacement than a stiff stereocilium.

8. Calculating the stiffness: The horizontal stiffness k_L of a stereocilium can be calculated from the horizontal force F_L exerted by the AFM sensor and the resulting horizontal displacement "x" using Hook's law for a linear spring:

$$k_L = \frac{F_L}{x}, \tag{1}$$

where F_L is the exerted force in horizontal direction causing a displacement "x" of the stereocilium. Assuming that friction is very low, F_L can be calculated from the measured vertical force F_N (*see* **Fig. 8**) using:

$$F_L = F_N \cdot \tan\alpha = k_{Cant} \cdot a \cdot \tan\alpha, \tag{2}$$

where α is the angle between the lateral face of the pyramidal-shaped AFM tip and the scan direction, "a" is the vertical deflection of the cantilever beam, and k_{Cant} is the spring constant of the AFM cantilever. For calculating the spring constant k_L

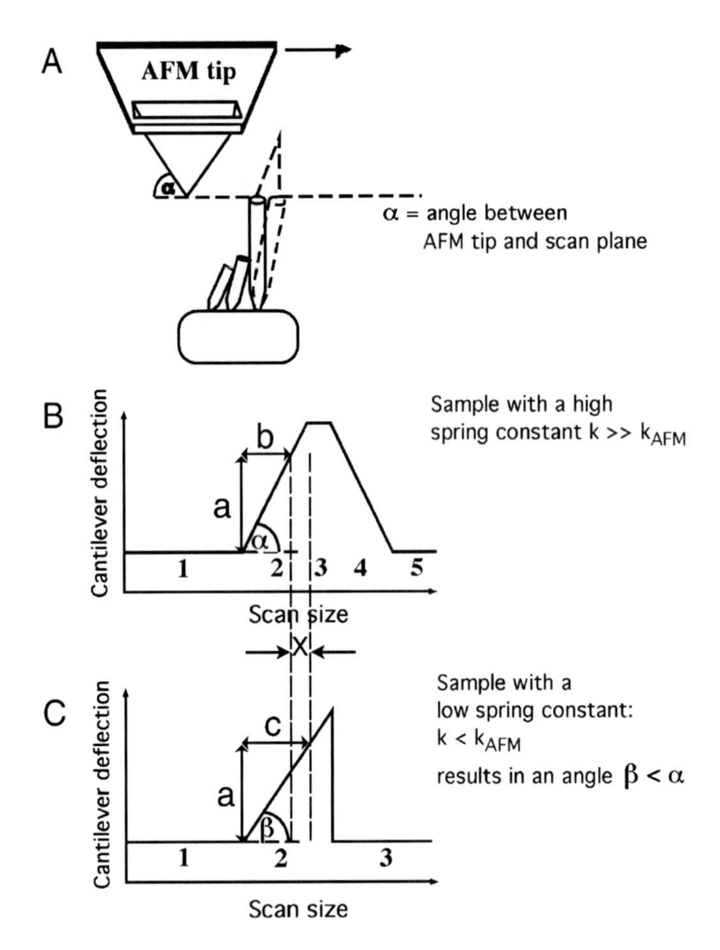

Fig. 8. Principle of stiffness measurement. **(A)** For experiments, atomic force microscopy (AFM) cantilevers with well-defined pyramidal-shaped tips were used. The angle between the lateral surface of the AFM tip and the scanning plane (for our AFM sensors, $\alpha = 55°$) is termed α. The horizontal force F_L is calculated from the vertical force F_N according to **Eq. 2**. **(B)** Expected experimental result for an AFM scan, performed on an infinitely stiff pin-like structure. The vertical cantilever deflection is plotted versus the horizontal scan size. At point 1, the AFM tip does not touch the specimen. At point 2, the cantilever deflection increases with scan size. The lateral tip surface is in contact with the stiff specimen, mapping the lateral plane of the AFM tip. At point 3, the AFM tip scans across the flat top of the specimen followed by a decrease in force where the AFM tip slips down the specimen (region 4). Finally, the AFM tip looses contact reaching its resting position (point 5). "*a*" indicates the cantilever deflection at a scan size "*b*" when being in contact with the stiff specimen.

in **Eq. 1**, we still have to assess the horizontal displacement "*x*" of the stereocilium. **Figure 8** reveals the experimental situation for an infinitely stiff pin-like sample and from an elastic stereocilium. The cantilever deflection is displayed scan size. "*b*" corresponds to the distance between the first point of contact (between AFM tip and stiff sample) and the horizontal position where the AFM tip reaches a vertical deflection "*a*." "*c*" represents the distance between the first point of contact (between AFM tip and elastic stereocilium) and the horizontal position where the AFM tip reaches the same vertical deflection "*a*." The horizontal displacement "*x*" of an elastic sample (stereocilium) can be calculated by subtracting distance "*b*" from distance "*c*." This subtraction accounts for the convolution of the tip shape with the measured topography of the sample. The tip shape is defined by the angle α between the scan direction and the lateral face of the tip. Distance "*b*" depends on α as follows:

$$b = \frac{a}{\tan \alpha}. \tag{3}$$

The vertical deflection of the elastic sample is assumed to be negligible for a horizontal displacement "*x*" being small compared with the length of the elastic object (stereocilium). Considering these assumptions, we can calculate the force constant k_L inserting the results of **Eq. 3** and **Eq. 2** into **Eq. 1**:

$$k_L = \frac{F_L}{c - b} = \frac{k_{Cant} \cdot a \cdot \tan \alpha}{c - (a/\tan \alpha)}. \tag{4}$$

Using the horizontal deflection method described in section 3.8, it is essential to know the exact angle α between the lateral face of the AFM tip and the scan plane. Therefore, double-beam cantilevers with pyramidal-shaped Si_3N_4 tips having a well-defined tip angle of 70° were used. In opposite to single-beam cantilevers, double-beam cantilevers scanned horizontally at right angle to the beam direction show only small buckling and torsion at forces being smaller than 2 nN.

Fig. 8. **(C)** Expected experimental situation for the same AFM scan recorded on an elastic stereocilium. Again point 1 marks the region of noncontact. At point 2, the tip hits the stereocilium resulting in an increasing cantilever deflection. In contrast to the situation displayed in **B**, the force exerted by the AFM tip results in a horizontal displacement of the elastic object. At the transition between points 2 and 3, AFM tip and specimen lose contact. The cantilever nearly instantaneously jumps back to the initial position, whereas the stereocilium horizontally moves back to the resting position on the left. "*c*" indicates the scan range starting from the first point of contact to the point where the cantilever reaches the same vertical deflection "*a*" as shown in **B**. "*x*" corresponds to the horizontal displacement of the elastic sample.

3.9. Simultaneous AFM/Patch-Clamp Recordings

The main purpose of this setup is to simultaneously apply AFM and patch clamp to mechanosensitive ion channels in living cells. Here, an experimental protocol is given emphasizing the major points that have to be considered.

1. Transfer and orientation of the sample: After the chamber is filled with extracellular solution, one has to wait for 10 min until thermal equilibrium is reached. A section of the organ of Corti mounted on a 10-mm glass coverslip is transferred into the liquid chamber. Under light microscopic control, hair bundles are oriented in a way that the axis of symmetry is aligned parallel to the scan direction and the patch-clamp pipette gets free access to the hair cell bodies.
2. Cleaning of hair cell bodies: For free access with the patch-clamp electrode to the hair cell bodies, supporting cells such as the Hensen cells have to be removed using a cleaning pipette. Pipettes have tip diameters ranging from 5 to 10 μm and are fabricated of 1-mm borosilicate glass. They are mounted in a holder of the patch-clamp headstage and positioned under light microscopic control underneath the cochlear hair cells. Supporting cells including the Deiters cells are carefully removed applying negative pressure to the cleaning pipette. This procedure is applied to just three to five cells in order to keep the formation of hair cells in a mechanically stable condition.
3. AFM adjustment: A new cantilever is mounted on the scanner of the AFM and attached to the three-axis translation stage. Under the light microscope, the AFM tip is coarsely approached to the cell of interest. The laser spot of the lever detection system is positioned onto the AFM cantilever surface using the XY-translation stage supporting the entire light microscope. During this procedure, relative position between AFM tip and hair cells is not changed. The angle of the laser diode is adjusted by means of a two-axis mirror mount such that the incoming laser beam is oriented perpendicular to the AFM cantilever surface. For an optimum adjusted laser beam, a shift of the focal plane of the objective does not change the signal on the detector.
4. Forming a seal: A Quartz glass pipette is filled from the backside with intracellular solution using a syringe connected to a microloader with a diameter of only 0.5 mm. The pipette is mounted on the holder of the patch-clamp headstage. A positive pressure is applied through silicon tubing preventing contamination of the pipette tip in the bath. After the pipette is transferred into the liquid, the serial resistance of the pipette is tested applying a voltage step of typically 2 or 5 mV between reference and headstage electrode. The coarse and fine approach of the pipette is stopped at a distance of 3–5 μm to the hair cell body. The pipette tip is carefully positioned underneath the cuticular plate of the hair cell. Owing to the positive pressure applied by the patch-clamp pipette, the membrane of the hair cell body starts to bend inward. At a typical membrane indentation of 1–2 μm, the pressure is rapidly released causing the membrane to snap into contact with the pipette tip. Before

breakthrough, the holding potential is switched to −80 mV roughly corresponding to the reversal potential of voltage-gated K^+ ion channels in OHCs. A small hole at the membrane patch is formed applying a slight negative pressure. The electrical contact of the patch-clamp electrode to the intracellular space is tested applying 10 incrementing voltage pulses starting from −80 to +10 mV. A characteristic pattern of outward current steps indicates opening of voltage-gated K^+ ion channels.

5. Approaching the AFM tip: A software protocol repetitively scanning the AFM tip in the same line is started. The hair bundle of interest is approached to the AFM tip simultaneously lifting the specimen chamber and the patch-clamp electrode. Vertical position of specimen chamber and patch-clamp electrode is manually controlled by means of a potentiometer at the front panel of the high-voltage amplifier driving both piezoelectric actuators. The feedback electronics in principle allows an automatic approach but takes the risk of damaging the hair bundle structure because of the limited speed of the feedback electronics. As the AFM tip makes contact to a stereocilium, a vertical displacement of the AFM cantilever can be monitored on the computer display.
6. Simultaneous AFM/patch-clamp recording: Displacement of a taller stereocilium by AFM toward the shorter stereocilia closes those transduction channels being open at rest (*see* **Fig. 9**), whereas displacement in the opposite direction opens the

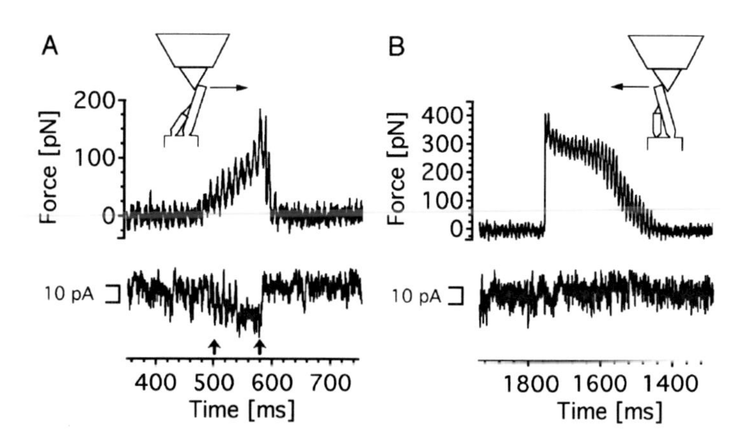

Fig. 9. Force and current simultaneously recorded for displacement of an individual stereocilium in excitatory **(A)** and inhibitory **(B)** direction. **(A)** A tall stereocilium of an outer hair is displaced by atomic force microscopy (AFM) scanning the AFM tip in the constant-height mode across the hair bundle. As the stereocilium is displaced by AFM, an inward current is detected (trace below). At about 550 ms, the current reaches saturation level. As the AFM tip loses contact with the stereocilium, the inward current synchronously reduces to the initial baseline level (greater than 600 ms). **(B)** For reversed scan in inhibitory direction, no inward current is detected.

transduction channels (*see* **Fig. 9**). Depending on the number of stereocilia being displaced, single-transduction channels can be opened under well-defined conditions. Variations in force change the open probability of ion channels. Measurements can be combined with local application of drugs such as aminoglycosides acting as transduction channel blockers.

4. Notes

1. Adsorbates and contamination on the AFM cantilever: A flat (low granularity) and clean gold-coated AFM cantilever surface is critical for proper detection of the cantilever deflection on the four-quadrant photodiode. If for example the laser spot hits a dust particle on the cantilever surface while scanning the AFM tip, a signal is detected on the photodiode although the AFM tip is not in contact with the sample. Therefore, AFM cantilevers have to be stored under clean conditions. Additional cleaning as for example in an ultrasonic cleaner is not recommended as it disrupts the cantilevers from the substrate. If necessary, the sensor surface is carefully cleaned before experiments using an air jet.
2. Detection through the objective: Owing to the limited working distance of 1.92 mm of the water-immersion objective, the laser beam used to detect the cantilever deflection has to be coupled through the objective. Detection from below, through the condenser of the optical microscope, was not chosen because of the quite inhomogeneous optical properties of the cell tissue disturbing the propagating laser beam.
3. Dust on the microscope objective: After experiments, the objective is removed from the extracellular liquid, which quiet often leads to the formation of salt crystals and dust sticking to the objective as the liquid film dries. Contaminations might disturb the laser signal on the detector and therefore have to be removed with a soft cloth or pure water before experiments.
4. Low-drift design of the pipette holder: For common pipette holders, patch-clamp electrodes show lateral drift as negative or positive pressure is applied. This was prevented integrating two instead of just one cone washer. The cone washers are located at both opponent ends of the holder, clamping the pipette at two different points. Thus, lateral movement is kept at a minimum level.
5. Synchronicity of sample and patch-clamp actuator during approach of the sample to the AFM tip. For patch-clamp experiments, it is essential that both patch-clamp headstage and sample stage are synchronously adjustable in vertical direction. As a seal is formed, relative movement between patch-clamp pipette and clamped hair cell has to be prevented. Any vibration or relative movement might rupture the membrane.
6. Mechanical vibrations are very critical for patch-clamp and AFM measurements. All cables and wires connected to the AFM head and patch-clamp setup are mechanically fixed on the granite support of the vibration isolation table.

7. Electrical noise: For single-channel recordings, electrical shielding of the piezoelectric tube of the AFM scanner is essential. The whole scanner is encapsulated in a copper cylinder connected to ground. Additionally, the AFM holder made of titanium is electrically connected to the potential of the reference electrode preventing DC voltage offsets.
8. Patch-clamp software "ScanClamp": The requirements for patch clamp and AFM are quiet different. This was the motivation to develop the custom-made software *ScanClamp* providing data acquisition and stimulation protocols adapted to both AFM and patch clamp. However, application to any other scanning probe technique is possible.
9. Constant-height constant-force mode: For a more precise control of the force acting between AFM tip and hair bundle, a feedback electronic in principle would be advantageous. However, tests have shown that the feedback cannot move the cantilever beam sufficiently fast at constant force without damaging the hair bundle structure. In particular, the steep slope of the rod-like stereocilia prevents the feedback electronic to follow the surface topology sufficiently fast. Hair bundles appear as trodden tussocks after investigations in constant-force mode. Therefore, the constant-height mode was chosen allowing gentle displacement of stereocilia without damaging the cytoskeletal structure.
10. Adsorbates on cochlear hair bundles (tectorial membrane): Quantitative analysis of the stereocilia stiffness works properly only for low friction between AFM tip and stereocilium. However, in some cases, pieces of the tectorial membrane sticking to the stereocilia were observed in the light microscope. In those cases, high attractive forces acting between AFM tip and stereocilium were detected. In few cases, attractive forces were so high that even at maximum retraction of the AFM cantilever it was not possible to get rid of the stereocilium. The experiment has to be stopped, and a new AFM cantilever is mounted to the holder.

Acknowledgments

I would like to thank S. Fink for providing **Fig. 9**, H.P. Zenner and F. Lehmann-Horn for providing laboratory space and equipment, K. Loeffler for technical assistance, J.K.H. Hoerber for supporting the development of the AFM/patch clamp setup, F. Grauvogel for providing **Figs. 4** and **5**, and A. Koitschev of the Department of Otolaryngology of the University of Tuebingen for proofreading the manuscript. This work was supported by a grant of the Deutsche Forschungsgemeinschaft to M.G. Langer (LA1227/1-3) and the Bundesministerium für Bildung und Forschung (BMBF) to M.G. Langer ("Bio-AFM"; 0312017A).

References

1. Hudspeth, A. J. (1989) How the ear's works work. *Nature* **341**, 397–404.
2. Corey, D. P., Garcia-Anoveros, J., Holt, J. R., Kwan, K. Y., Lin, S. Y., Vollrath, M. A., Amalfitano, A., Cheung, E. L., Derfler, B. H., Duggan, A., Geleoc, G. S., Gray, P. A., Hoffman, M. P., Rehm, H. L., Tamasauskas, D., and Zhang, D. S. (2004) TRPA1 is a candidate for the mechanosensitive transduction channel of vertebrate hair cells. *Nature* **432**, 723–30.
3. Sobkowicz, H. M., Bereman, B., and Rose, J. E. (1975) Organotypic development of the organ of Corti in culture. *J Neurocytol* **4**, 543–72.
4. Russell, I. J., and Richardson, G. P. (1987) The morphology and physiology of hair cells in organotypic cultures of the mouse cochlea. *Hear Res* **31**, 9–24.
5. Binnig, G., and Smith, D. P. E. (1986) Single-tube three-dimensional scanner for scanning tunneling microscopy. *Rev Sci Instrum* **57**, 1688–89.
6. Siegel, J., Witt, J., Venturi, N., and Field, S. (1995) Compact large-range cryogenic scanner. *Rev Sci Instrum* **66**, 2520–32.
7. Langer, M. G., Öffner, W., Wittmann, H., Flösser, H., Schaar, H., Häberle, W., Pralle, A., Ruppersberg, J. P., and Hörber, J. K. H. (1997) A scanning force microscope for simultaneous force and patch-clamp measurements on living cell tissues. *Rev Sci Instrum* **68**, 2583–90.
8. Meyer, G., and Amer, N. M. (1988) Novel optical approach to atomic force microscopy. *Appl Phys Lett* **53**, 1045–47.
9. Langer, M. G., Fink, S., Koitschev, A., Rexhausen, U., Horber, J. K., and Ruppersberg, J. P. (2001) Lateral mechanical coupling of stereocilia in cochlear hair bundles. *Biophys J* **80**, 2608–21.
10. Alexander, S., Hellemans, L., Marti, O., Schneir, J., Ellings, V., Hansma, P. K., Longmire, M., and Gurleey, J. (1989) An atomic-resolution atomic-force microscope implemented using an optical lever. *J Appl Phys* **65**, 164–67.
11. Langer, M. G., Koitschev, A., Haase, H., Rexhausen, U., Horber, J. K., and Ruppersberg, J. P. (2000) Mechanical stimulation of individual stereocilia of living cochlear hair cells by atomic force microscopy. *Ultramicroscopy* **82**, 269–78.

10

Synaptic Connectivity in Engineered Neuronal Networks

Peter Molnar, Jung-Fong Kang, Neelima Bhargava, Mainak Das, and James J. Hickman

Summary

In this study, we have demonstrated a method to organize cells in dissociated cultures using engineered chemical clues on the culture surface and determined their connectivity patterns. Although almost all elements of the synaptic transmission machinery between neurons or between neurons and muscle fibers can be studied separately in single-cell models in dissociated cultures, the difficulty of clarifying the complex interactions between these elements makes random cultures not particularly suitable for specific studies. Factors affecting synaptic transmission are generally studied in organotypic cultures, brain slices, or in vivo where the cellular architecture generally remains intact. However, by utilizing engineered neuronal networks, complex phenomenon such as synaptic transmission can be studied in a simple, functional, cell culture-based system. We have utilized self-assembled monolayers (SAMs) and photolithography to create the surface templates. Embryonic hippocampal cells, plated on the resultant patterns in serum-free medium, followed the surface clues and formed the engineered neuronal networks. Basic electrophysiological methods were applied to characterize the synaptic connectivity in these engineered two-cell networks.

Key Words: Engineered networks; hippocampal cultures; synaptic connectivity; EPSC; action potential; SAM; photolithography; serum-free.

1. Introduction

The behavior of the cells (attachment, proliferation, and differentiation function) is determined by their internal programs and by extracellular clues. The origin of these extracellular signals can be the culture medium,

From: *Methods in Molecular Biology, vol. 288: Patch-Clamp Methods and Protocols*
Edited by: P. Molnar and J. J. Hickman © Humana Press Inc., Totowa, NJ

other cells, or the culture surface. Recent developments in cell biology and fabrication technology have enabled the engineering of active surfaces to present these signaling molecules to the cells in a controlled manner ***(1–5)***. Using functionalized self-assembled monolayers (SAMs) in combination with advanced surface patterning methods, the inherent differentiation and self-organizing programs utilized by the neurons can be controlled and guided to form directed networks ***(6–8)***.

Synaptic transmission is a complex process, even between similar cell types ***(9)***, regulated by several presynaptic and postsynaptic mechanisms ***(10)***. It plays an important role in memory formation and several neurodegenerative diseases ***(11–13)***. An in vitro functional system that would enable high-throughput testing of drugs on synaptic transmission in a controlled environment would be high significance.

2. Materials

2.1. Surface Modification

1. Glass coverslips (Thomas Scientific, Swedesboro, NJ, USA, cat. no. 6661F52, 22×22 mm no. 1).
2. Plasma cleaner (Harrick, Ithaca, NY, USA, cat. no. PDC-32G).
3. Trimethoxysilylpropyldiethylenetriamine (DETA, United Chemical Technologies, Bristol, PA, USA, cat. no. T2910KG).
4. Toluene (Fisher, Pittsburgh, PA, USA, cat. no. T2904).
5. Molecular sieves, 4A (Sigma, St. Louis, MO, USA, cat. no. 334294).
6. Goniometer (KSV Instruments, Monroe, CT, USA, cat. no. Cam 200).
7. X-ray photoelectron spectroscope (Kratos Axis 165) equipped with Al Kα X-ray source.

2.2. Photolithography

1. CleWin layout editor (WieWeb, The Netherlands).
2. Chrome/Quartz mask (Bandwidth Foundry Pty Ltd, Australia).
3. Ar/F laser (LPX200i, Lambda Physik, USA) combined with a beam homogenizer (Microlas, USA).
4. Tridecafluoro-1,1,2,2-tetrahydroctyl-1-trichlorosilane (13F, Gelest, Morrisville, PA, USA).
5. Chloroform (Sigma, cat. no. 439142).
6. $PdCl_4$ solution: 10 mg $PdCl_4$ and 1.75 g NaCl/50 ml water, pH adjusted to 1.0 with concentrated HCl.
7. Dimethylamine borane (DMAB) solution: 1.7 g DMAB/50 ml water.
8. Copper bath solution: 3 g copper sulfate, 14 g sodium potassium tartrate, 2 g of NaOH, and 1 ml formaldehyde (37.2%) in 100 ml distilled water.

2.3. Embryonic Hippocampal Cultures

1. Dissecting medium: 500 ml Hibernate E (BrainBits, USA), 10 ml B27 supplement (Invitrogen, cat. no. 17504), 5 ml glutamax (Invitrogen, Carlsbad, CA, USA, cat. no. 35050-061), and 5 ml antibiotic/antimycotic (Invitrogen, 15240-062).
2. Culture medium 2: 500 ml neurobasal E (Invitrogen, cat. no. 21103-049), 10 ml B27, 5 ml glutamax, and 5 ml antibiotic/antimycotic.
3. Culture medium 1: culture medium 2 plus 25 μM glutamate (Sigma, cat. no. A8875).
4. Stereo microscope.
5. Laminar flow hood.
6. CO_2 incubator.
7. Hematocytometer Hausser Scientific, USA.
8. Inverted cell-culture microscope.
9. Trypan blue (Fluka, USA).

2.4. Dual Patch-Clamp Recordings

1. Zeiss Axioscope 2 FS plus upright microscope.
2. PCS-5000 piezoelectric 3D micromanipulators (×2, Burleigh Quebec, Canada).
3. Gibraltar platform (Burleigh).
4. Vibration isolation table (TMC Peabody, MA, USA).
5. Multiclamp 700B patch-clamp amplifier (Axon Sunnyvale, CA).
6. Digidata 1320 A/D converter (Axon).
7. pClamp 9.0 software.
8. Sutter P97 pipette puller (Sutter, USA).
9. Borosilicate glass pipettes (BF150-86-10, Sutter).
10. pH meter, balance, osmometer (Fiske USA).
11. Extracellular solution: neurobasal E medium, pH is adjusted to 7.3 with 4-(2-hydroxyethyl)-1-piperazineethane sulfonic acid (HEPES), Sigma.
12. Intracellular solution: 140 mM K-gluconate, 1 mM ethylene glycol tetraacetic acid (EGTA), Sigma, 2 mM $MgCl_2$, 2 mM Na_2ATP, 10 mM HEPES; pH = 7.2; 276 mOsm. In some experiments, 1% Alexa Fluor 488 (Molecular Probes/invitrogen, cat. no. A20100) dye was added.

3. Methods

3.1. Surface Modification

The glass coverslip surface was modified with DETA SAMs ***(14,15)***.

1. Glass coverslips were cleaned using an O_2 plasma cleaner for 30 min at 400 mTorr (*see* **Note 1**).
2. The SAM was formed by the reaction of the cleaned surface with a 0.1% (v/v) mixture of the organosilane (DETA) in freshly distilled toluene (*see* **Note 2**).

3. The coverslips were heated with the mixture to just below the boiling point of the toluene, rinsed with toluene, reheated to just below the boiling temperature, and then oven dried.
4. Surfaces were characterized by contact angle measurements using an optical contact angle goniometer and by X-ray photoelectron spectroscopy by monitoring the N 1s peak (*see* **Note 3**) ***(16)***.
5. For contact angle measurements, a static, sessile drop (5 μl) of deionized water was applied to the surface. Three measurements were taken and averaged. The stable contact angles (40.64 ± 2.9 [mean ± SD]) throughout the study period indicated a high reproducibility and quality of the DETA coatings.
6. For the X-ray photoelectron spectroscopy (XPS analysis), the pass energy of the analyzer and the take-off angle of photoelectron were set at 40 eV and 90°. Survey scans as well as high-resolution scans for fluorine (1s), oxygen (1s), nitrogen (1s), carbon (1s), and silicon (2p) were performed on each sample. Based on the high-resolution scans, elemental composition percentages were calculated for each element ***(15)***.

3.2. Photolithography (see Note 4)

DETA patterns were created on the surface by ablation using a 193-nm excimer laser and a quartz photomask. The ablated region was then backfilled with a second SAM (13F) ***(8)***. The patterns (amino groups on the surface) were visualized by electroless copper deposition.

1. Masks were designed using the Clewin program (*see* **Note 5**) and written by a commercial vendor (Bandwidth Foundry Pty Ltd).
2. DETA was ablated by the exposure to ultraviolet (UV) light through the photomask (193 nm, $50 mJ/mm^2$) (*see* **Note 6**, **Fig. 1**).
3. Patterns were backfilled by dipping the ablated coverslips in a 13F solution in chloroform (0.1% v/v) (*see* **Note 7**). 13F is not permissive for cellular growth.
4. For the copper deposition, the coverslips were immersed in the $PdCl_4^{-2}$ solution for 15 min. They were rinsed thoroughly with water and immersed for 15 min in DMAB solution. The samples were again rinsed thoroughly with water and then immersed in the copper bath solution for 5 min (*see* **Note 8**).

3.3. Embryonic Hippocampal Cultures

The hippocampal cultures were prepared from E17 rat embryos according to the animal protocol accepted by the IACUC of University of Central Florida.

1. Rats were euthanized in a precharged CO_2 chamber.
2. The hippocampal tissue was dissected from the embryos and placed in ice-cold dissociating medium.

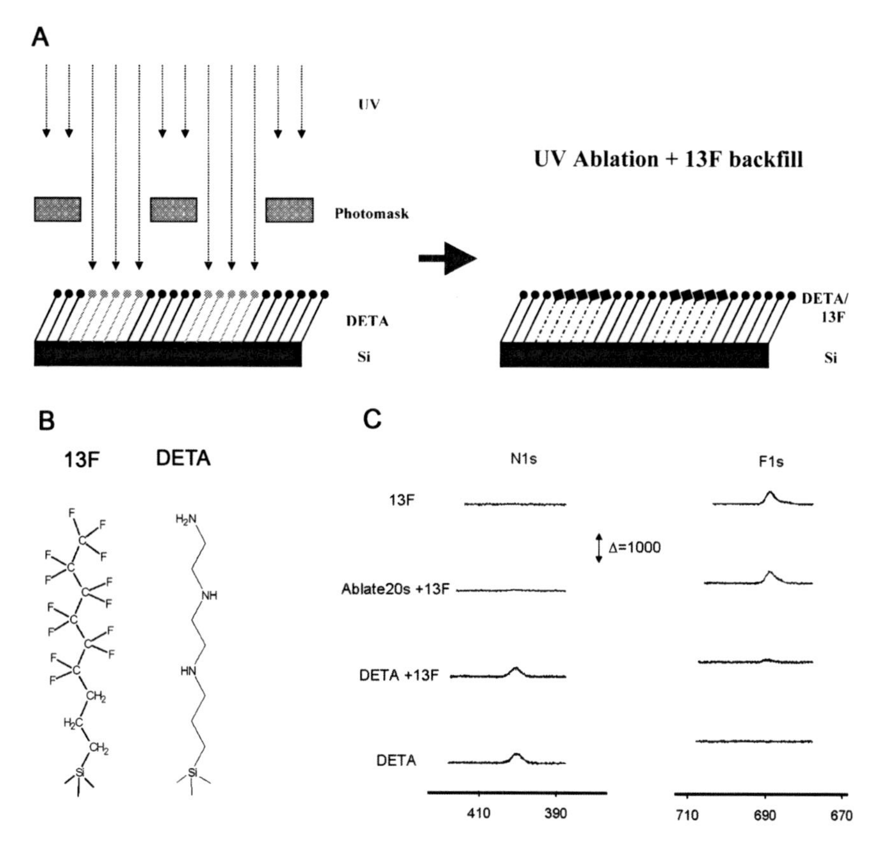

Fig. 1. Photolithographic patterning of self-assembled monolayers. **(A)** Trimethoxysilylpropyldiethylenetriamine (DETA) coverslips were ablated through a photomask followed by backfilling the ablated areas by 13F. **(B)** Chemical structure of DETA and 13F. **(C)** High-resolution XPS spectra of the surfaces at different stages of the patterning process. The area under the peaks was proportional to the amount of the given element on the surface. DETA coverslips (N 1s peak, no F 1s peak) were ablated and backfilled with 13F. Ablate 20s + 13F shows that the N 1s peak disappeared and were replaced by the F 1s peak of 13F. DETA + 13F shows some incorporation of 13F into the DETA monolayer. x-axis: binding energy, y-axis: arbitrary intensity unit.

3. The tissue was triturated using a sterile Pasteur pipette (*see* **Note 9**).
4. The dissociated tissue was centrifuged at 300 *g* for 2 min at 4°C.
5. The pellet was resuspended in 2 ml culture medium 1.
6. Cells were counted in a hemocytometer and plated on the patterns at a density of 100 cells/mm^2.
7. Half of the medium was changed every 4 days using culture medium 2.

3.4. Dual Patch-Clamp Recordings

Dual patch-clamp recordings were performed on the neurons plated on the patterns. Synaptic connectivity was verified by evoking action potentials in one of the cells in current-clamp mode, whereas recording of the excitatory postsynaptic currents (EPSCs) was done in voltage-clamp mode (*see* **Fig. 2**).

1. On the 14th day of the culture, the patterned coverslips were transferred to a recording chamber on the stage of a Zeiss Axioscope 2 FS plus upright microscope. Experiments were performed at room temperature.
2. Electrodes were pulled from borosilicate glass and filled with the intracellular solution. The resistance of the electrodes was 6–10 MΩ.
3. Dual current clamp–voltage clamp experiments were performed on two apparently connected neurons on the patterns.
4. Two filled electrodes were placed into the two headstages; positive pressure (about 2 cc from a 10-cc syringe) was applied to both pipettes before touching the extracellular solution; the electrodes were brought close to the target cells; pipette offset was compensated at both channels; cells were touched with the tip of the electrodes (seal position) under visual control; −5 mV seal test was applied; pipette capacitances were compensated; −70 mV holding was applied on both channels (voltage-clamp mode); gigaseal was formed; the cell membrane was ruptured by the application

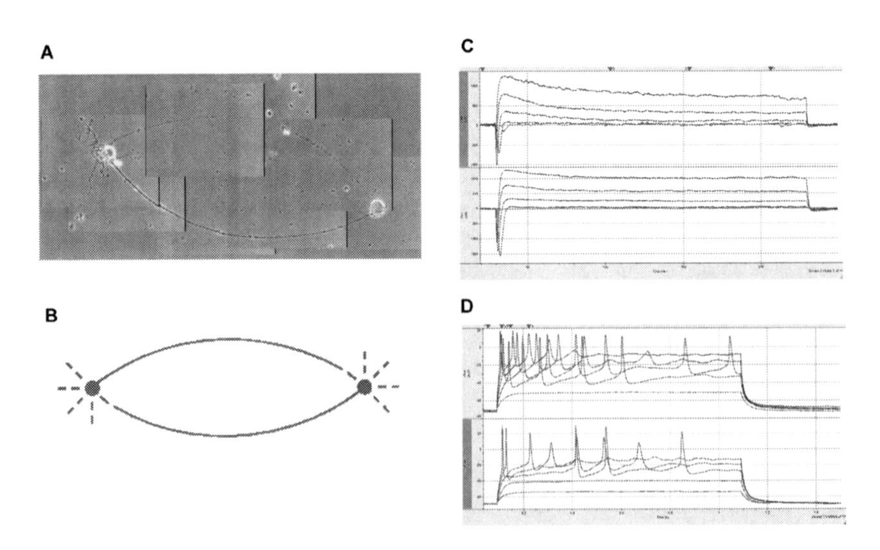

Fig. 2. Functional engineered networks. (**A**) Embryonic hippocampal cells complied to the surface patterns. (**B**) Mask design for two-cell networks. The diameter of the soma-attachment area is 25 μ*M*. (**C**) Ionic currents recorded from the patterned cells in voltage-clamp mode. (**D**) Action potentials were evoked by current injections.

of short suction pulses; resting membrane potential was measured in $I = 0$ mode; whole cell capacitance and resistance were compensated.

5. Voltage step protocols were applied in parallel to both cells. Signals were filtered at 2 kHz and digitized at 20 kHz. Passive membrane properties (membrane resistance and capacitance) and ionic currents (voltage-dependent sodium and potassium) were measured.

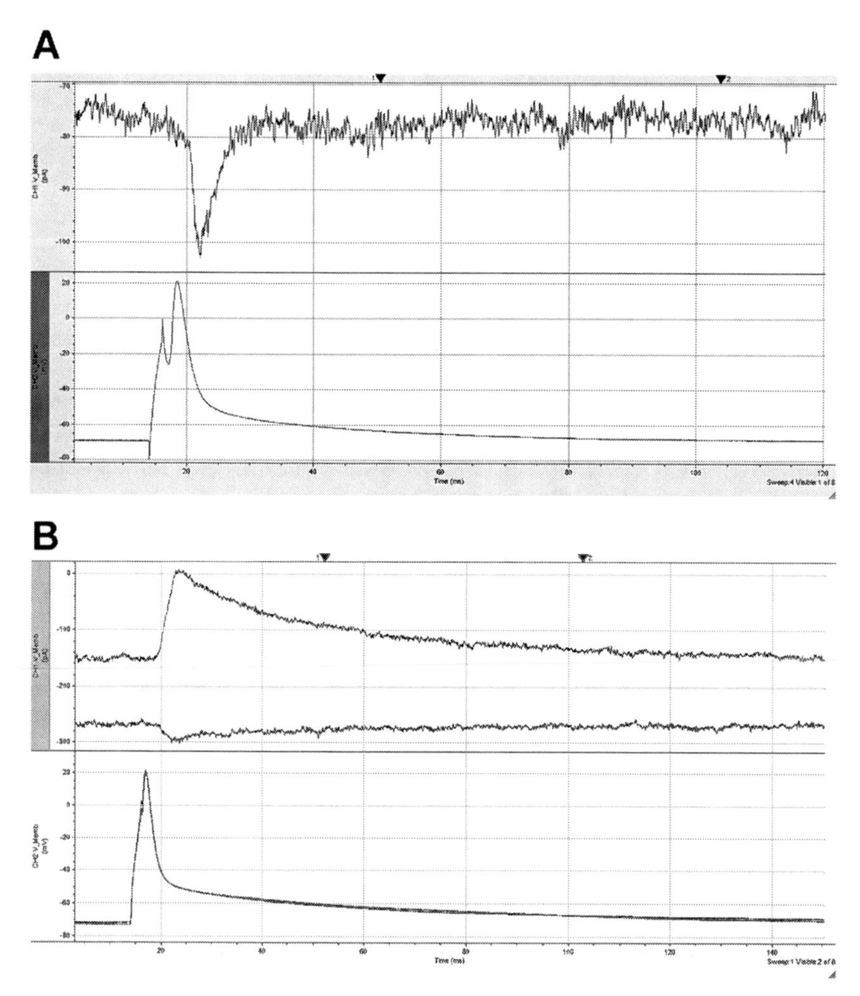

Fig. 3. Synaptic transmission in engineered networks. Action potentials were evoked by current injection in one of the cells, whereas postsynaptic currents were recorded from the other cell. **(A)** Action Potential (AP) lower trace and excitatory postsynaptic current recorded at −70 mV holding—upper trace. **(B)** AP—lower trace and Inhibitory Post Synaptic Current (IPSC) recorded at −70 and −30 mV holding—upper trace. Note the reversal of the current.

6. The action potential threshold was measured in current-clamp mode with increasing current injections.
7. Synaptic transmission was measured by evoking action potentials in one cell and recording EPSCs from the other cell in voltage-clamp mode. EPSCs were recorded at −70, −30, 0, and 20 mV holding potential to discriminate between glutamatergic and GABAergic synapses. (The reversal potential for chloride currents is about −50 mV in our experimental conditions, whereas reversal of AMPA receptor-mediated currents is about 0 mV.) (*see* **Fig. 3**.)
8. In some experiments, the cells were filled with a fluorescent dye through the patch pipette for consequent visualization with a confocal microscope

4. Notes

1. Instead of plasma cleaning of the coverslips, acid cleaning (30 min in cc H_2SO_4) could also be used.
2. It is very important (and difficult) to control the amount of water in the solvent. The water content of the toluene and the surface determine the speed of the reaction. At high water content, the organosilane can polymerize. We distilled the toluene and used molecular sieves to control the water content.
3. In our experience, DETA forms a monolayer on the surface when the ratio of the areas below the N 1s and Si 2p peaks is about 1500 ± 500.
4. If the laser is not available, several other methods can be used for patterning, for example, micro contact printing or ink-jet printing ***(17)***.
5. For the photomask creation, several other programs can be used such as AutoCAD. Neuronal polarity can be determined by spatial clues ***(8)***.
6. The mask should be tightly pressed to the substrate. The UV wavelength and output power is important ***(18)***.
7. The timing of the 13F backfill is important, because 13F incorporates to the DETA layer. Backfilling should be optimized to result in a minimal incorporation and maximal coverage of the ablated surface. Incorporation can be tested by XPS.
8. In the last step of the copperless deposition, the timing is difficult and the coverslip must be checked regularly and removed from the solution when it is ready.
9. Gentle tituration of the tissue without bubble formation is important.

Acknowledgments

This work was supported by NIH Career Award K01 EB03465 and DOE grant DE-FG02-04ER46171.

References

1. Bhadriraju, K. and C.S. Chen, Engineering cellular microenvironments to cell-based drug testing. *Drug Discovery Today*, 2002; 7(11): p. 612–620.
2. Singhvi, R., et al., Engineering cell shape and function. *Science*, 1994; 264(5159): p. 696–698.

3. Chen, C.S., et al., Micropatterned surfaces for control of cell shape, position, and function. *Biotechnology Progress*, 1998; 14(3): p. 356–363.
4. Das, M., et al., Long-term culture of embryonic rat cardiomyocytes on an organosilane surface in a serum-free medium. *Biomaterials*, 2004; 25(25): p. 5643–5647.
5. Park, T.H. and M.L. Shuler, Integration of cell culture and microfabrication technology. *Biotechnology Progress*, 2003; 19(2): p. 243–253.
6. Ravenscroft, M.S., et al., Developmental neurobiology implications from fabrication and analysis of hippocampal neuronal networks on patterned silane-modified surfaces. *Journal of the American Chemical Society*, 1998; 120(47): p. 12169–12177.
7. Stenger, D.A., et al., Surface determinants of neuronal survival and growth on self-assembled monolayers in culture. *Brain Research*, 1993; 630(1–2): p. 136–147.
8. Stenger, D.A., et al., Microlithographic determination of axonal/dendritic polarity in cultured hippocampal neurons. *Journal of Neuroscience Methods*, 1998; 82(2): p. 167–173.
9. Craig, A.M. and H. Boudin, Molecular heterogeneity of central synapses: afferent and target regulation. *Nature Neuroscience*, 2001; 4(6): p. 569–578.
10. MacDermott, A.B., L.W. Role, and S.A. Siegelbaum, Presynaptic ionotropic receptors and the control of transmitter release. *Annual Review of Neuroscience*, 1999; 22: p. 443–485.
11. Zimmermann, M., Pathobiology of neuropathic pain. *European Journal of Pharmacology*, 2001; 429(1–3): p. 23–37.
12. Yamada, K.A., Modulating excitatory synaptic neurotransmission: potential treatment for neurological disease? *Neurobiology of Disease*, 1998; 5(2): p. 67–80.
13. Albuquerque, E.X., et al., Modulation of nicotinic receptor activity in the central nervous system: a novel approach to the treatment of Alzheimer disease. *Alzheimer Disease and Associated Disorders*, 2001; 15: p. S19–S25.
14. Das, M., et al., Electrophysiological and morphological characterization of rat embryonic motoneurons in a defined system. *Biotechnology Progress*, 2003; 19: p. 1756–1761.
15. Kang, J.F., et al. Patterned neuronal networks for robotics, neurocomputing, toxin detection and rehabilitation. In *24th Army Conference*, 2004, Orlando, FL.
16. Schaffner, A.E., et al., Investigation of the factors necessary for growth of hippocampal neurons in a defined system. *Journal of Neuroscience Methods*, 1995; 62(1–2): p. 111–119.
17. Kane, R.S., et al., Patterning proteins and cells using soft lithography. *Biomaterials*, 1999; 20(23–24): p. 2363–2376.
18. Hickman, J.J., et al., Rational pattern design for in-vitro cellular networks using surface photochemistry. *Journal of Vacuum Science & Technology. A, Vacuum, Surfaces, and Films: An Official Journal of the American Vacuum Society*, 1994; 12(3): p. 607–616.

11

Modeling of Action Potential Generation in NG108-15 Cells

Peter Molnar and James J. Hickman

Summary

In order to explore the possibility of identifying toxins based on their effect on the shape of action potentials, we created a computer model of the action potential generation in NG108-15 cells (a neuroblastoma/glioma hybrid cell line). To generate the experimental data for model validation, voltage-dependent sodium, potassium, and high-threshold calcium currents, as well as action potentials, were recorded from NG108-15 cells with conventional whole-cell patch-clamp methods. Based on the classic Hodgkin–Huxley formalism and the linear thermodynamic description of the rate constants, ion channel parameters were estimated using an automatic fitting method. Utilizing the established parameters, action potentials were generated using the Hodgkin–Huxley formalism and were fitted to the recorded action potentials. To demonstrate the applicability of the method for toxin detection and discrimination, the effect of tetrodotoxin (a sodium channel blocker) and tefluthrin (a pyrethroid that is a sodium channel opener) was studied. The two toxins affected the shape of the action potentials differently, and their respective effects were identified based on the predicted changes in the fitted parameters.

Key Words: Action potential; computer modeling; Hodgkin–Huxley; linear thermodynamic description; drug detection; NG108-15; parameter fitting.

1. Introduction

Action potential generation and the shape of the action potential depends on the status of several ion channels located in a cell's membrane, which are regulated by receptors and intracellular messenger systems ***(1–3)***. Changes in the extracellular or intracellular environment (receptor activation and gene

From: *Methods in Molecular Biology, vol. 288: Patch-Clamp Methods and Protocols*
Edited by: P. Molnar and J. J. Hickman © Humana Press Inc., Totowa, NJ

expression) can be reflected in an alteration of spontaneous firing properties such as the frequency and firing pattern ***(1,4–6)*** of excitable cells and in the changes of the action potential shape ***(7–12)***.

Experimental data-driven computer modeling has been an excellent tool to integrate our knowledge concerning elements of a complex biological system as well as to draw conclusions about the behavior of the complex system under different experimental conditions. These predictions can be correlated again with the experimental data. One typical example is the modeling of the electrophysiological behavior of an excitable cell. We have detailed knowledge about individual ion channels as well as extensive knowledge about the behavior of the whole cell, but, we know relatively little about the interaction and modulation of ionic currents that shape the action potentials. One of the most complex single-cell models created simulates the electrophysiological behavior of human cardiac myocytes ***(12–14)***. This model was validated based on electrophysiological experiments under physiological and pathophysiological conditions ***(15)*** and used for a deeper understanding of behavior of ion channels in disease models.

The most commonly used mathematical formalism that describes the action potential generation in excitable cells was developed by Hodgkin and Huxley in 1952 ***(16)***. In our studies, we have used this formalism, although for the description of the ion channels, we utilized the linear thermodynamic approach, which, in our opinion, was more general and does not require "guessing" the form of the functions describing the voltage dependence of the state parameters ***(17,18)***.

We used the NG108-15 neuroblastoma/glioma cell line in our studies ***(10)***, because these cells do not form synapses in culture, thus they are ideal single-cell sensors ***(19,20)***.

2. Materials

2.1. NG108-15 Cultures

1. Culture medium: 90% Dulbecco's modified Eagle's medium (DMEM, Invitrogen, USA) supplemented with 10% fetal bovine serum (Invitrogen) and 1% HAT supplement (Invitrogen).
2. Differentiating medium: DMEM + 2% B27 supplement (Invitrogen).
3. Poly-D-lysine (PDL) solution: 5 mg PDL (Sigma, USA, cat. no. P7405) in 500 ml water and sterile filtered.

2.2. Patch-Clamp Recording of Ionic Currents and Action Potentials

1. Extracellular solution: 140 m*M* NaCl, 3.5 m*M* KCl, 2 m*M* $MgCl_2$, 2 m*M* $CaCl_2$, 10 m*M* glucose, 10 m*M* 4-(-2-hydroxyethyl)-1-piperazineethane sulfonic acid (HEPES), pH = 7.34.

2. Extracellular solution for the recording of potassium currents: extracellular solution +1 μ*M* tetrodotoxin (Alomone Labs, Israel, cat. no. T550).
3. Extracellular solution for the recording of sodium currents: 50 m*M* NaCl, 100 m*M* Tetra Ethyl Ammonium (TEA)-cl 5 m*M* CsCl, 1 m*M* $CaCl_2$, 1 m*M* $CoCl_2$, 1 m*M* $MgCl_2$, 10 m*M* glucose, 10 m*M* HEPES, pH = 7.34.
4. Extracellular solution for the recording of calcium currents: 100 m*M* NaCl, 30 m*M* TEA-Cl, 10 m*M* $CaCl_2$, 2 m*M* $MgCl_2$, 10 m*M* glucose, 10 m*M* HEPES, 1 m*M* tetrodotoxin, pH = 7.34.
5. Intracellular solution for action potential and potassium channel measurement: 130 m*M* K-gluconate, 2 m*M* $MgCl_2$, 1 m*M* Ethylene glycol tetraacetic acid (EGTA), Sigma , 15 m*M* HEPES, 5 m*M* Adenosine 5'-triphosphate (ATP), pH = 7.2, osmolarity = 276 mOsm.
6. Intracellular solution for sodium channel measurement: 130 m*M* CsF, 10 m*M* NaCl, 10 m*M* TEA-Cl, 2 m*M* $MgCl_2$, 1 m*M* EGTA, 10 m*M* HEPES, 5 m*M* ATP.
7. Intracellular solution for calcium channel measurement: 120 m*M* CsCl, 20 m*M* TEA-Cl, 2 m*M* $MgCl_2$, 1 m*M* EGTA, 10 m*M* HEPES, 5 m*M* ATP. For selecting L-type calcium channels, 1 μ*M* ωCTxGVIA (Tocris, UK) was added.
8. Pipette puller (Sutter USA, cat. no. P97).
9. Glass pipettes (Sutter, cat. no. BF150-86-10).
10. Vibration isolation table with Faraday cage (TMC, USA).
11. Microscope Axioskop 2 FS plus (Zeiss, Germany).
12. Patch-clamp amplifier Multiclamp 700A (Axon, USA), A/D converter Digidata 1322A (Axon), patch-clamp software pClamp 8 (Axon).

2.3. Obtaining the Parameters Describing Ion Channel Currents and Action Potentials in NG108-15 Cells

1. Matlab software (MathWorks, USA).

2.4. Effect of Drugs on Action Potential Shape and on Ion Channel Parameters

1. Tetrodotoxin (TTX) (Alomone Labs).
2. Tefluthrin (Riedel-de Haën, Germany).

3. Methods

3.1. NG108-15 Cultures

1. The NG108-15 cell line (passage number 16) was obtained from Dr. M.W. Nirenberg (NIH). Cells were stored frozen in liquid N_2 in 1-ml vials (1 million cells/vials).
2. Cell stock was grown in a T-75 flask in culture medium at 37°C with 10% CO_2 (*see* **Note 1**).
3. Glass coverslips were cleaned utilizing concentrated nitric acid for 30 min, rinsed three times with water, and sterilized by submerging in 100% alcohol for 20 min.

4. Coverslips were placed in six-well plates and were incubated in PDL solution for 1 h in the 37°C incubator.
5. After confluence, the culture medium was replaced by 6 ml differentiating medium. The cell layer was dislodged by knocking the flask on the table. Cells were dissociated by tituration using a 5-ml pipette.
6. NG108-15 cells were plated at a density of 40,000 cells/35-mm culture dish in 2 ml differentiating medium on the PDL-coated coverslips and were cultured for 1 week at 37°C with 5% CO_2.

3.2. Patch-Clamp Recording of Ionic Currents and Action Potentials

1. Coverslips were transferred into a chamber on the microscope stage, which was continuously perfused with the appropriate extracellular solution. Experiments were performed at room temperature.
2. Glass pipettes, pulled with the electrode puller, were filled with the appropriate intracellular solution and had a resistance of 4–6 MΩ.
3. Signals were filtered at 2 kHz and digitized at 20 kHz.
4. Sodium and potassium currents were measured in voltage-clamp mode using 10 mV steps from a −85 mV holding potential. To record high-threshold calcium currents, a −40 mV holding was used.
5. Whole-cell capacitance and series resistance were compensated, and a p/6 protocol was used.
6. Action potentials were evoked with short (2 ms) current injections in current-clamp mode either at resting membrane potential or at a −85 mV holding potential. Data were saved in text format and imported into Matlab for further analysis.
7. Tip potential was calculated by the built-in routine of the pClamp program and was compensated by subtracting 15 mV from the membrane potential values.

3.3. Obtaining the Parameters Describing Ion Channel Currents in NG108-15 Cells

1. A computer program was created in Matlab to fit the parameters to the recorded data according to the following equations. Basically, the experimental data were given as total ionic current versus time (note that sodium, potassium, and calcium currents were measured independently in separate experiments) and membrane potential time.
2. Total ionic current (sodium I_{Na}, potassium I_K, calcium I_{Ca}, and leakage channels I_l included):

$$\begin{aligned} I_{\text{ionic}} = I_{\text{Na}} + I_K + I_{\text{Ca}} + I_l &= \bar{g}_{\text{Na}} m^3 h (V - V_{\text{Na}}) + \bar{g}_K n^4 (V - V_K) \\ &+ \bar{g}_{\text{Ca}} e^3 (V - V_{\text{Ca}}) + \bar{g}_l (V - V_l), \end{aligned}$$

where $\bar{g}_{Na}$, $\bar{g}_K$, $\bar{g}_{Ca}$ and V_{Na}, V_K, V_{Ca} are parameters (maximum conductances of the channels and reversal potentials, respectively) and m, n, h, and e are the state variables.

3. The dynamics of the state variables were

$$\frac{\mathrm{d}m}{\mathrm{d}t} = \frac{m_\infty - m}{\tau_\mathrm{m}},$$

where m_∞, n_∞, h_∞, and e_∞ are the steady-state values of the state variables and τ_m, τ_n, τ_h, and τ_e are their voltage-dependent time constants, respectively.
4. The voltage dependence of the time constants and the steady-state state parameters were given according to the general thermodynamic formalism, an example was given for the m state parameter:

$$m_\infty = \frac{1}{1 + \exp^{-(zF/RT)(V_\mathrm{m} - V_{1/2})}}$$

and

$$\tau_\mathrm{m} = \frac{A}{\exp^{[(zF/RT)\xi](V_\mathrm{m} - V_{1/2})} \cos h\left[(zF/2RT)\left(V_\mathrm{m} - V_{1/2}\right)\right]},$$

where z, $V_{1/2}$, A, and ξ are fitting parameters and V_m represents the membrane potential. As it can be seen from these equations, $V_{1/2}$ corresponds to the half activation/inactivation potential of the channel and A is linearly related to the activation or inactivation time constant. The meanings of z and ξ are not that obvious: z is related to the number of moving charges during the opening or closing of the channel, whereas ξ describes the asymmetric position of the moving charge in the cell membrane.
5. Sodium, potassium, and calcium channel-mediated current data, which were recorded in voltage-clamp mode at different membrane potentials (5 mV increments and −40 to +30 mV range), were imported into Matlab. Based on the previously listed equations, simulated current traces were calculated (*see* **Fig. 1**).
6. An error function was generated based on the difference between the recorded and the simulated traces through the whole voltage range (*see* **Notes 2** and **3**).
7. Matlab's fminsearch routine was used to optimize the parameters to minimize the error function (*see* **Note 4**).
8. Parameters obtained from different cells were averaged ($n = 4$–6) and considered as initial values for the action potential modeling.

3.4. Fitting Parameters to the Experimentally Recorded Action Potentials

1. Effect of the currents on the membrane potential:

$$\frac{\mathrm{d}V}{\mathrm{d}t} = \frac{I_\mathrm{external} - I_\mathrm{ionic}}{C_\mathrm{M}},$$

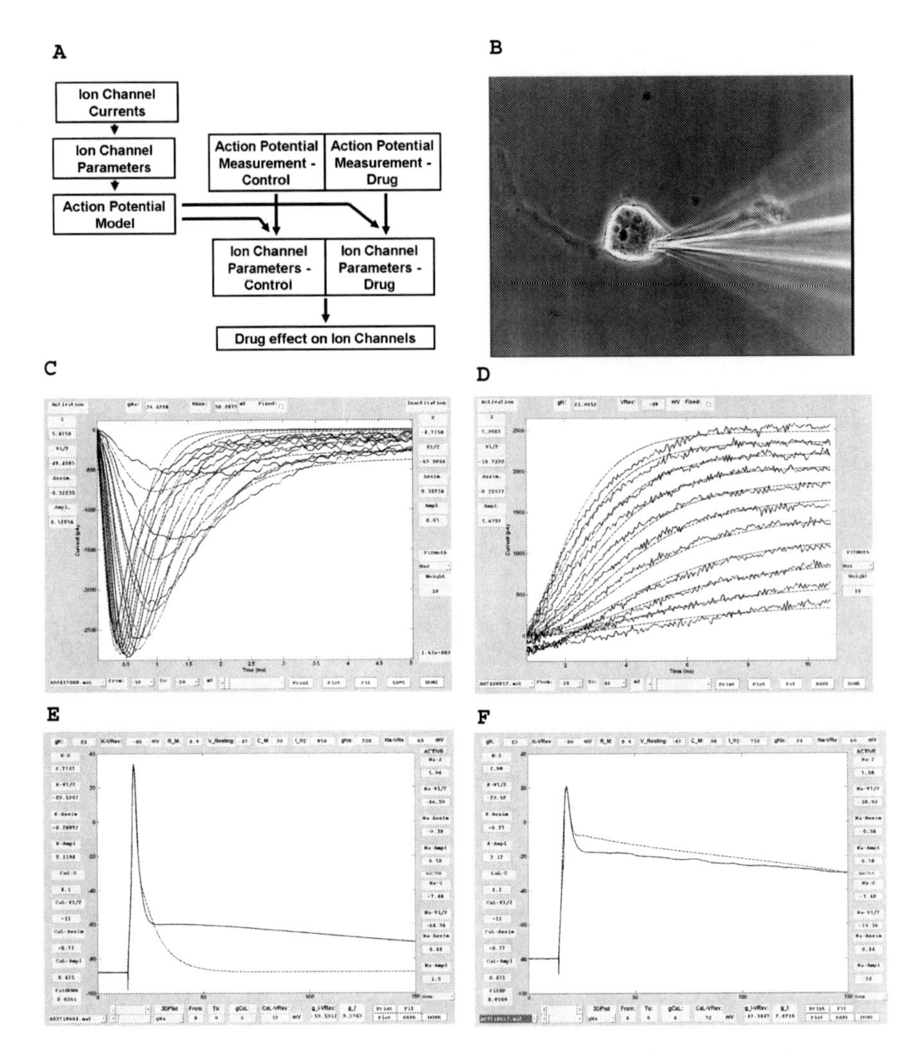

Fig. 1. Determination of the ion channel parameters obtained from action potential shapes. **(A)** Ion channel parameters are determined from action potential recordings based on a computer model of action potential generation and parameter fitting. Drug effects could be quantified based on changes of ion channel parameters. **(B)** Phase-contrast picture of an NG108-15 cell with a patch electrode. **(C)** Measured sodium channel currents (solid line) with simulated current traces (dotted line). Parameters for the simulation were determined by an automatic fitting routine. **(D)** Measured potassium channel currents (solid line) with simulated current traces (dotted line). **(E)** Control action potential (solid line) with simulated action potential trace (dotted line). **(F)** Action potential shape after 1 μ*M* tefluthrin (solid line) and simulated action potential trace (dotted line).

Table 1
Effect of TTX and Tefluthrin on the Action Potential Parameters

			Activation				Inactivation			
Channel	g	V_{Rev}	z	$V_{1/2}$	ξ	A	z	$V_{1/2}$	ξ	A
Control	**550**	60	6.02	−45.99	−0.38	**0.6**	−7.48	−64.36	0.53	**0.37**
TTX	**16.5**	60	6.02	45.99	−0.39	**1.30**	−7.48	−64.36	0.51	**6.00**
Control	**220**	60	5.98	**−46.93**	−0.38	0.58	−7.48	**−64.36**	0.44	**1.5**
Tefluthrin	**24**	60	5.98	**−58.93**	−0.38	0.58	−7.48	**−74.36**	0.44	**70**

Representative recordings of three to four experiments with similar results. Only sodium channel parameters are shown, other ion channel parameters did not change. The boldface values represent more than 10% change.

where $I_{external}$ is the externally injected current used to depolarize the membrane and evoke the action potential in current-clamp mode.

2. The following parameters were obtained from the patch-clamp recordings and used in the modeling: membrane resistance, resting membrane potential, membrane capacitance, and injected current. The maximum conductance of the leakage current (g_l) was calculated from the ionic conductances and resting membrane potential.
3. The differential equation previously listed and the differential equations for the state parameters formed a first-order coupled differential equation system, which was solved with Matlab's ODE23 solver.
4. Parameters determining the simulated curve were fitted to the experimental data using the fminunc routine.

3.5. Effect of Drugs on Action Potential Shape and on Ion Channel Parameters

1. Drugs were applied by perfusing the experimental chamber with extracellular medium containing the drug.
2. Action potentials were recorded before and 10 min after the drug administration.
3. Ion channel parameters were calculated for the control and drug-modified action potentials by fitting the simulated action potentials to the recorded experimental data (*see* **Table 1**).
4. Drug effects were quantified as percentage changes in the ion channel parameters obtained before and after drug administration.

4. Notes

1. Cells also grow well at 5% CO_2.
2. Curves were fitted after an initial 0.1 ms delay to eliminate the effect of experimental artifacts.
3. To quantify the difference between the fitted curves and the recorded data, the following error functions were implemented: Maximum error: $E_{Max} = Max\{Abs[R(t_n) - S(t_n)]\}$, where $R(t_n)$ is the recorded value and $S(t_n)$ is the simulated data at time t_n; least square: $E_{Lsquare} = \sum_n [R(t_n) - S(t_n)]^2$; weighted least square: $E_{WLsquare} = E_{Lsquare}$ if $t_n < 30\,\text{ms}$ and $E_{WLsquare} = 5 \times E_{Lsquare}$ if $t_n \geq 30\,\text{ms}$. After several trials, it was concluded that simulations that used the weighted least square error function gave the most satisfactory results because the other error functions occasionally obtained a noninactivating sodium-current component in the simulated data.
4. Fminunc was used to find a minimum of a scalar function (the error function) for several variables, starting at an initial estimate. This method is generally referred to as unconstrained nonlinear optimization. Because it finds only local minimums, it is very important to start the optimization as close to the final result as possible. In our case, the averaged ion channel parameters obtained from voltage-clamp experiments served the initial values for the parameter estimations.

Acknowledgments

This work was supported by NIH Career Award K01 EB03465 and DOE grant DE-FG02-04ER46171.

References

1. Gross, G.W., et al., Odor, drug and toxin analysis with neuronal networks in vitro: extracellular array recording of network responses. *Biosensors & Bioelectronics*, 1997; 12(5): p. 373–393.
2. Gross, G.W., et al., The use of neuronal networks on multielectrode arrays as biosensors. *Biosensors & Bioelectronics*, 1995; 10(6–7): p. 553–567.
3. Morefield, S.I., et al., Drug evaluations using neuronal networks cultured on microelectrode arrays. *Biosensors & Bioelectronics*, 2000; 15(7–8): p. 383–396.
4. Amigo, J.M., et al., On the number of states of the neuronal sources. *Biosystems*, 2003; 68(1): p. 57–66.
5. Chiappalone, M., et al., Networks of neurons coupled to microelectrode arrays: a neuronal sensory system for pharmacological applications. *Biosensors & Bioelectronics*, 2003; 18: p. 627–634.
6. Xia, Y., K.V. Gopal, and G.W. Gross, Differential acute effects of fluoxetine on frontal and auditory cortex networks in vitro. *Brain Research*, 2003; 973(2): p. 151–160.
7. Akay, M., E. Mazza, and J.A. Neubauer, Non-linear dynamic analysis of hypoxia-induced changes in action potential shape in neurons cultured from the rostral ventrolateral medulla (RVLM). *The FASEB Journal: Official Publication of the Federation of American Societies for Experimental Biology*, 1998; 12(4): p. 2881.
8. Clark, R.B., et al., Heterogeneity of action-potential wave-forms and potassium currents in rat ventricle. *Cardiovascular Research*, 1993; 27(10): p. 1795–1799.
9. Djouhri, L. and S.N. Lawson, Changes in somatic action potential shape in guinea-pig nociceptive primary afferent neurones during inflammation in vivo. *Journal of Physiology-London*, 1999; 520(2): p. 565–576.
10. Mohan, D.K., P. Molnar, and J.J. Hickman, Toxin detection based on action potential shape analysis using a realistic mathematical model of differentiated NG108-15 cells. *Biosensors & Bioelectronics*, 2006; 21(9): p. 1804–1811.
11. Muraki, K., Y. Imaizumi, and M. Watanabe, Effects of noradrenaline on membrane currents and action-potential shape in smooth-muscle cells from guinea-pig ureter. *Journal of Physiology-London*, 1994; 481(3): p. 617–627.
12. Nygren, A., et al., Mathematical model of an adult human atrial cell – the role of K+ currents in repolarization. *Circulation Research*, 1998; 82(1): p. 63–81.
13. Bernus, O., et al., A computationally efficient electrophysiological model of human ventricular cells. *American Journal of Physiology. Heart and Circulatory Physiology*, 2002; 282(6): p. H2296–H2308.

14. Nygren, A., L.J. Leon, and W.R. Giles, Simulations of the human atrial action potential. *Philosophical Transactions of the Royal Society of London. Series A: Mathematical Physical and Engineering Sciences*, 2001; 359(1783): p. 1111–1125.
15. Winslow, R.L., et al., Computational models of the failing myocyte: relating altered gene expression to cellular function. *Philosophical Transactions of the Royal Society of London. Series A: Mathematical Physical And Engineering Sciences*, 2001; 359(1783): p. 1187–1200.
16. Hodgkin, A.L. and A.F. Huxley, A quantitative description of membrane current and its application to conduction and excitation in nerve. *Journal of Physiology*, 1952; 117: p. 500–544.
17. Destexhe, A. and J.R. Huguenard, Nonlinear thermodynamic models of voltage-dependent currents. *Journal of Computational Neuroscience*, 2000; 9: p. 259–270.
18. Weiss, T.F., *Cellular Biophysics*, 1996. Cambridge, MA: MIT Press.
19. Kowtha, V.C., et al., Comparative electrophysiological properties of NG108-15 cells in serum-containing and serum-free media. *Neuroscience Letters*, 1993; 164(1–2): p. 129–133.
20. Ma, W., et al., Neuronal and glial epitopes and transmitter-synthesizing enzymes appear in parallel with membrane excitability during neuroblastoma x glioma hybrid differentiation. *Brain Research. Developmental Brain Research*, 1998; 106: p. 155–163.

12

Whole-Cell Voltage Clamp on Skeletal Muscle Fibers With the Silicone-Clamp Technique

Sandrine Pouvreau, Claude Collet, Bruno Allard, and Vincent Jacquemond

Summary

Control of membrane voltage and membrane current measurements are of strong interest for the study of numerous aspects of skeletal muscle physiology and pathophysiology. The silicone-clamp technique makes use of a conventional patch-clamp apparatus to achieve whole-cell voltage clamp of a restricted portion of a fully differentiated adult skeletal muscle fiber. The major part of an isolated muscle fiber is insulated from the extracellular medium with silicone grease, and the tip of a single microelectrode connected to the amplifier is then inserted within the fiber through the silicone layer. This method represents an alternative to the traditional vaseline-gap isolation and two or three microelectrode voltage-clamp techniques. This chapter reviews the main benefits of the silicone-clamp technique and provides detailed insights into its practical implementation.

Key Words: Skeletal muscle; voltage clamp; silicone grease; mammalian muscle; ion channels; excitation–contraction coupling.

1. Introduction

Whole-cell voltage clamp on isolated adult skeletal muscle fibers was first achieved in the late 1960's *(1)* and has since then been mastered in several groups with either two or three microelectrode techniques *(2–5)* or various vaseline-gap isolation methods *(6–9)*. Of course each method has its own advantages and drawbacks, but overall, none of those could be readily

From: *Methods in Molecular Biology, vol. 288: Patch-Clamp Methods and Protocols*
Edited by: P. Molnar and J. J. Hickman © Humana Press Inc., Totowa, NJ

qualified of easy to implement and use. Furthermore, some specific difficulties inherent to the two previously described lines of approach were worth circumvent; for instance, microelectrodes suffer from their invasive features, whereas gap-isolation techniques necessitate a specific experimental chamber within which the transfer and mounting of a single isolated fiber can be a tricky demanding task. There was thus room for a more accessible way to do voltage clamp on skeletal muscle, especially when having in mind the relative simplicity and easiness of practice of the whole-cell patch-clamp technique. Furthermore, until the recent years, there were only scarce data from mammalian isolated fibers, largely because of the quite challenging microdissection step in this type of muscle preparation. In this respect, the alternative possibility of using enzymatically isolated fibers ***(10)*** was appealing. Still, owing to their large size and to the extent of the transverse tubular membrane system, skeletal muscle fibers have to be left off the list of the cell types suitable for conventional whole-cell patch clamping. We initially had the idea that whole-cell voltage clamp could then be achieved by sealing the tip of a patch-clamp pipette on an electrically insulated restricted portion of a muscle fiber, sort of a combination between a gap-isolation system and the patch-clamp technique. While attempting to work this out, we found that silicone grease could be used as a nontraumatic electrically insulating agent on fibers isolated by collagenase treatment of a muscle, opening up the possibility to work on a short portion of such an isolated intact fiber. We also found that the inside of a glass micropipette could be given access to the cytoplasmic compartment of the fiber through the silicone-insulating material. These two small tricks have had important practical outcomes. First, enzymatically isolated muscle fibers could, for the first time, be voltage clamped by a classical gap-isolation method using silicone instead of vaseline ***(11,12)***. Second, voltage clamp could also be achieved with a patch-clamp amplifier on a short silicone-free end portion of a fiber, simply with the tip of a micropipette inserted within the fiber through the silicone layer ***(13)***; this is what we have called the silicone voltage-clamp technique. This configuration is much more stable and less damageable to the preparation than sealing the tip of a patch pipette on the silicone-free end portion of fiber. Finally, microinjection can be performed within the silicone-embedded portion of fiber without fearing the detrimental consequences of membrane damage. This makes intracellular loading of a dye, a buffer, or any other compound of interest easy. The simplicity of the silicone voltage-clamp technique makes it readily combinable with an extracellular perfusion system, with fluorescence detection under either conventional or confocal microscopy and with single-channel activity measurements using an additional pipette in cell-attached mode

(14). The different types of membranous and intracellular signals that can be routinely collected under silicone voltage-clamp conditions were reviewed by Collet et al. *(15)*.

2. Materials

All solutions are prepared with deionized water and stored at 4°C until use unless otherwise specified.

2.1. Isolation of the Muscle Fibers

1. Fibers are isolated from the hind limb *flexor digitorum brevis* and/or interosseal muscles from mouse.
2. Tyrode solution: 140 m*M* NaCl, 5 m*M* KCl, 2.5 m*M* $CaCl_2$, 1 m*M* $MgCl_2$, 10 HEPES, pH 7.2.
3. Collagenase (type 1, Sigma, USA).

2.2. Preparation of a Muscle Fiber for the Voltage Clamp

1. Silicone grease (Pâte 70428; Rhodia Silicones, France) (*see* **Note 1**).
2. Culture medium supplemented with fetal bovine serum (FBS 10%) (M199; Eurobio, France).

2.3. Electrophysiology

1. The technique makes use of a conventional patch-clamp setup.
2. The composition of the extracellular solution is designed according to the aim of the experiment; care should be taken to avoid conditions into which large currents may produce membrane voltage escape. For routine measurements of voltage-activated intracellular calcium transients in the silicone-free end portion of fiber, we use the following: 140 m*M* tetraethylammonium-methanesulfonate, 2.5m*M* $CaCl_2$, 2 m*M* $MgCl_2$, 10 m*M* HEPES, 2 m*M* tetrodotoxin, pH 7.2; for measurements of calcium current, the solution also contains 1 m*M* of 4-amino-pyridine to further block potassium conductances; for measurements of intramembrane charge movement, the $CaCl_2$ concentration is lowered to 0.25 m*M*, and the solution also contains 0.5 m*M* $CdCl_2$ and 0.3 m*M* $LaCl_3$ to eliminate the calcium current.
3. Internal solution: 120 m*M* K-glutamate, 5 m*M* Na_2ATP, 5 m*M* sodium phosphocreatine, 5.5 m*M* $MgCl_2$, 5 m*M* glucose, 10 m*M* HEPES, pH 7.2. Aliquots of 1.5 ml of this solution are stored at −20°C.

3. Methods

3.1. Isolation of the Muscle Fibers

Intact Isolated muscle fibers can be easily obtained by enzymatic treatment of small muscles from mouse. The procedure is particularly efficient with either the *flexor digitorum brevis* or the interosseal muscles.

1. Mice are killed by cervical dislocation.
2. Muscles are removed and incubated at 37°C for 60 min in typically 3 ml Tyrode solution containing 6 mg collagenase (*see* **Note 2**).
3. Muscles are then rinsed with collagenase-free Tyrode and can be stored at 4°C until use. They can be used up to 24 h following the dissociation.
4. Isolated fibers are obtained by gently sucking a collagenase-treated muscle a few times in and out of the cut disposable blue tip of a Pipetman (or of the tip of a Pasteur pipette), within the experimental chamber (see section 3.2.1). This process yields tens of viable, healthy fibers from a single round of trituration of one muscle.

3.2. Preparation of a Muscle Fiber for the Voltage Clamp

1. For simple voltage-clamp experiments, we use a standard 35-mm culture dish as experimental chamber. For experiments combining voltage-clamp and fluorescence measurements, we use a homemade chamber with a glass coverslip bottom. In all cases, the bottom of the chamber is first coated with a thin layer of silicone (*see* **Note 3**).
2. The chamber is then filled with culture medium (supplemented with FBS, *see* **Subheading 2.**). This solution is necessary to preserve the integrity of the fibers on first contact with the silicone. After approximately 10 min, it can be safely replaced by the experimental solution (*see* **Note 4**). Notice that isolated fibers do not adhere to the silicone material.
3. The next step consists in embedding the major part of a fiber with silicone so as to leave only a short (typically 50–100 μm long) emerging portion in contact with the extracellular solution. Obviously, this is a critical step of the process; although the first time it may seem hazardous, with practice, it gets extremely trivial and can be achieved within a couple of minutes (*see* **Note 5**). In our hands, this step is routinely achieved in the following way:

    a. The tip of a glass micropipette is broken so as to yield a diameter of approximately 50–100 μm.
    b. This pipette is backfilled with silicone; this is done by applying pressure through a crude device consisting in a thin plastic capillary tube inserted within the lumen of the pipette and linked on the other end to a standard 5-ml medical syringe filled with silicone.
    c. The silicone-filled pipette is then mounted on the holder of a pressure microinjection device (Picospritzer II; General Valve Corp., USA) held on a micromanipulator. The gross tip of the pipette is then positioned close to the median portion of a given fiber, in touch with the silicone layer covering the bottom of the chamber (*see* **Fig. 1B**). At that point, care should be taken that no air bubble remains within the pipette tip.
    d. A constant pressure is then applied in the system (typically 10–20 psi), so that the silicone flows out of the pipette tip, at a rate sufficiently low to allow control of the process.

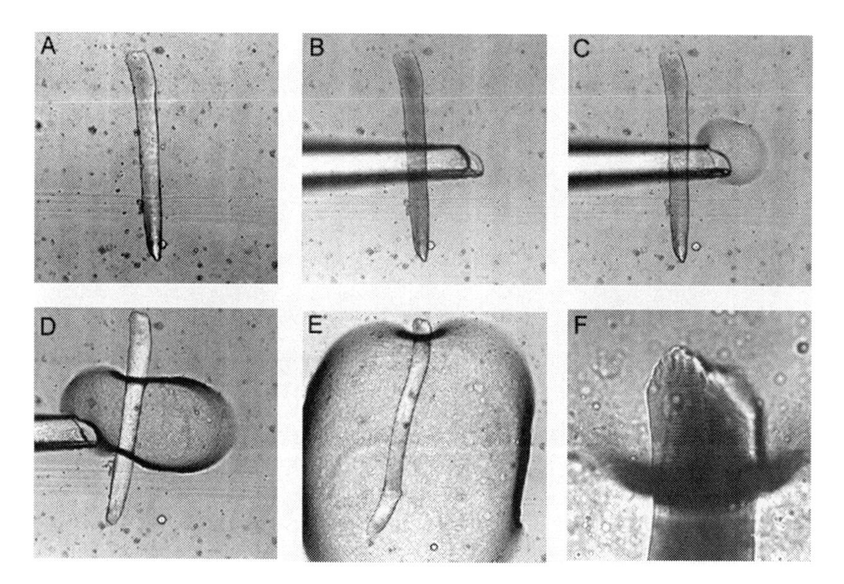

Fig. 1. Silicone embedding of a single isolated muscle fiber. Images of a muscle fiber were taken during the consecutive steps of the process. In panels **(A–E)** and **(F)**, image side size is 709 and 113 μm, respectively. See text for details.

e. Once the flowing out silicone has merged with the layer on the bottom of the chamber (*see* **Fig. 1C**), the pipette is rapidly moved up (keeping the silicone flowing out), then laterally above the fiber and then down on the opposite side of the fiber so as to make a bridge of silicone covering the fiber (*see* **Fig. 1D**). This is repeated several times along the fiber until only a short portion remains out of the silicone (*see* **Fig. 1E, F**). Once this is completed, any compound of interest can be microinjected within the portion of fiber embedded in silicone (*see* **Note 6**).

3.3. Electrophysiology

1. The voltage-clamp pipettes are pulled so as to yield a typical resistance of 1–2 MΩ when filled with the internal solution (*see* **Note 7**).
2. Positive pressure is maintained in the pipette until the tip is inserted within the silicone layer coating the fiber. This makes the resistance go way up (hundreds of megaohms); pressure is then released, and the desired holding voltage value is applied.
3. The pipette is then moved further down until the tip gets in touch with the membrane of the fiber. In order for the tip to go through the membrane, a single or couple of brief taps on the supporting table is usually sufficient. Access can in many cases be

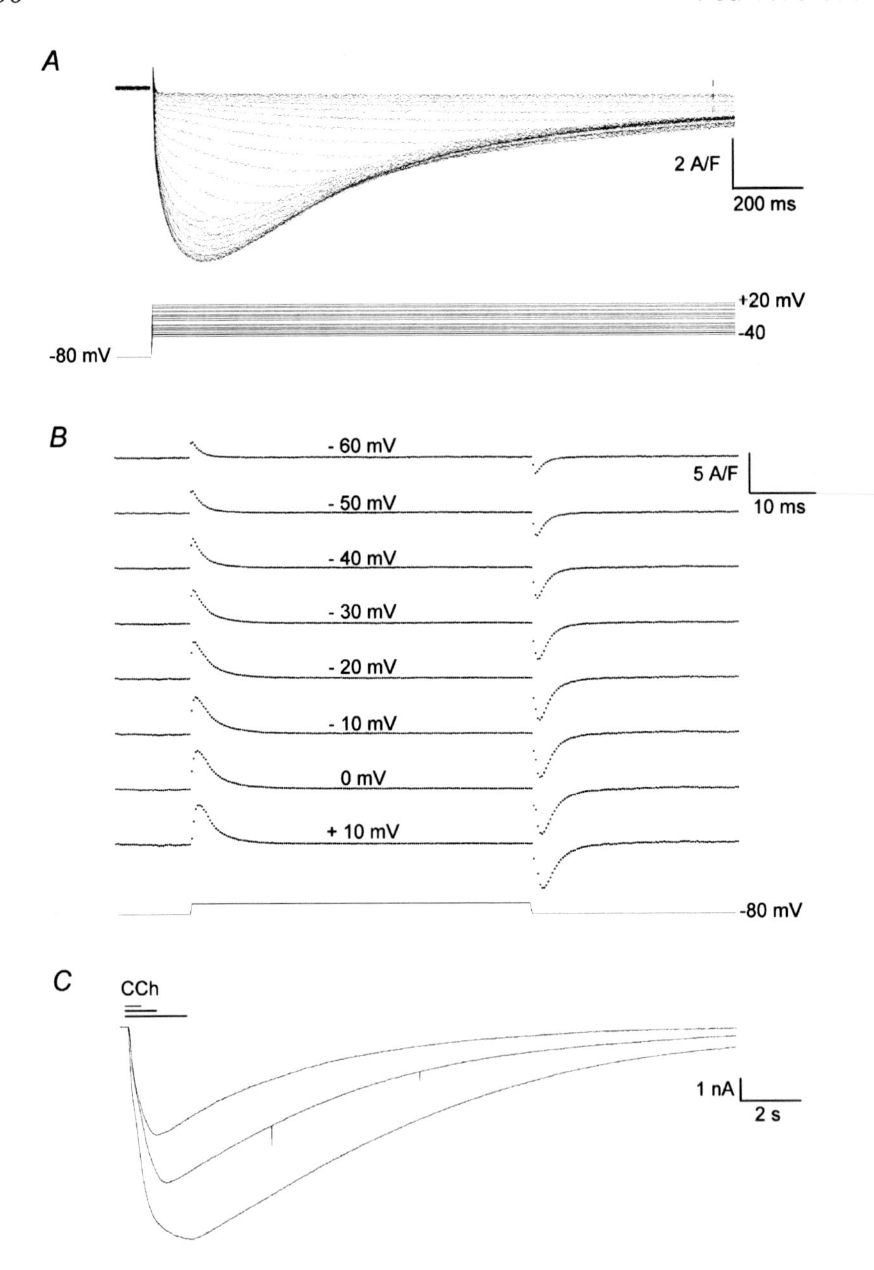

Fig. 2. Examples of membrane current records from silicone voltage-clamped fibers under different conditions. **(A)** Calcium current records obtained in response to depolarizing pulses to the indicated values; the linear resistive component of the current has been subtracted. **(B)** Intramembrane charge movement records in response to

further improved by moving the tip of the pipette within the fiber and/or by giving a few more gentle taps on the table once the tip is in.

4. Following the initial jump in current produced upon insertion of the tip, the holding current usually tends to decrease toward a lower value with a time constant of a few minutes; we thus routinely wait approximately 10 min for the current to stabilize, adjust the series resistance compensation, and start the experiment. The impalement site is usually located at a distance of 100–150 μm from the silicone-free end portion of fiber.

Figure 2 shows examples of membrane current traces obtained from silicone voltage-clamped fibers under various experimental conditions. **Figure 2A** shows L-type calcium current traces obtained in response to depolarizing pulses of increasing amplitude with a 2-mV increment. **Figure 2B** shows intramembrane charge movement traces obtained in response to depolarizations to the indicated voltages. Linear capacitive and resistive components of the current have been subtracted. **Figure 2C** shows membrane current traces recorded from a fiber, the silicone-free end portion of which included the endplate area. Inward currents were elicited by puffing carbamylcholine (CCh) onto the endplate region for the indicated periods of time.

Notes

1. A compound called Chemplex 825 (NFO Technologies, USA) has also been used with success. The quality of the silicone is extremely important. Special attention has to be paid to its viscosity. If the silicone is too viscous, it can be very hard to get a good insulation of the silicone-embedded main portion of fiber. On the contrary, if the silicone is too fluid, the silicone-treated fibers end up spontaneously popping out of the silicone after a few minutes. Although we always purchased the same silicone product (Pâte 7048, *see* **Subheading 3.**), we occasionally had a clear difference in viscosity in certain batches. Also one batch exhibited a surprisingly elevated level of intrinsic fluorescence upon ultraviolet (UV) excitation, hardly compatible with intracellular calcium measurements using UV excitable dyes.

Fig. 2. depolarizing pulses to the indicated values; linear resistive and capacitive components of the current have been subtracted. **(C)** Membrane currents through endplate nicotinic receptors at −80 mV. The silicone-free end portion of fiber included the endplate area; carbamylcholine (CCh) was puffed onto the endplate region using a pipette containing 10 m*M* CCh diluted in the extracellular solution. The pipette was connected to the Picospritzer apparatus; its tip was placed in the vicinity of the endplate and pressure pulses of the indicated duration were delivered. The extracellular solution contained 130 m*M* $MgSO_4$, 10 m*M* $MgCl_2$, 100 m*M* mannitol, 10 m*M* PIPES, pH 7.2.

2. The period of time during which the muscles are incubated within the collagenase-containing solution can be more precisely adjusted (within a range of 45 to 90 min) so as to achieve the best quality of the cells according to the following gross criteria. If the duration of incubation is too short, fibers hardly if any dissociate upon mechanical trituration of the muscles. Conversely, if the incubation lasts too long, fibers dissociate very easily but are then usually more fragile and may be more likely to die during the subsequent steps of the process.
3. For this, a small amount of the material is first spread all over the bottom of the chamber with a plastic or glass rod. This silicone layer is then made as thin as possible and uniform by sliding the edge of a square coverslip back and forth against and along the bottom of the chamber.
4. If one uses a standard extracellular solution as for instance Tyrode, most fibers exhibit a dramatic and irreversible contracture on first contact with the silicone. Why silicone is toxic to the fibers and how this is prevented by the presence of culture medium are not entirely clear. Notice that FBS alone was reported to protect enzymatically isolated muscle fibers mounted in a double silicone-gap device ***(12)***.
5. This step can be carried out in the presence of either the FBS-supplemented culture medium or the experimental extracellular solution provided that the fibers first spent approximately 10 min in the presence of the culture medium. Although the Picospritzer injection device is particularly convenient for embedding the fiber within silicone, the process can also be achieved using any crude manually controlled pressure delivery system.
6. There is no specific trick to be successful with the microinjection under these conditions. The tip of the pipette is inserted through the silicone within the embedded end portion of fiber. Pressure is raised until local swelling is perceptible at the impalement site. In some cases, the injection can be facilitated by gently moving the pipette tip back and forth within the fiber. Once the injection is complete, pressure is turned off and the pipette removed. A period of at least 30 min is then left to allow for equilibration of the injected solution within the entire myoplasmic volume. For best control of the injection and of the equilibration of the injected compound within the myoplasm, a colorimetric or fluorescent dye can be used in the injected solution ***(16,17)***.
7. In early experiments ***(13)***, the voltage-clamp micropipettes had a much smaller tip; they were filled with a solution containing 3 *M* K-acetate and 20 m*M* KCl and thus exhibited a similar resistance (1–3 MΩ) as the ones used later. One advantage was then that the insertion of the tip within the silicone-insulated part of fiber was much easier and immediate. However, we found that during long-lasting experiments performed under these conditions, the voltage-clamped end portion of the fiber tended to end up swelling, which is the reason why this procedure was abandoned.

Acknowledgments

This work was supported by the Centre National de la Recherche Scientifique, the Université Claude Bernard Lyon 1, and the Association Française contre les Myopathies.

References

1. Adrian, R. H., Chandler, W. K., and Hodgkin, A. L. (1966) Voltage clamp experiments in skeletal muscle fibres. *J. Physiol.* **186**, 51P–52P.
2. Adrian, R. H., Chandler, W. K., and Hodgkin, A. L. (1970) Voltage clamp experiments in striated muscle fibres. *J. Physiol.* **208**, 607–644.
3. Chandler, W. K., Rakowski, R. F., and Schneider, M. F. (1976) A non-linear voltage dependent charge movement in frog skeletal muscle. *J. Physiol.* **254**, 245–283.
4. Heistracher, P. and Hunt, C. C. (1969) The relation of membrane changes to contraction in twitch muscle fibres. *J. Physiol.* **201**, 589–611.
5. Adrian, R. H. and Marshall, M. W. (1977) Sodium currents in mammalian muscle. *J. Physiol.* **268**, 223–250.
6. Ildefonse, M. and Rougier, O. (1972) Voltage-clamp analysis of the early current in frog skeletal muscle fibre using the double sucrose-gap method. *J. Physiol.* **222**, 373–395.
7. Hille, B. and Campbell, D. T. (1976) An improved vaseline gap voltage clamp for skeletal muscle fibers. *J. Gen. Physiol.* **67**, 265–293.
8. Kovacs, L. and Schneider, M. F. (1978) Contractile activation by voltage clamp depolarization of cut skeletal muscle fibres. *J. Physiol.* **277**, 483–506.
9. Kovacs, L., Rios, E., and Schneider, M. F. (1983) Measurement and modification of free calcium transients in frog skeletal muscle fibres by a metallochromic indicator dye. *J. Physiol.* **343**, 161–196.
10. Bekoff, A. and Betz, W. J. (1977) Physiological properties of dissociated muscle fibres obtained from innervated and denervated adult rat muscle. *J. Physiol.* **271**, 25–40.
11. Szentesi, P., Jacquemond, V., Kovacs, L., and Csernoch, L. (1997) Intramembrane charge movement and sarcoplasmic calcium release in enzymatically isolated mammalian skeletal muscle fibres. *J. Physiol.* **505**, 371–384.
12. Woods, C. E., Novo, D., DiFranco, M., and Vergara, J. L. (2004) The action potential-evoked sarcoplasmic reticulum calcium release is impaired in mdx mouse muscle fibres. *J. Physiol.* **557**, 59–75.
13. Jacquemond, V. (1997) Indo-1 fluorescence signals elicited by membrane depolarization in enzymatically isolated mouse skeletal muscle fibers. *Biophys. J.* **73**, 920–928.
14. Jacquemond, V. and Allard, B. (1998). Activation of Ca^{2+}-activated K^+ channels by an increase in intracellular Ca^{2+} induced by depolarization of mouse skeletal muscle fibres. *J. Physiol.* **509**, 93–102.

15. Collet, C., Pouvreau, S., Csernoch, L., Allard, B., and Jacquemond, V. (2004) Calcium signaling in isolated skeletal muscle fibers investigated under "silicone voltage-clamp" conditions. *Cell Biochem. Biophys.* **40**, 225–236.
16. Csernoch, L., Bernengo, J. C., Szentesi, P., and Jacquemond, V. (1998) Measurements of intracellular Mg^{2+} concentration in mouse skeletal muscle fibers with the fluorescent indicator mag-indo-1. *Biophys. J.* **75**, 957–967.
17. Bernengo, J. C., Collet, C., and Jacquemond, V. (2001) Intracellular Mg^{2+} diffusion within isolated rat skeletal muscle fibers. *Biophys. Chem.* **89**, 35–51.

13

Determination of Channel Properties at the Unitary Level in Adult Mammalian Isolated Cardiomyocytes

Romain Guinamard

Summary

This chapter describes methods to investigate mammalian cardiac channel properties at the single-channel level. Cell isolation is performed from adult heart by enzymatic digestion using the Langendorff apparatus. Isolation proceeding is suitable for rabbit, rat, and mouse hearts. Also, isolation of human atrial cardiomyocytes is described. In freshly isolated cells or cells maintained in primary culture, the single-channel variants of the patch-clamp technique (cell-attached, inside-out, and outside-out) are used to investigate channel properties. Biophysical properties such as conductance and ionic selectivity are determined. Also, regulations by extracellular and intracellular mechanisms are investigated. To illustrate the study, the author provides an example by the characterization of a calcium-activated nonselective cation channel (TRPM4).

Key Words: Single channel; patch clamp; cell isolation; mammalian cardiomyocytes; Langendorff; calcium-activated nonselective cation channel; TRPM4.

1. Introduction

To study ionic currents in cardiac cells, the first step is to obtain freshly isolated cardiomyocytes that can be maintained in culture. Isolated ventricular myocytes have the advantages by providing a uniform population of single cell-type free of neuronal and humoral influences in a controllable environment. These cells are suitable for electrophysiological studies but also are used in metabolic, receptor and ligand binding, pharmacological, or immunological studies. The highest difficulties come by the facts that adult cardiomyocytes are

From: *Methods in Molecular Biology, vol. 288: Patch-Clamp Methods and Protocols*
Edited by: P. Molnar and J. J. Hickman © Humana Press Inc., Totowa, NJ

highly differentiated then do not multiply anymore and are strongly associated together in the cardiac tissue, thus complicating cell dissociation and culture.

Concerning laboratory mammals, adult cardiomyocyte dissociation is achieved using the Langendorff method, described by the end of the nineteenth century ***(1)***. This technique allows the perfusion of the isolated heart by retrograde flow into the aorta with different buffers and enzymatic solutions that reach the coronary vascular system. Enzymes induce the degradation and losing of the extracellular and collagen matrixes that hold the myocytes in the myocardium. As the aortic valve prevents the retrograde entrance of solutions, the ventricles do not fill with the perfusate and then do not perform pressure-volume work, preventing cell injury because of ischemia. After isolation, cells can be used as freshly isolated cells or maintained in culture to follow their modifications in controlled conditions. The technique that gives routinely high yields of calcium-tolerant myocytes from different parts of the heart appears to be convenient and reliable for providing electrophysiologically useful cardiac single cells. It can be commonly used by any investigator and easily applied to the heart dissociation of small animals ***(2)***.

Dissociated cells are suitable for both whole-cell and single-channel electrophysiological recordings, including the physiological cell-attached configuration and both inside-out and outside-out excised configurations. Although whole-cell recordings provide currents from thousands of channels, single-channel proceedings give currents from a unique channel and thus provide informations at the molecular level. If experiments are well conduct, single-channel results can be considered as free of contamination by other currents. Thus, it is possible to identify precisely a channel type through its biophysical properties such as single-channel conductance, rectification, open and close time constants, ionic selectivity, and voltage dependency. According to these results, a rectification observed at the whole-cell level will be attributed to a variation of single-channel conductance and/or open probability upon voltage. In addition, the intracellular and extracellular channel modulators can be easily investigated as solutions bathing both side of the membrane can be rapidly changed in the excised configurations. Also, the cell-attached configuration, considered as the more physiological condition, allows the description of intracellular pathways involved in channel regulation. These approaches are suitable for most of channel types with sufficiently large (few pS) single-channel conductance to allow individual events detection.

This chapter focuses on the proceedings to obtain freshly isolated adult cardiomyocytes using the Langendorff apparatus, their surviving in culture, and the use of the patch-clamp technique to investigate channel properties at the unitary level. In the patch-clamp section, neither the patch-clamp setup nor the softwares used for analysis are described as both are basic apparatus. However, the proceedings to characterize channel biophysical and regulatory properties are described. To illustrate the study, the author provides an example by the characterization of a calcium-activated nonselective cation channel (TRPM4).

2. Materials

2.1. Cell Dissociation and Culture

2.1.1. Langendorff Apparatus

To perform cell dissociation, we used the isolated perfused heart preparation, according to the Langendorff technique *(1)* that is suitable for most warm-blooded animals including rabbit, mouse, rat, ferret, and guinea pig. The aim of the technique is to perfuse the coronary vessels with solutions that allow the digestion of the ventricular extracellular matrix. Simplistic apparatus could easily be home made. However, companies provide complete setups. The Langendorff apparatus is drawn in **Fig. 1** and composed of the following.

1. A water bath to warm solutions at 37°C.
2. A system to oxygenate the solutions (95% O_2 + 5% CO_2 is normally used).
3. Tubulures that go from stocks of solutions through the water bath to a faucet which allow to select the perfused solution.
4. A bubble trap.

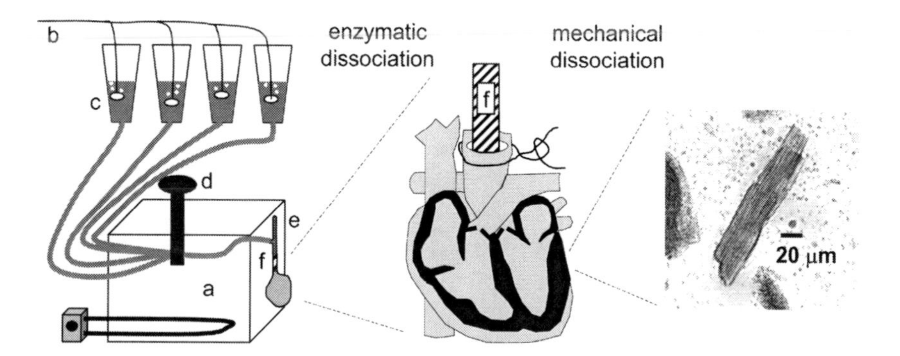

Fig. 1. Cardiomyocytes isolation using a Langendorff apparatus. The apparatus is composed of (a) water bath, (b) oxygenation system, (c) tank for solutions and tubulures, (d) faucet, (e) bubble trap, and (f) cannula. Introduction of the cannula into the aorta is schematized in the central magnification. Right panel is a phase-contrast micrograph of a freshly isolated adult rat ventricular cardiomyocyte.

5. An aortic cannula. It can be made in stainless steel or polyethylene tube with section and length that depend on the animal. The external diameter is typically similar to or slightly larger than that of the aorta, for example, 2–4 cm in length and 2–3 mm in diameter for a rat heart. Several small circumferential grooves are machined into the distal end of the cannula to prevent the aorta from slipping off.
6. A chamber to protect the heart.

2.1.2. Solutions Used for Cell Isolation

The complete compositions of solutions used for cell isolation are given in **Table 1**. Five solutions are successively used.

1. A classical Tyrode solution.
2. A nominally calcium-free Tyrode solution containing 0.1 m*M* ethyleneglycol-bis-N,N,N′, N′-tetraacetic acid (EGTA), so-called Tyrode 0 Ca^{2+}.
3. A saline solution containing 100 IU/ml collagenase (Worthington, USA, type II), a low calcium concentration, bovine serum albumin (BSA), and phenol red, so-called collagenase solution.
4. A solution similar to the previous one but free of enzyme, so-called washing solution.
5. A Kraft-Brühe solution, so-called KB *(3)*.

Table 1
Ionic Solutions Used for Rat Cardiomyocyte Isolation Using a Langendorff Apparatus

Solution	Tyrode	Ca-free Tyrode	Collagenase	Wash	KB
NaCl (μ*M*)	140	140	140	140	
KCl (μ*M*)	5.4	5.4	5.4	5.4	70
$CaCl_2$ (μ*M*)	1.8		0.07	0.07	0.08
$MgCl_2$ (μ*M*)	1.8	1.8	1.8	1.8	
EGTA (μ*M*)		0.1			0.5
Taurine (μ*M*)		20	20	20	20
BSA (%)			0.1	0.1	
Glucose (μ*M*)	10	10	10	10	10
Phenol red (mg/ml)			15	15	
HEPES (μ*M*)	10	5	5	5	10
Collagenase (IU/ml)			100		
pH	7.4	7.2	7.2	7.2	7.2
Osmolarity (mOsm)	300	290	290	290	240

BSA, bovine serum albumin; EGTA, ethyleneglycol-bis-N,N,N′, N′-tetraacetic acid.
In addition, the Kraft-Brühe (KB) solution contains 20 m*M* KH_2PO_4, 5 m*M* $MgSO_4$, 5 m*M* Na_2-ATP, 5 m*M* K-glutamate, and 5 m*M* creatine.

2.1.3. Materials for Cell Culture

The medium used for cardiomyocytes culture is M199 (Invitrogen, USA) supplemented with 0.2% BSA, 10^{-7} *M* insulin (Sigma, USA), 1% antibiotics (100 IU/ml penicillin G and 50 IU/ml streptomycin, Sigma), and 10 μ*M* cytosine 1-β-D-arabinofuranoside (Sigma), an inhibitor of cell proliferation, including fibroblasts.

Cells are seeded in 35-mm Petri dishes preincubated for 30 min with laminin (20 μg/ml, Sigma).

2.2. Patch Clamp

1. Patch-clamp setup: The patch-clamp setup used for the electrophysiological recordings includes an inverted microscope, a perfusion system, a patch-clamp amplifier connected to a headstage, and a digital audio tape recorder or a computer for data storage.
2. Perfusion system: The perfusion system can differ for cell-attached and cell-free configuration experiments. For cell-attached recordings, it is preferred a unique perfusion tube into the recording chamber connected to all the experimental solution tanks. At the opposite, for the cell-free configurations, it is useful to have a superfusion tube for each solution. That system minimizes blending of the different solutions by putting the pipette at the entrance of the concern tube.
3. Patch pipettes: Patch pipettes are made from borosilicate microhematocrits glass tubes pulled in two stages and coated with the hydrophobic silicon elastomer Sylgard 184 (Dow Corning, Belgium) to reduce pipette capacitance and noise. After application on the pipette, Sylgard is polymerized using a heated wire. Also the tip of the pipettes is heat polished using a "microforge" before filling with pipette solution (*see* **Note 1**). Good seals and recordings are achieved with a pipette tip resistance of 7–10 MΩ (in 140 m*M* NaCl solution).
4. Solutions for patch-clamp experiments: For electrophysiological recordings, the standard 140 m*M* NaCl solution used for bath and pipette contains the following: 140 m*M* NaCl, 4.8 m*M* KCl, 1.2 m*M* $MgCl_2$, 10 m*M* glucose, and 10 m*M* HEPES. Pipette and bath solutions usually contain 1 and 1.8 m*M* $CaCl_2$, respectively.

3. Methods

3.1. Cell Preparation

After animal anesthesia and excision, the heart is perfused in a retrograde way. Therefore, five steps have to be performed for cells preparation: (1) washing of the preparation to clear the blood, (2) perfusion of the tissue with a calcium-free solution to weaken intercellular adhesion, (3) enzymatic digestion of extracellular matrix, (4) mechanical agitation to free the cells from the tissue, and (5) cell seeding. *See* **refs. *2*** and ***4*** for original proceedings.

3.1.1. Animal Preparation

The proceeding is suitable for most laboratory mammals but is described here for rats. Adult animals of either sex are used; however, better results are obtained from young animals (approximately 1–2 months old weighing 0.25–0.3 kg) (*see* **Note 2**).

Animals are injected with heparin (5000 IU/kg, intraperitoneal). Anesthesia can be performed with ether 10 min after injection of heparin. However, as ether is highly flammable and an irritant to the animal and experimenter, it will be preferable inhalation of methoxyflurane or intraperitoneal injection of 0.3 g/kg hydroxychloral. If the last proceeding is applied, it is possible to inject heparin and hydroxychloral at the same time.

3.1.2. Heart Excision

Once the animal is anesthetized, a transabdominal incision is performed. The thorax is opened, and the thoracic cage is reflected over the animal head, exposing the heart. The aorta is pinched with forceps as high as possible, and aorta, vena cava, and pulmonary vessels are incised. Immediately after excision, the heart is immersed in an ice-cold Tyrode solution to stop beating and limit any ischemic injury during the period between the excision and the restoration of vascular perfusion.

3.1.3. Aorta Cannulation

The cannula is mounted on a 10-ml syringe containing Tyrode solution (at 4°C). One must be sure that no air bubbles are trapped in the cannula, to avoid the chance of air emboli. Then, the cannula is introduced into the aorta. At this point, it should be taken care that the position of the cannula is not too low to impede the aortic valves or the coronary ostia. The aorta is fixed on the cannula using a thread or a clamp. Then, the syringe content is perfused gently into the coronary vessels to empty any traces of blood. At that time, any surplus tissue can be trimmed away, and the cannula is removed from the syringe and mounted on the Langendorff apparatus. Once again, a great care should be taken to avoid any air bubble go into the heart. Isolation success highly depends on the time from heart excision to recirculation that should be reduced to only few minutes.

3.1.4. Heart Perfusion

1. The heart is retrogradely perfused with the 37°C Tyrode solution (about 2–3 min) at a perfusion rate of approximately 2 ml/min. Owing to warming, the heart recovers its spontaneous activity.

2. Tyrode solution is switched to the calcium-free Tyrode solution for 4 min which causes total Ca^{2+} depletion, cessation of heartbeat, and intercellular cements flimsiness (*see* **Note 3**).
3. Then follows perfusion for 10–20 min with the low calcium solution supplemented with 100 IU/ml collagenase (Worthington, type II). By the time of collagenase action, the heart becomes flaccid and its color lightens (*see* **Note 4**).
4. Once flaccid, the heart is rinsed 2–5 min with the washing solution to washout collagenase and stop digestion.
5. The heart is removed from the cannula and placed in a Petri dish containing KB solution at room temperature (*see* **Note 5**).

3.1.5. Mechanical Isolation

1. At this point, the right or left ventricle is dissected and finely minced with iridectomy scissors in the KB solution.
2. Using a fire-polished Pasteur pipette, pieces are gently triturated by up and down movements to release isolated cells.
3. The material was filtered through a 200-μ*M* mesh nylon gauze to remove undissociated pieces and let in a 50-ml tube for 15 min to allow cell sedimentation.
4. The supernatant that contains solution and fibroblasts is removed except 5 ml. Then, increasing volumes of culture medium are reintroduced every 5 min to double the initial volume (e.g., 3 μl; 10 μl; 30 μl, 0.1 ml; 0.3 ml; 1 ml; 3 ml) (*see* **Note 6**). Cells are let 15 min for sedimentation.
5. The supernatant is removed, and cells are resuspended in the culture medium containing 1.8 m*M* Ca^{2+}.

3.1.6. Cell Culture

Culture method is derived from the protocols used by Jacobson and Piper *(5)* and Volz et al. *(6)*.

1. Cells are seeded in Petri dishes (35 mm) preincubated for 30 min with laminin (20 μg/ml, Sigma, dissolved in sterile water and stored in freeze aliquots) to facilitate cell adhesion and incubated for 2 h at 37°C for plating.
2. The culture medium is changed once. At this point, cells are ready for electrophysiological experiments or cell culture (*see* **Note 7**).
3. If cells have to be maintained in culture, the medium is renewed every 2 days while dishes are kept at 37°C with a constant moist 95% air + 5% CO_2 (*see* **Note 8**).

Isolation and culture of cardiomyocytes can also be done from atrial specimens, including human atrial appendages (*see* **Note 9**) *(7)*.

3.2. Patch-Clamp Experiments

Proceeding for patch-clamp recordings, including ionic solutions, highly depends on the type of channel that is expected (*see* **Note 10**). In the following sections 3.2.1 to 3.2.5, we will take the example of a nonselective cation channel in mammalian cardiomyocytes. Nonselective cation channels are a family of channels hard to investigate in native cells as several members coexist in the same cell, and no specific pharmacology is known. That is particularly true for the TRPM4 channel, a calcium-activated nonselective cation channel (NSC_{Ca}) widely expressed in tissues including heart ***(8)***. Recordings of this channel at the whole-cell level are compromised by the presence of a large variety of other cationic channels, in particular in the heart. Thus, an approach at the single-channel level was necessary to investigate biophysical and regulatory properties of this channel in native cardiomyocytes. The study was done on several preparations such as rat ventricular myocytes and human atrial myocytes. In these preparations, an NSC_{Ca} was sharply described at the single-channel level. Its properties are similar to those described for the TRPM4 protein expressed in cell lines. Thus, the work showed a functional expression of such channel in mammalian cardiomyocytes, participating to the understanding of their role in heart physiology and occurrence of arrhythmias ***(7,9,10)***.

3.2.1. Recording Dish Preparation

In most of the studies, the bath solution is the physiological saline solution (so-called 140 μM NaCl solution) described in the **Subheading 2.,** with a pH adjusted to 7.4. Before patching, the culture medium is removed and the bath solution is added gently to avoid cell wasting (*see* **Note 11**). Several solution changes are necessary to totally remove culture medium components such as BSA which will prevent seal formation.

Experiments can be carried out at room temperature that do not modify cell surviving by the time of experiment or at 37°C using a microchamber coupled to a cooler/heater with a temperature controller.

3.2.2. Seal Formation

1. Before introducing the patch pipette into the bath, a small positive pressure is applied in the pipette. This pushes back all impurities present at the surface of the dish. The patch amplifier is switched to the Vtrack mode.
2. The pipette resistance is estimated using a voltage pulse and measure of the resulted current.
3. Liquid junction potential is adjusted to zero.
4. The pipette is approached to the cell under microscopic observation, and then the final approach is achieved following pipette resistance variations. When the cell membrane is touched, the pipette resistance increases by few megaohms, then the suppression of the positive pressure gives a second weak increase in resistance. At that time, a small suction is applied in the pipette, inducing a drop of resistance to

1 GΩ. Then, a cell-attached seal is reached, the suction is suppressed, and the patch amplifier switched to the voltage-clamp mode. Seal formation and duration highly depend on cell quality. Thus, depending on the dissociation, seal formation success is in the range of 50–10% of trials.

3.2.3. Cell-Attached Recordings

1. In the cell-attached configuration ***(11)***, the cell remains intact and so is its resting membrane potential (*Em*) that is not known. Thus, the applied potential ($\Delta Vm =$ −pipette potential) corresponding to the difference between bath and pipette potentials is superimposed on the spontaneous cell membrane potential for the patch membrane. In these conditions, the real potential applied to the patch membrane (*Vm*) corresponds to the sum of ΔVm and *Em*. This has to be taken in account for protocols and interpretation of the results (*see* **Note 12**).
2. The first approach to investigate channel properties is the realization of a current/voltage relationship (I/V). This can be done by long voltage steps (30 s to few minutes) with an increase of 10–20 mV after each sweep from −100 to +100 mV. Channel openings and closings are observed as rectangular changes in current traces. In some cases, voltage-ramp protocols can be used.
3. Current changes are analyzed with classical patch-clamp analysis software providing histogram amplitude and thus unitary current values and open probabilities (*see* **Fig. 2**). These data are used for I/V construction and estimation of reversal potential, unitary conductance, and variation of open probability in function of voltage (*see* **Fig. 3**). Also, kinetics studies can be achieved, providing open and close lifetime distribution.
4. Investigation of channel selectivity in the cell-attached configuration will be more qualitative than quantitative as neither *Em* nor the precise intracellular ionic composition is known. However, an approach can be the realization of I/V with different ionic solutions into the pipette. For example, replacing a part (120 m*M*) of Cl^- in the pipette by gluconate shifts the I/V, changing reversal potential, in case of a chloride channel. The same can be done for cations, replaced by N-methyl-D-glucamine (NMDG). The disadvantage is that these experiments are done on different patch membranes. Lastly, Unknown *Em* can be shifted to zero by bathing the cells in a 145 m*M* KCl-rich solution that totally depolarizes them. **Figure 3** illustrates such recordings from a nonselective cation channel.
5. Channel regulation can be investigated in cell-attached configuration by perfusing a molecule in the bath. In that case, the channel is not directly accessible; so, active molecules have to permeate cell membrane before reaching the channel or to interact with an extracellular receptor that modifies channel activity. Then, channel modulation and particularly washing can be slow. An example is provided in **Fig. 3**.
6. Although delicate to use, some apparatus were designed to allow changes in pipette solution during a recording. This can be useful to investigate channel properties in the cell-attached configuration ***(12)***.

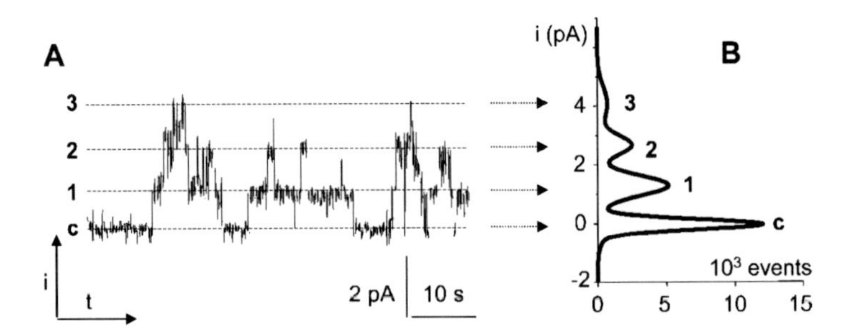

Fig. 2. Single-channel trace and amplitude histogram construction. Single-channel current trace recorded from a rat ventricular cardiomyocyte. **(A)** Changes in current (i) recorded during a constant voltage pulse. Three identical channels are present in the patch membrane, labeled "c" for the closed state and 1–3 indicating the current opening level for the three channels. **(B)** Corresponding amplitude histogram. Each peak corresponds to one level of current. Then, fitting data points to a Gaussian distribution allows the estimation of the single-channel current amplitude for the corresponding voltage and also estimation of the channel open probability using ratios from areas under curve for each level.

3.2.4. Inside-Out Recordings

1. The inside-out configuration ***(11)*** is obtained by lifting up the pipette after the cell-attached seal was obtained.
2. Free of the cell, the pipette can be approached very close to the perfusion system. As the intracellular side is now on the bath, the pH of perfused solutions is adjusted to 7.2.
3. Recordings of single-channel traces resemble those of the cell-attached recordings (*see* **Fig. 4**). Realization of I/V allows the determination of single-channel conductance, rectification properties, and channel dependency on voltage, that have to be compared with the properties observed in the cell-attached configuration. However, it is possible that channels appear only in the inside-out configuration or inversely.
4. As solutions bathing both sides of the membrane are now controlled, the ionic channel selectivity can be sharply investigated by replacing ions in the perfusing solution. I/V reversal potentials are used to estimate permeability ratios, according to the Goldman–Hodgkin–Katz equation ***(13)*** (*see* **Note 13**).
5. During analysis, the membrane potentials have to be corrected for the liquid junction potentials arising from changes in bathing solutions at the inner surface of the membrane patch. Variation in liquid junction potentials can be measured beforehand with a pipette containing 2.7 *M* KCl by measuring the zero-current voltage deflections induced by NaCl-substituted solutions ***(14)***. Junction potentials can also be calculated according to Henderson equation by the JPCalc program ***(15,16)*** with Clampex software (Axon Instruments, USA).
6. In the inside-out configuration, direct channel activation can be investigated by adding molecules or modifying ionic concentrations in the perfusing solutions.

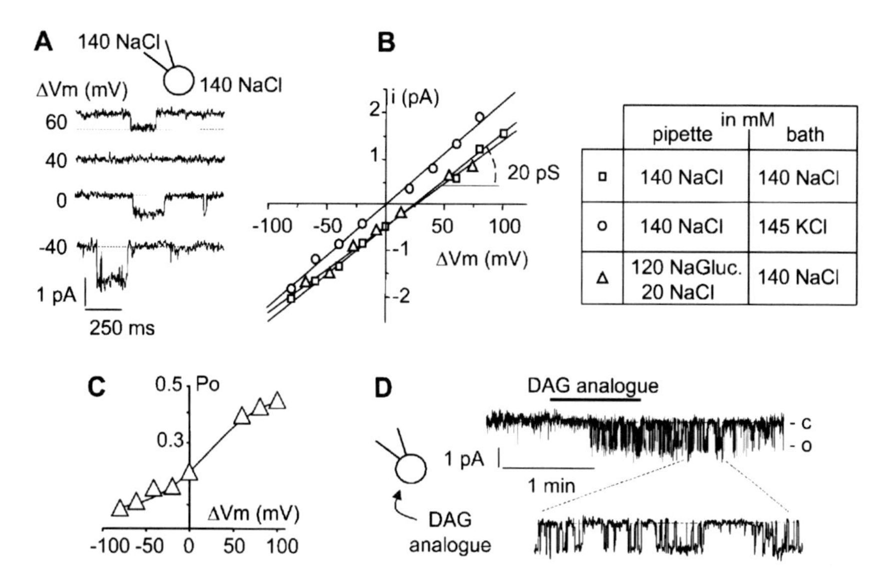

	pipette (in mM)	bath
□	140 NaCl	140 NaCl
○	140 NaCl	145 KCl
Δ	120 NaGluc. 20 NaCl	140 NaCl

Fig. 3. Properties of a nonselective cation channel in the cell-attached configuration. **(A)** Single-channel tracings recorded at various voltages from a cell-attached patch of an adult rat cardiomyocyte after 8 days in culture. Pipette and bath contained the standard 140 μM NaCl solution. ΔVm corresponds to the clamp potential between the pipette and the bath. **(B)** Channel conductance and selectivity. Current–voltage (I–V) relationships in ionic conditions as indicated in the right table. In the control condition (140 μM NaCl in pipette and bath, squares), the I/V is linear with a slope conductance of 20 pS and a reversal potential at 24 mV. This corresponds to the opposite of resting membrane potential *(Em)* of such cell ***(9)***. Thus, it indicates that ions going through the channel are at the equilibrium when the membrane potential ($Vm = \Delta Vm + Em$) is zero. That is, the case only for nonselective cation channels if one considers that the cell contains approximately 140 μM K^+ and 5 μM Na^+. Depolarizing the cell using a bathing solution with KCl instead of NaCl (circles) annuls *Em*, thus $Vm = \Delta Vm$. In these conditions, the I/V reversal potential is close to zero, which confirms the previous statement. In addition, replacing most of the Cl^- in the pipette by gluconate (triangles) does not modify reversal potential, indicating that Cl^- does not permeate through the channel and thus that it is a cationic nonselective channel. **(C)** Voltage dependence. Open probability (Po), determined at various voltages using amplitude histograms, is plotted against imposed voltage. Fitting the points indicates that the channel is activated by membrane depolarization. **(D)** Channel regulation. During a recording, the bath is perfused with a diacylglycerol analogue ($\Delta Vm = -40\,mV$) ***(10)***. It produces a long activation of the channel. c and o indicate the close and open states of the channel, respectively.

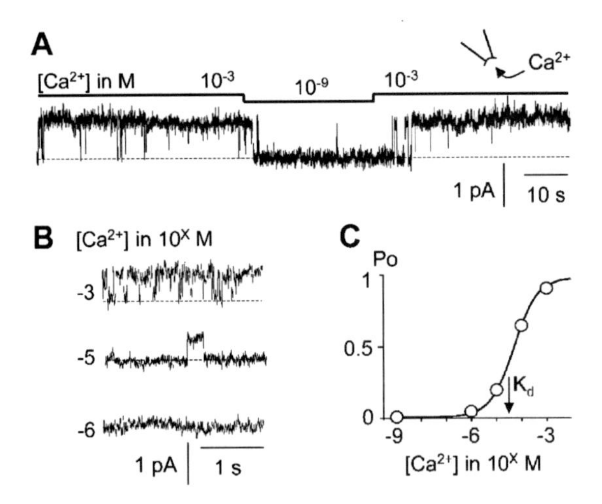

Fig. 4. Inside-out recordings. The effect of $[Ca^{2+}]_i$ on cardiac calcium-activated nonselective cation channel (NSC_{Ca}) activity is investigated on rat ventricular cardiomyocytes in the inside-out configuration *(9)*. Pipette and bath contained the standard 140 μM NaCl solution. $Vm = +40\,mV$. **(A)** Single-channel tracing showing the rapid and reversible change in channel activity when $[Ca^{2+}]_i$ is strongly reduced. **(B)** Current tracings from the same patch membrane with variable $[Ca^{2+}]_i$. **(C)** Corresponding values of open probability (Po) in function of $[Ca^{2+}]_i$. Fitting of the Hill equation to the experimental data points yields apparent dissociation constant (K_d) and Hill coefficient.

Contrary to the cell-attached configuration, the intracellular pathways are now totally removed and only direct effects can be observed. However, it is conceivable that proteins associated to the channel are present. **Figure 4** illustrates the direct regulation of the cardiac NSC_{Ca} by intracellular calcium.

3.2.5. Outside-Out Recordings

1. The outside-out configuration is obtained after the formation of the giga seal in the cell-attached configuration. A strong suction in the pipette breaks the membrane, and thus, the whole-cell configuration is reached. This is detectable by the appearance of capacitive components at the onset and end of the pulse, on the oscilloscope. The pipette is then lifted up to reach the outside-out configuration.
2. Investigations of channel properties are identical to those described in the **Subheading 3.2.4**. However, active molecules will now act at the outside of the channel as the extracellular side of the membrane faces the perfusion system.

4. Notes

1. Although it is feasible to pull several pipettes in advance, it is recommended to polish the pipette at the time of patching.
2. It is possible to use older animals until 1 year old, but the success of cell dissociation is reduced due to the high level of fibroblasts and extracellular matrix.
3. When reintroduced in 37°C Tyrode solution, heartbeat should be regular, otherwise the perfusion is not efficient and then the dissociation process compromised. Also, when calcium is removed, the heartbeat should stop within a minute and should not start again, indicating that the perfusion is correct.
4. The endpoint of collagenase perfusion is one of the most important points of the process. Unfortunately, the time of perfusion highly depend on the age of animal, quality of perfusion, and efficiency of the collagenase. If the heart is not sufficiently digested, very few isolated cells are obtained. On the contrary, a strong digestion deteriorates cell membrane. One trick is to remove a small sample of the ventricle during the perfusion of the collagenase solution, quickly triturate with a Pasteur pipette in KB medium, and examine, using a microscope, to see whether some cells are already isolated, indicating time to stop the digestion. Sometimes the heart recovers spontaneous beating in the collagenase solution, indicating that the solution is contaminated with calcium. That is injurious for the following of the experiment.
5. The high K^+ concentration of the KB solution that depolarizes the cells helps their protection during mechanical dissociation.
6. This proceeding allows a gentle restoration of calcium concentration. In a calcium-free medium, the cells are highly calcium sensitive and any strong variation of calcium is noxious.
7. Freshly isolated cells can be used rapidly after dissociation proceeding. If so, it is not necessary to reintroduce calcium in the solution, using the culture medium. A Tyrode solution can be used. However, owing to the traumatic injury induced by cell-dissociation proceeding, it is recommended to let cells recover few hours before using for experiments.
8. One should take care of the differentiating process that occurs on cardiomyocytes in culture. After 3–4 days, the morphology of the ventricular cells changes and also is their protein expression. The freshly isolated rod-shaped myocytes progress to a rounded shape; after 1 week, most of the cells are largely flattened and considerably bigger with apparent multinucleation and then show spontaneous activity. Interestingly, these dedifferentiated cells have similarities with cardiomyocytes obtained after cardiac hypertrophy ***(4,5,9,17)***.
9. *See* Benardeau et al. ***(18)*** and Guinamard et al. ***(7)*** for complete proceedings. Right atrial appendages can be obtained from patients undergoing cardiac surgery. Specimens are finely minced with iridectomy scissors in Kreps buffer containing 30 μ*M* 2.3 butanedione monoxime (BDM) and 0.5 μ*M* EGTA. Digestion is performed in Kreps additioned with 0.5% BSA and 6 IU/ml protease type XXIV

(Invitrogen, USA), then 165 IU/ml collagenase type V (Invitrogen, USA) for 15 min. After a further two to four 10-min steps in the presence of collagenase, samples are triturated with a Pasteur pipette. Cells are filtered and gradually resuspended in Dulbecco's modified Eagle's medium (DMEM) culture medium supplemented with 1% antibiotics (100 IU/ml penicillin G and 50 IU/ml streptomycin), 1 n*M* insulin, and 10% fetal calf serum. Cells suitable for patch-clamp experiments are seeded and maintained in culture as described in section 3.1.6 for rat cardiomyocytes.

10. For single-channel recordings, it is important to minimize the number of different types of currents present in the patch membrane. A first approach to eliminate undesirable currents is the use of known specific inhibitors. However, one has to be sure that these molecules do not interfere with the expected channel. A second approach is the use of solutions with specific ionic compositions, for example, lacking the ion permeating through undesirable channels.
11. A simple Pasteur pipette can be used on well-attached cells. However, if cells are isolated for only a few hours, it can be useful to use a perfusion and suction system during few minutes that has a lower flow than a manually operated Pasteur pipette.
12. One should consider that in the cell-attached configuration, the junction potential that was corrected before sealing disappears after sealing. Thus, this has to be taken in account during the analysis if bath and pipette solutions have different compositions.
13. A useful method is to use solutions containing low NaCl concentrations in the perfusing system (e.g., 14 m*M* NaCl with 256 m*M* sucrose added to maintain osmolarity) and a 140 m*M* NaCl standard solution in the pipette. Thus, Na^+ and Cl^- have opposite equilibrium potentials. In these conditions, if several types of currents are present, the unexpected one can be eliminated by doing measurements at the reversal potential for its permeating ion.

Acknowledgments

The author is indebted to Dr. Daniel Potreau, Pr. Patrick Bois, and Dr. Jacques Teulon for helpful comments on this manuscript and teaching in both tissue preparation and patch-clamp proceedings.

References

1. Langendorff, O. (1895) Untersuchungen am uberlebenden Saugethierherzen. *Pflugers Arch.* **61**, 291–332.
2. Bouron, A., Potreau, D., Besse, C., and Raymond, G. (1990) An efficient isolation procedure of Ca-tolerant ventricular myocytes from ferret heart for applications in electrophysiological studies. *Biol. Cell.* **70**, 121–127.
3. Isenberg, G., and Klockner, U. (1982) Calcium tolerant ventricular myocytes prepared by pre-incubation in a "KB medium". *Pflugers Arch.* **395**, 6–18.

4. Fares, N., Gomez, J.P., and Potreau, D. (1996) T-type calcium current is expressed in dedifferentiated adult rat ventricular cells in primary culture. *C. R. Acad. Sci. III* **319**, 569–576.
5. Jacobson, S.L., and Piper, H.M. (1986) Cell cultures of adult cardiomyocytes as models of the myocardium. *J. Mol. Cell. Cardiol.* **18**, 661–678.
6. Volz, A., Piper, H.M., Siegmund, B., and Schwartz, P. (1991) Longevity of adult ventricular rat heart muscle cells in serum-free primary culture. *J. Mol. Cell. Cardiol.* **23**, 161–173.
7. Guinamard, R., Chatelier, A., Demion, M., Potreau, D., Patri, S., Rahmati, M., and Bois, P. (2004) Functional characterization of a Ca^{2+}-activated non-selective cation channel in human atrial cardiomyocytes. *J. Physiol.* **558**, 75–83.
8. Launay, P., Fleig, A., Perraud, A.L., Scharenberg, A.M., Penner, R. and Kinet, J.P. (2002) TRPM4 is a Ca^{2+}-activated nonselective cation channel mediating cell membrane depolarization. *Cell* **109**, 397–407.
9. Guinamard, R., Rahmati, M., Lenfant, J., and Bois, P. (2002) Characterization of a Ca^{2+}-activated nonselective cation channel during dedifferentiation of cultured rat ventricular cardiomyocytes. *J. Membr. Biol.* **188**, 127–135.
10. Guinamard, R., Chatelier, A., Lenfant, J., and Bois, P. (2004) Activation of the Ca^{2+}-activated nonselective cation channel by diacylglycerol analogues in rat cardiomyocytes. *J. Cardiovasc. Electrophysiol.* **15**, 342–348.
11. Hamill, O.P., Marty, A., Neher, E., Sakmann, B., and Sigworth, F.J. (1981) Improved patch-clamp techniques for high-resolution current recording from cell-free membrane patches. *Pflugers Arch.* **391**, 85–100.
12. Tang, J.M., Wang, J., Quandt, F.N., and Eisenberg, R.S. (1990) Perfusing pipettes. *Pflugers Arch.* **416**, 347–350.
13. Hille, B. (1984) Selective permeabilities. In *Ionic Channels of Excitable Membranes*. Sinauer Associated Inc. Publishers, Sunderland, MA.
14. Nastuk, W.I., and Hodkin, A.L. (1950) The electrical activity of single muscles fibers. *J. Cell. Comp. Physiol.* **35**, 39–73.
15. Barry, P.H. (1994) JPCalc, a software package for calculating liquid junction potential corrections in patch-clamp, intracellular, epithelial and bilayer measurements and for correcting junction potential measurements. *J. Neurosci. Methods* **51**, 107–116.
16. Ng, B., and Barry, P.H. (1995) The measurement of ionic conductivities and mobilities of certain less common organic ions needed for junction potential corrections in electrophysiology. *J. Neurosci. Methods* **56**, 37–41.
17. Hefti, M.A., Harder, B.A., Eppenberger, H.M., and Schaub, M.C. (1997) Signaling pathways in cardiac myocyte hypertrophy. *J. Mol. Cell. Cardiol.* **29**, 2873–2892.
18. Benardeau, A., Hatem, S.N., Rucker-Martin, C., Le Grand, B., Mace, L., Dervanian, P. *et al.* (1996) Contribution of Na^{+}/Ca^{2+} exchange to action potential of human atrial myocytes. *Am. J. Physiol.* **271**, H1151–H1161.

14

Electrophysiological Properties of Embryonic Stem Cells During Differentiation Into Cardiomyocyte-Like Cell Types

Antoni C. G. van Ginneken and Arnoud C. Fijnvandraat

Summary

The method described here to differentiate mouse embryonic stem (ES) cells into cardiomyocytes is adapted from Maltsev et al. and results in a high percentage of spontaneously beating cardiomyocyte-like cells. In order to determine to what extent the differentiating ES cells resemble true cardiomyocytes, the cells were electrophysiologically characterized during differentiation, using the whole-cell variant of the patch-clamp technique. Action potentials (APs) and membrane currents were recorded and analyzed off-line to determine electrophysiological changes during development.

Key Words: Stem cells; differentiation; HM-1; cardiomyocyte; electrophysiology; current clamp; voltage clamp.

1. Introduction

There is a continuing interest in stem cell therapy as a means to repair damaged tissue. For repairing infarcted cardiac ventricular tissue, one would like to inject stem cells in the infarcted area in the hope that they will differentiate into true cardiomyocytes. These stem cells not only need to develop a cardiomyocyte-like contractile apparatus, but also their electrophysiology should be comparable with that of the ventricular myocyte, that is, they preferably should have a stable resting potential.

From: *Methods in Molecular Biology, vol. 288: Patch-Clamp Methods and Protocols*
Edited by: P. Molnar and J. J. Hickman © Humana Press Inc., Totowa, NJ

We questioned to what extent mouse embryonic stem (ES) cells can differentiate into cell types, comparable with those in the intact murine heart. We used an embryonic mouse stem cell line that readily differentiates into cardiomyocyte-like cells (ES cell-derived cardiomyocytes [ESDCs]). These ESDCs were allowed in vitro to differentiate and were electrophysiologically characterized. We showed that a great variety in electrophysiological phenotypes exists and that during differentiation the phenotype shifts toward cells with a short action potential (AP), a fast upstroke velocity, and a fast beating rate.

2. Materials

2.1. Maintenance Culture of Undifferentiated Mouse ES Cells

1. Culture flasks 25 ml (BD Falcon, Franklin Lakes, NJ, USA, cat. no. 353018).
2. ES cell line hamster macrophage-1 (HM-1) *(1)*.
3. Irradiated primary cultures of mouse embryonic fibroblasts (MEFs) (*see* **Notes 1** and **2**).
4. Phosphate-buffered saline (PBS) 7.2 (1×) (Invitrogen Europe, Paisley, Scotland, cat. no. 20012050).
5. Cell-dissociation medium: 191 ml PBS + 5 ml 40 μ*M* EDTA + 2 ml chicken serum (Gibco/Invitrogen, Carlsbad, CA USA, cat. no. 16110-033) +2 ml Trypsin 2.5% (Gibco/Invitrogen 25090-028).
6. Gelatin for coating: (10×) 1% gelatin (Sigma-Aldrich, St. Louis, MO USA, cat. no. G1890) in water.
7. Complete medium: 500 ml Glasgow Minimal Essential Medium (GMEM) (Flow Laboratories, Irvine, Great Britian, cat. no. 12-302-54), supplemented with 5 ml 200 μ*M* L-glutamine (100×) (Gibco/BRL, Gaithersburg MD USA, cat. no. 25030-024), 5 ml 100 μ*M* sodium pyruvate (Gibco/BRL, cat. no. 11360-039), 5 ml non-essential amino acids (100×, Gibco/BRL, cat. no. 11140-035), and 50 ml fetal calf serum (FCS, Gibco/BRL or Bodinco, Bodinco, Alkmaar, Holland, cat. no. 39454) (*see* **Note 3**).
8. Leukemia-inhibiting factor (LIF) (1000×) 10^6 U/ml in complete medium of Esgro-LIF (Gibco/BRL, cat no. 13275-029). Working concentration 10^3 U/ml.
9. Complete medium + beta +LIF: 100 ml complete medium +0.1 ml beta-mercaptoethanol (1000×) +0.1 ml LIF (1000×).

2.2. Differentiation of Mouse ES Cells

1. Disposable Petri dish 100 × 2 mm (Falcon 3003 dish Oxnard, CA USA, cat. no.).
2. Twenty-four-well culture plate, 15.6 mm diameter, 1.9 cm^2, 3.4 ml (Costar, Sigma-Aldrich, cat. no. CLS3524).
3. Differentiation medium: Iscove's Modified Dulbecco's Medium (Gibco/BRL, cat. no. 12440053), supplemented with 20% FCS (v/v), 1× amino acids, and 1× penicillin/streptavidin (both from Gibco/BRL).
4. Filter Millex-GV 0.22 μ*M* (Millipore, Billerica, MA USA, cat. no. SLGV025BS).

5. For coating of wells: 1% agar dissolved in warm differentiation medium and sterilized by pressure filtering through Millex-GV 0.22-μ*M* filter.

2.3. Preparation of Single Cells for Electrophysiology

1. Glass coverslips, 12–18 mm diameter (SPI Supplies, USA).
2. AAS solution for coating coverslips: dilute 5 ml 3-aminopropyltriethoxysilane [(AAS) also known as TESPA] (Sigma, cat. no. A3648 [100 ml]) to 250 ml with acetone.
3. Glass Petri dish 90 mm diameter.
4. Eppendorf tubes, 1.5 ml (Eppendorf, Hamburg, Germany, cat. no. 022364111).
5. Disposable culture dishes 35 mm diameter (BD Falcon plastics, cat. no. 353001).

2.4. Preparation of Electrophysiological Measurements

1. Inverted microscope (Nikon Diaphot, Tokyo, Japan) equipped with fixed stage.
2. Micromanipulator Narashige M01-103, mounted on fixed stage of microscope.
3. Perspex perfusion chamber in which coverslips can be mounted as bottom of the chamber.
4. Syringe (50 ml) with three-way tap.
5. Syringe (1 ml).
6. Outer diameter (1 mm) plastic tubing.
7. Injection needle, connected to a vacuum bottle.
8. Extracellular solution for superfusion: 140 m*M* NaCl, 5.4 m*M* KCl, 1.8 m*M* $CaCl_2$, 1.0 m*M* $MgCl_2$, 5.5 m*M* glucose, and 5.0 m*M* HEPES (pH set at 7.4 with NaOH).
9. Intracellular solution for filling pipettes: 140 m*M* K-gluconate, 10 m*M* KCl, and 5.0 m*M* HEPES (pH set at 7.2 with KOH).
10. Electrode glass capillaries (1 mm OD with filament, World Precision Instruments, Inc. Sarasota FA USA, cat. no. TW100F-3).

3. Methods

All culturing activities should be done in a laminar flow cabinet, using sterile equipment and sterile solutions. If not sterile, solutions should be filtered through a Millex-GV 0.22-μ*M* filter.

Mouse ES cells of the line HM-1 *(1)* are cultured in undifferentiated state as described in section 3.1 *(2)*. Cardiac differentiation is evoked by culturing the cells in aggregates called embryoid bodies (EBs), using the hanging-drop assay, essentially as described by Maltsev et al. *(3)* with some small modifications. After 3 days of incubation in hanging drops, EBs are brought in a floating culture in 24-well plates coated with agar to prevent attachment of the EBs. At various times during this floating culture, EBs are dissociated into single cells. These cardiomyocyte-like cells are transferred to coverslips and allowed to attach overnight.

The next day, the coverslips are mounted in a perfusion chamber on the stage of microscope. A gravity-fed perfusion system is used. Excess solution is removed by suction with an injection needle, placed above the surface of the solution. After the start of perfusion, the electrophysiological properties of the cells are determined by whole-cell measurements of APs and membrane currents (*see* **Fig. 1**).

3.1. Maintenance Culture of Undifferentiated Mouse ES Cells

1. Prepare coating solution: 1 ml gelatin solution in 9 ml PBS.
2. Coat tissue-culture flasks or wells with this coating solution for 30 min at room temperature.

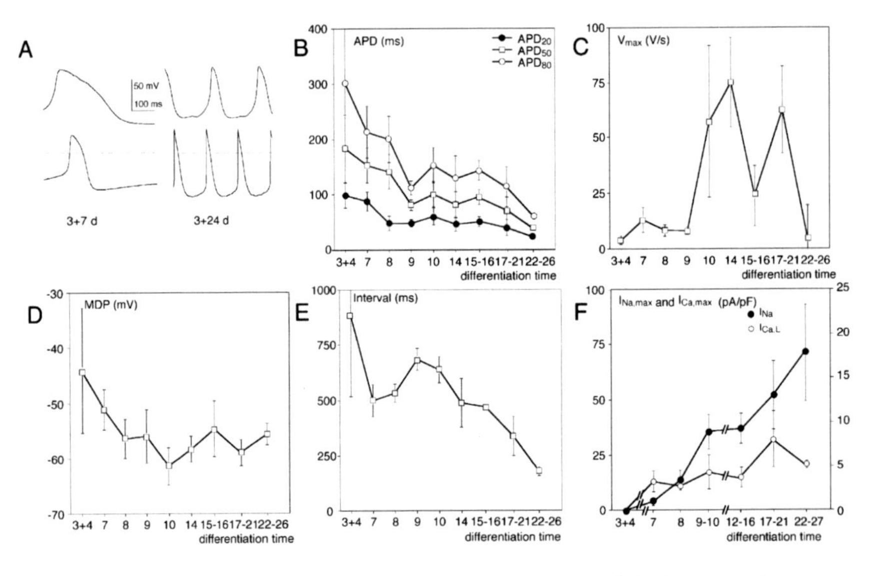

Fig. 1. Electrical properties of embryonic stem cell-derived cardiomyocytes (reprinted from **ref. 4**, copyright 2003, with permission from European Society of Cardiology). **(A)** Representative action potentials (APs) recorded early (3 + 7 days, left) and late (3 + 24 days, right) during differentiation. **(B–E)** Changes in average AP characteristics during differentiation. Number of measured cells and culture days after 3 days in a hanging drop (age/*n*): 4/2; 7/14; 8/15; 9/10; 10/9; 14/18; 15–16/6; 17–21/9; 22–26/21. During differentiation, the following parameters were measured: AP duration (APD) at 20, 50, and 80% of depolarization (**B**), upstroke velocity (V_{max}, **C**), maximal diastolic max potential (MDP, **D**), spontaneous beating interval (**E**), and density of peak sodium current (I_{Na}, left scale) and L-type calcium current ($I_{Ca,L}$, right scale) (**F**).

3. Thaw irradiated MEF cells at 37°C, spin cells down through 10 ml complete medium (5 min at 161g (1200 RPM)), and seed cells onto gelatin-coated surfaces to obtain a confluent monolayer.
4. Thaw frozen ES cells at 37°C and spin down through 10 ml complete medium (+beta-mercaptoethanol + LIF) (5 min at 161g (1200 rpm)).
5. Resuspend ES cells and seed onto MEF cell layer. Add 3 ml medium.
6. Gas flasks with 5% CO_2/95% air, close lid but leave room for air to enter.
7. Store flasks in incubator at 37°C, 5% CO_2.
8. Refresh medium the next day. Thereafter, refresh medium when buffer is exhausted (indicated by orange-yellow color of medium), usually after 3 days.
9. Reseed the cells when cells tend to grow in a confluent layer or when medium is yellow 1 day after refeeding (usually 3–5 days of culturing).
10. Wash cells with small volume of PBS.
11. Trypsinize with 1 ml cell-dissociation medium (10×) for 10 min at 37°C.

    - Detach cells by careful shaking and tapping.
    - Add 1 ml medium and resuspend to single cells with 10-ml pipette.
    - Reseed cells 1:3–1:7.

12. ES cells can be frozen in complete medium + beta-mercaptoethanol + LIF + 10% dimethyl sulfoxide (DMSO).

3.2. Differentiation of Mouse ES Cells

1. Dissociate ES cells with 1 ml TVP (10×) for 10 min at 37°C.
2. Count cells in hemocytometer.
4. Dilute cell suspension to 600 cells/20 μl in ISCOVE's MDM culture medium.
5. Resuspend cells and use pipette to place 20 μl drops on the lid of a Petri dish.
6. Fill dish with 20 ml PBS to prevent evaporation of the drops.
7. Turn lid over and put on dish. Allow cells to differentiate into EBs during 3 days (*see* **Note 4**).
8. Heat agar gel in a water bath until liquid. Prepare 24-well plates by pipetting 1 ml agar solution/well and allow the agar to solidify.
9. Add 5 ml differentiation medium/well.
10. Transfer EBs to well plate, one EB/well.
11. Change differentiation medium every other day.

3.3. Preparation of Single Cells for Electrophysiology

1. Pour 20 ml AAS solution in glass Petri dish.
2. Dip coverslips in solution for 5 min.
3. Allow coverslips to dry in the air and store them in a closed 90-mm Petri dish before use.
4. Transfer EBs from well to 1.5-ml Eppendorf tube, using a 100-μl pipette. Remove excess medium.

5. Add about 100 μl TVP medium/EB.
6. Dissociate EBs during 20 min at 37°C and triturate every 10 min with a 100-μl pipette.
7. Fill 35-mm culture dishes with 2.5 ml differentiation medium and immerse AAS-coated coverslip in medium.
8. Transfer 50–100 μl cell suspension onto the middle of the coverslip.
9. Carefully transfer culture dishes to incubator and allow the cells to attach overnight.

3.4. Preparation of Electrophysiological Measurements

1. Mount coverslip with cells attached to it in cell chamber (*see* **Note 5**).
2. Fill 50-ml syringe (tap closed) with extracellular solution.
3. Use 1-ml tubing to connect tap to the inlet of perfusion chamber.
4. Start superfusion with pre-warmed extracellular solution. Use 1-ml syringe on extra outlet of the tap to prime the tubing. Place suction needle just above the surface of the solution in the perfusion chamber.
5. Pull small-tipped pipettes and fill with intracellular solution.
4. Mount pipette in electrode holder.

3.5. Electrophysiology

1. Approach cell carefully, as the cells are very flat (*see* **Notes 6** and **7**).
2. Stop movement of pipette as soon as a slight increase in resistance is seen.
3. Apply suction by mouth until resistance increases to about 500 MΩ.
4. Increase holding potential to −40 mV and wait for seal to improve (*see* **Note 8**).
5. Increase suction shortly, until whole-cell configuration is obtained.
6. Determine cell capacitance (*see* **Note 9**).
7. Compensate for C_m and R_m.
8. Record APs as spontaneous activity as driven by 2-ms rectangular current pulses applied through the patch pipette.
9. Record membrane currents elicited by clamp steps to 200 ms and 15–10 mV increment in amplitude using a holding potential of −90 mV (recording of I_{Na}) as well as −50 mV (recording of $I_{Ca,L}$) (*see* **Note 10**).

4. Notes

1. MEF cells are made by trypsinizing 14-day mouse embryo carcasses (viscera and head removed). The cells are irradiated with 2500 rads to prevent them form further dividing. MEFs can be frozen in complete medium +20% DMSO and stored in liquid nitrogen for later use.
2. MEF and ES cells should be tested for mycoplasm contamination.
3. Sera should be tested for optimal growth for ES cells, before larger quanta of the sera are ordered.
4. After 3 days, the EBs should beat spontaneously.

5. Use a thin silicone rubber gasket between perspex ring and coverslip to prevent leakage.
6. It is very helpful to use a video camera mounted on the microscope. This makes it more easy to manipulate the recording pipette while simultaneously monitoring the development of seal resistance on the oscilloscope.
7. ESDCs tend to be very flat. Therefore, mechanical drift of the pipette should be minimized. Such drift may be caused by the stiffness of the cable connecting the headstage to the clamp amplifier. Therefore, this cable should have enough freeplay during long recordings. Also changes in temperature may cause significant vertical movements of the bottom.
8. During seal formation, seal resistance often improves when the pipette potential is made slightly negative (−40 mV). This also improves stability of the recording after establishing whole-cell configuration.
9. On some voltage-clamp amplifiers, cell capacitance can be read from a capacity compensation dial. Alternatively, capacitance can be calculated from the change in slope during short current pulses: $C_m = I_m \cdot (\Delta t/\Delta V)$.
10. We measured currents immediately after the recording of APs. As recording time was limited, no specific blockers were used to discriminate between various currents. Instead, we used a depolarized holding potential to inactivate I_{Na}.

Acknowledgments

We would like to thank C. Verhoek-Pockock and Dr. M.A. van Roon for their helpful suggestions.

References

1. Magin T.M., McWhir J., Melton D.W. (1992) A new mouse embryonic stem cell line with good germ line contribution and gene targeting frequency. *Nucleic Acids Research*, **20**, 3795–3796.
2. Fijnvandraat A.C., de Boer P.A.J., Lekanne Deprez R.H., Moorman A.M. (2002) Non-radioactive in situ detection of mRNA in ES cell-derived cardiomyocytes and in the developing heart. *Microscopy Research and Technique*, **58**, 387–394.
3. Maltsev V., Wobus A.M., Rohwedel J., Bader M., Hescheler J. (1994) Cardiomyocytes differentiated in vitro from embryonic stem cells developmentally express cardiac-specific genes and ionic currents. *Circulation Research*, **75**, 233–244.
4. Fijnvandraat A.C., van Ginneken A.C.G., de Boer P.A.J., Ruijter J.M., Christoffels V.M., Moorman A.F.M., Lekanne Deprez R.H. (2003) Cardiomyocytes derived from embryonic stem cells resemble cardiomyocytes of the embryonic heart tube. *Cardiovascular Research*, **58**, 399–409.

15

Hybrid Neuronal Network Studies Under Dynamic Clamp

Alan D. Dorval II, Jonathan Bettencourt, Theoden I. Netoff, and John A. White

Summary

Even a complete understanding of the biophysical properties driving neuronal behavior would be insufficient to explain the interactions between neurons, neuronal assemblies, and brain regions. Exploring interactions between small numbers of synaptically coupled neurons in vitro can provide insight into the in vivo activity of neuronal assemblies. However, pairs of synaptically coupled neurons are notoriously difficult to find in vitro, and trying to study networks of more than two neurons is nearly impossible. The advent of the dynamic-clamp technique enables researchers to generate hybrid networks of neurons in which living neurons are synaptically coupled through computationally generated synapses. In this chapter, we provide an overview of the components of a dynamic-clamp system. We detail how to use dynamic clamp to construct simple neuronal networks from living neurons as well as hybrid networks including both living and in silico neurons.

Key Words: Neuronal networks; hybrid networks; dynamic clamp; conductance clamp; synchronization.

1. Introduction

The term dynamic clamp is typically used interchangeably with conductance clamp because the first experiments utilizing the technique were, in fact, clamping conductances (*1,2*). A versatile dynamic-clamp system, however, can actually clamp simultaneously many neuronal characteristics as a function of time. A dynamic-clamp system can clamp current or voltage (superseding the

From: *Methods in Molecular Biology, vol. 288: Patch-Clamp Methods and Protocols*
Edited by: P. Molnar and J. J. Hickman © Humana Press Inc., Totowa, NJ

classic techniques of current and voltage clamp), conductance, or more abstract measures such as neuronal firing rate or membrane potential variance. This chapter details both conductance and firing rate clamps.

During a standard dynamic-clamp experiment, some calculation typically performed by a computer simulates the biophysical process of some membrane conductance in the neuron. In the context of this chapter that biophysical process is the conductance change that occurs at a synapse following a presynaptic action potential. Conceptually, each neuron we consider is a hybrid entity consisting a biological cell body and a computational synapse. The simplest network, a neuron synapsing onto itself, requires only one hybrid neuron. For all other networks of interest, this hybrid neuron must be connected to at least one other neuron: either a second patched neuron or a computational model neuron. In the subsequent sections, we explain how to construct a few simple neuronal networks from one or two cells. Straightforward extrapolation to larger networks is left to the reader.

2. Materials

This chapter assumes that the reader is familiar with the more general techniques of patch clamping. In this list of materials, **Subheadings 2.1.** and **2.2.** are admittedly brief collections of the absolute essential materials required to prepare and patch-clamp neurons in slice. Subsequent sections focus on materials more specific to dynamic clamp. **Subheading 2.3.** lists the required components of a dynamic-clamp system, with particular focus given to digital dynamic-clamp systems run on personal computers. **Subheading 2.4.** delineates the required actions that a dynamic clamp must be capable of performing to generate hybrid neuronal networks. Depending on the dynamic-clamp system, these modules may need to be written prior to experimentation or they may already exist as part of the dynamic clamp itself.

2.1. Standard Tissue Preparation Equipment

1. Air-tight anesthesia chamber with isoflurane.
2. Appropriately sized animal guillotine.
3. Surgical scissors, scoopula, scalpel.
4. Horizontal serial slicer (e.g., vibratome) with walled slice chamber.
5. Filter paper, fine artist's paint brush.
6. Bubble chamber with tubing and oxygen.
7. Artificial cerebrospinal fluid (ACSF: 126 m*M* NaCl, 1.25 m*M* NaH_2PO_4, 2.0 m*M* $MgSO_4$, 26 m*M* $NaHCO_3$, 10 m*M* dextrose, 3 m*M* KCl, and 2 m*M* $CaCl_2$) (*see* **Note 1**).

2.2. Patch-Clamp Hardware

1. Infrared differential interference contrast microscope (e.g., Axioskop 2FS+; Zeiss, Germany).
2. Stage chamber (e.g., RC-26GLP; Warner Instruments, USA) with temperature controller (e.g., TC-344B; Warner Instruments, USA) and influx/efflux tubing for ACSF.
3. One micromanipulator (e.g., MP-225; Sutter Instruments, USA) for each living neuron in the desired hybrid network.
4. One or more current-clamp amplifier (e.g., Multiclamp 700B; Molecular Devices, USA) with headstage. Many current-clamp amplifiers come equipped with two channels: you must have one channel for each living neuron in the desired network. Headstages should be attached to the micromanipulators.
5. Artificial neuroplasm: 135 m*M* K-gluconate, 2 m*M* $MgCl_2$, 2 m*M* NaCl, 1.25 m*M* ethyleneglycol-bis-N, N, N′, N′-tetraacetic acid (EGTA), 10 m*M* HEPES, 2 m*M* Tris-ATP, and 0.4 m*M* Tris-GTP (*see* **Note 1**).
6. Borosilicate glass electrodes (e.g., 1.0/0.75 mm OD/ID) drawn on a horizontal puller (e.g., P-97; Sutter Instruments, USA) to approximately 1 μm diameter tip, 3–6 MΩ when filled with artificial neuroplasm. Electrodes should be loaded onto headstages.

2.3. Dynamic-Clamp System

The exact equipment needed depends on the dynamic-clamp system being used, but the conceptual framework is always the same (*see* **Fig. 1A**). We focus on a software-based dynamic clamps (*3,4*) that do not require additional computers, digital signal processing (DSP) boards, analog circuitry, or any other specialty hardware boxes.

1. Personal computer, preferably running a hard real-time operating system (OS) (*see* **Note 2**).
2. Multipurpose A/D–D/A board installed in the computer (e.g., PCI-6036E or PCI-6052E; National Instruments, USA) (*see* **Note 3**). Voltage outputs from the current-clamp amplifier channels should be connected to A/D inputs (AI) on the board. Current inputs to the current-clamp amplifier should be connected to D/A outputs (AO) on the board (*see* **Note 4**).
3. Dynamic-clamp software installed on the computer (*see* **Note 5**).

2.4. Dynamic-Clamp Modules

1. The ability to detect spikes from a neuron's membrane potential in real time (*see* **Fig. 1B**).
2. The ability to simulate conductances and present currents to neurons (*see* **Fig. 1C**).
   a. The ability to control arbitrary neuronal outputs, such as firing rate (*see* **Fig. 1D**).

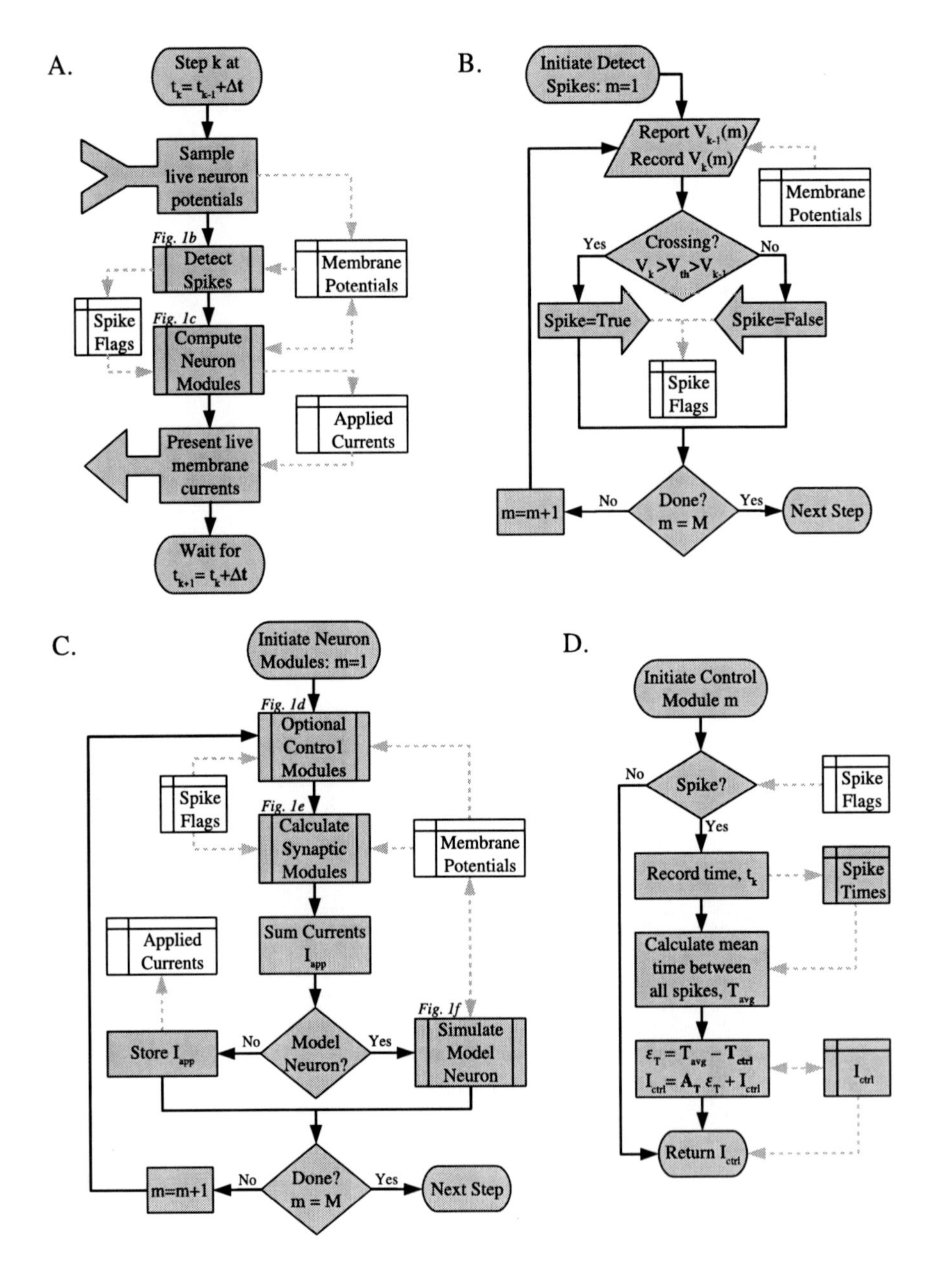

A.
Step k at
$t_k = t_{k-1} + \Delta t$
Sample live neuron potentials
Fig. 1b
Detect Spikes
Membrane Potentials
Spike Flags
Fig. 1c
Compute Neuron Modules
Applied Currents
Present live membrane currents
Wait for
$t_{k+1} = t_k + \Delta t$
B.
Initiate Detect Spikes: m=1
Report $V_{k-1}(m)$
Record $V_k(m)$
Membrane Potentials
Yes
Crossing?
$V_k > V_{th} > V_{k-1}$
No
Spike=True
Spike=False
Spike Flags
m=m+1
No
Done?
m = M
Yes
Next Step
C.
Initiate Neuron Modules: m=1
Fig. 1d
Optional Control Modules
Spike Flags
Fig. 1e
Calculate Synaptic Modules
Membrane Potentials
Sum Currents
I_{app}
Applied Currents
Store I_{app}
No
Model Neuron?
Yes
Fig. 1f
Simulate Model Neuron
m=m+1
No
Done?
m = M
Yes
Next Step
D.
Initiate Control Module m
No
Spike?
Spike Flags
Yes
Record time, t_k
Spike Times
Calculate mean time between all spikes, T_{avg}
$\varepsilon_T = T_{avg} - T_{ctrl}$
$I_{ctrl} = A_T \varepsilon_T + I_{ctrl}$
I_{ctrl}
Return I_{ctrl}

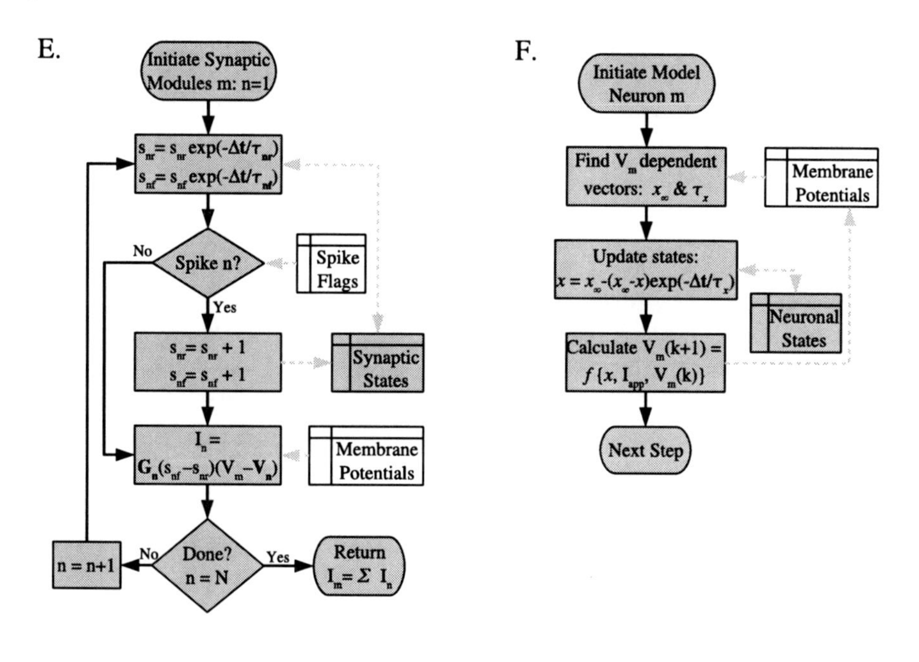

Fig. 1. Flowcharts depicting dynamic clamp and module activity. White boxes denote shared memory facilities that are accessible to other modules. **(A)** Overview of dynamic-clamp procedure for generating hybrid networks. The time step Δt is a parameter and ideally would be at most 1/10 the smallest time constant in the network. **(B)** The process of detecting spikes in a network with M neurons. The spike detection threshold V_{th} could vary across cells of different types. **(C)** The process of calculating the currents provided to each of the M neurons, some biological and some computational. **(D)** A simple proportional control algorithm used to control neuronal firing rate. The parameter T_{ctrl} is the desired time between consecutive spikes; A_T is the proportionality gain constant. **(E)** A process that calculates the synaptic current summed from the N cells in the network that synapse on M neurons. Synapses are modeled as a difference-of-exponential function with time constant parameters τ_{nr} and τ_{nf}. Each synapse has a maximum conductance of G_n and associated reversal potential V_n. All four parameters can vary across the $n = 1$ to N synapses. **(F)** A process that simulates a conductance-based model neuron with Hodgkin–Huxley style gating variables, although any class of model neuron could be used. From the hybrid network perspective, what is important is that the membrane potential V_m is updated each time step.

b. The ability to use spike times from one neuron and a synapse model to drive another neuron (*see* **Fig. 1E**).
c. The ability to generate a computational model neuron and interface it through model synapses with living neurons (*see* **Fig. 1F**).

3. Methods

Subheadings 3.1. and **3.2.** are brief descriptions of how to get healthy whole-cell-patched neurons. More thorough descriptions can be found throughout this book. As a general patch-clamping tool, however, the dynamic-clamp technique and this chapter, in particular, can be beneficial for physiologists, even those not interested in network behavior. For example, an electrophysiologist may desire a certain baseline behavior from a neuron (e.g., a specific firing rate) that under normal slice conditions would be achieved by an aging graduate student constantly adjusting the background applied current. Most dynamic-clamp systems can be programmed to perform these tedious real-time adjustments. We illustrate a simple firing rate control algorithm here (*see* **Fig. 1D**), though in practice more sophisticated algorithms should be considered.

With one or more healthy cells patched and the synaptic and neuronal models coded into the dynamic-clamp computer (*see* **Subheading 2.4.**), any of many neuronal networks can be, and have been, created ***(5)***. This chapter provides instructions for how to create some of the simplest networks (*see* **Fig. 2**): a lone neuron with a dynamically clamped firing rate (*see* **Subheading 3.3.**); a single neuron with a self-mediating computational Type A gamma-amino butyric acid ($GABA_A$) autapse (*see* **Subheading 3.4.**); a network of two biological neurons driving each other with computational alpha-amino-3-hydroxy-5-methyl-4-isoxaxolepropionic acid (AMPA) synapses (*see* **Subheading 3.5.**); and a network of one biological and one model neuron driving each other with opposing synapse types (*see* **Subheading 3.6.**).

Because the details of implementing model synapses and neurons vary across dynamic-clamp systems, we will enumerate their implementation in conceptual terms applicable to all systems. Each numbered step will be subdivided into alphabetically labeled steps that describe the syntactic requirements for the Real-Time eXperimental Interface (RTXI). These specifics are only completely applicable to the RTXI system but likely provide guidance for any dynamic-clamp implementation.

3.1. Prepare Healthy Tissue Slices

1. Place animal into sealed chamber with isoflurane vapor until rat is anesthetized.
2. Decapitate animal with guillotine. Open skull with scissors.

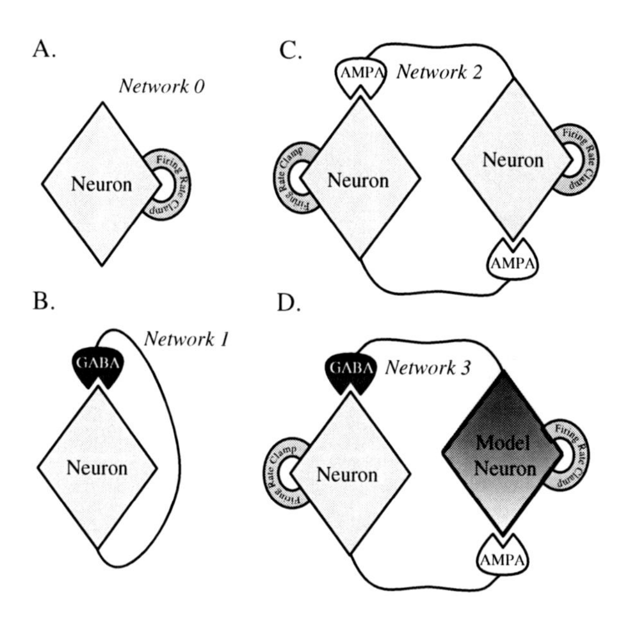

Fig. 2. Depictions of the example networks described in **Subheading 3.** **(A)** Network 0: a single neuron with a clamped firing rate. **(B)** Network 1: a single neuron with an inhibitory autapse. **(C)** Network 2: two firing rate clamped neurons with mutually excitatory synapses. **(D)** Network 3: one firing rate clamped biological neuron with an excitatory projection to a firing rate clamped computational neuron that extends an inhibitory projection back to the biological neuron.

3. With scoopula, remove brain and place in 0 °C ACSF.
4. With scalpel, isolate chunk(s) of tissue containing the brain region(s) of interest. Glue tissue to the bottom of a walled slice chamber with cyanoacrylate glue. Fill chamber with 0 °C ACSF.
5. Attach chamber to vibratome. Cut 300–400-μm thick slices. Place tissue into chamber containing ACSF bubbled with 95% O_2 and 5% CO_2.
6. Allow tissue to recover for greater than or equal to 1 h before transferring to microscope stage chamber.

3.2. Patch Neuron(s)

1. With a microscope using dual interference contrast microscopy, select a target neuron in slice tissue.
2. With micromanipulator, place electrode tip against the cell body.
3. While measuring electrode resistance, apply light negative pressure to form a gigaohm seal between electrode and neuron.
4. Apply negative pressure to pop through the membrane.

5. Balance bridge to compensate for access resistance and capacitance.
6. Repeat these steps for each living neuron in the desired hybrid network.

3.3. Network 0: Clamp Neuronal Firing Rate

1. Configure input/output channels.
   a. Select ***Controls → System Control***.
   b. Select ***Input Channel → Analog Input 0***. Set: *Range = −0.1 to 0.1*; *Gain =10V*. (All gains should be set as appropriate for the amplifier used.)
   c. Select ***Output Channel → Analog Output 0***. Set: *Range = −10 to 10; Gain =10mA*.
2. Initiate spike-detect module.
   a. Select ***Controls → Load Plugin***.
   b. Select ***spike_detect.so.*** Set: *Spike Detection Threshold* $= -30mV$.
3. Initiate firing rate control module.
   a. Select ***Controls → Load Plugin***.
   b. Select ***spike_rate_control.so.*** Set: *Target Interspike Interval = 100ms*.
4. Connect the modules.
   a. Connect ***Analog Input 0 → spike_detect: Vin***.
   b. Connect ***spike_detect: state → spike_rate_controller: state***.
   c. Connect ***spike_rate_controller: Iapp → Analog Output 0***.
5. Observe the neuronal firing rate approach target frequency.

3.4. Network 1: Single Neuron with Inhibitory Autapse

1. Configure input/output channels.
   a. Select ***Controls → System Control***.
   b. Select ***Input Channel → Analog Input 0***. Set: *Range = −0.1 to 0.1; Gain =10V*.
   c. Select ***Output Channel → Analog Output 0***. Set: *Range = −10 to 10; Gain =10mA*.
2. Initiate spike detect module.
   a. Select ***Controls → Load Plugin***.
   b. Select ***spike_detect.so.*** Set: *Spike Detection Threshold* $= -30mV$.
3. Initiate GABA synapse module.
   a. Select ***Controls → Load Plugin***.

b. Select ***synapse_biexp.so.*** Set: *Rising Time Constant = 2ms; Falling Time Constant = 6ms; Amplitude = 1 nS; Vsyn = −70mV*.

4. Connect the modules.

a. Connect ***Analog Input 0 → spike_detect: Vin***.
b. Connect ***spike_detect: state → synapse_biexp: state***.
c. Connect ***synapse_biexp: Iapp → Analog Output 0***.

5. Observe membrane potential hyperpolarization following each spike!

3.5. Network 2: Periodic Network of Two Mutually Excitatory Neurons

1. Configure input/output channels.

a. Select ***Controls → System Control***.
b. Select first ***Input Channel → Analog Input 0***. Set: *Range = −0.1 to 0.1; Gain =10V*.
c. Select second ***Input Channel → Analog Input 1***. Set: *Range = −0.1 to 0.1; Gain =10V*.
d. Select first ***Output Channel → Analog Output 0***. Set: *Range = −10 to 10; Gain =10mA*.
e. Select second ***Output Channel → Analog Output 1***. Set *Range = −10 to 10; Gain =10mA*.

2. Initiate spike detect modules.

a. Select ***Controls → Load Plugin***.
b. Select ***spike_detect.so***. Set: *Spike Detection Threshold = −30mV*.
c. Select ***Controls → Load Plugin***.
d. Select ***spike_detect.so***. Set: *Spike Detection Threshold = −30mV*.

3. Initiate firing rate control module.

a. Select ***Controls → Load Plugin***.
b. Select ***spike_rate_control.so***. Set: *Target Interspike Interval = 100ms*.
c. Select ***Controls → Load Plugin***.
d. Select ***spike_rate_control.so***. Set: *Target Interspike Interval = 100ms*.

4. Initiate AMPA synapse modules.

a. Select ***Controls → Load Plugin***.
b. Select ***synapse_biexp.so***. Set: *Rising Time Constant = 1ms; Falling Time Constant = 3ms; Amplitude = 1 nS; Vsyn = 0mV*.
c. Select ***Controls → Load Plugin***.
d. Select ***synapse_biexp.so***. Set: *Rising Time Constant = 1ms; Falling Time Constant = 3ms; Amplitude = 1 nS; Vsyn = 0mV*.

5. Connect the modules.
   a. Connect ***Analog input 0 → first spike_detect: Vin***.
   b. Connect ***Analog input 1 → second spike_detect: Vin***.
   c. Connect ***first spike_detect: state → first spike_rate_controller: state***.
   d. Connect ***first spike_detect: state → first synapse_biexp: state***.
   e. Connect ***second spike_detect: state → second spike_rate_controller: state***.
   f. Connect ***second spike_detect: state → second synapse_biexp: state***.
   g. Connect ***first synapse_biexp: Iapp → Analog Output 1***.
   h. Connect ***second synapse_biexp: Iapp → Analog Output 0***.
6. Observe network synchronize, or not (Fig. 3a).

3.6. Network 3: Periodic Biological-computational Hybrid Network

1. Configure input/output channels.
   a. Select ***Controls → System Control***.
   b. Select ***Input Channel → Analog Input 0***. Set: *Range = −0.1 to 0.1; Gain = 10V.*
   c. Select ***Output Channel → Analog Output 0***. Set: *Range = −10 to 10; Gain = 10mA.*
2. Initiate model neuron module.
   a. Select ***Controls → Load Plugin***.
   b. Select ***neuron.so***.
3. Initiate spike detect modules.
   a. Select ***Controls → Load Plugin***.
   b. Select ***spike_detect.so.*** Set: Spike Detection Threshold = *−30mV*.
   c. Select ***Controls → Load Plugin***.
   d. Select ***spike_detect.so***. Set: *Spike Detection Threshold* = −30*mV*.
4. Initiate firing rate control module.
   a. Select ***Controls → Load Plugin***.
   b. Select ***spike_rate_control.so***. Set: *Target Interspike Interval = 100ms.*
   c. Select ***Controls → Load Plugin***.
   d. Select ***spike_rate_control.so***. Set: *Target Interspike Interval = 100ms.*
5. Initiate synapse modules.
   a. Select ***Controls → Load Plugin***.
   b. Select ***synapse_biexp.so***. Set: *Rising Time Constant = 1ms; Falling Time Constant = 3ms; Amplitude = 1 nS; Vsyn = 0mV.*
   c. Select ***Controls → Load Plugin***.
   d. Select ***synapse_biexp.so***. Set: *Rising Time Constant = 2ms; Falling Time Constant = 6ms; Amplitude = 1 nS; Vsyn = −70mV.*

6. Connect the modules.

   a. Connect ***Analog Input 0 → first spike_detect: Vin***.
   b. Connect ***Neuron: Vm → second spike_detect: Vin***.
   c. Connect ***first spike_detect: state → first spike_rate_controller: state***.
   d. Connect ***first spike_detect: state → first synapse_biexp: state***.
   e. Connect ***second spike_detect: state → second spike_rate_controller: state***.
   f. Connect ***second spike_detect: state → second synapse_biexp: state***.
   g. Connect ***first synapse_biexp: Iapp → Neuron: Iapp***.
   h. Connect ***second synapse_biexp: Iapp → Analog Output 0***.

7. Observe network phase slip (**Fig. 3b**).

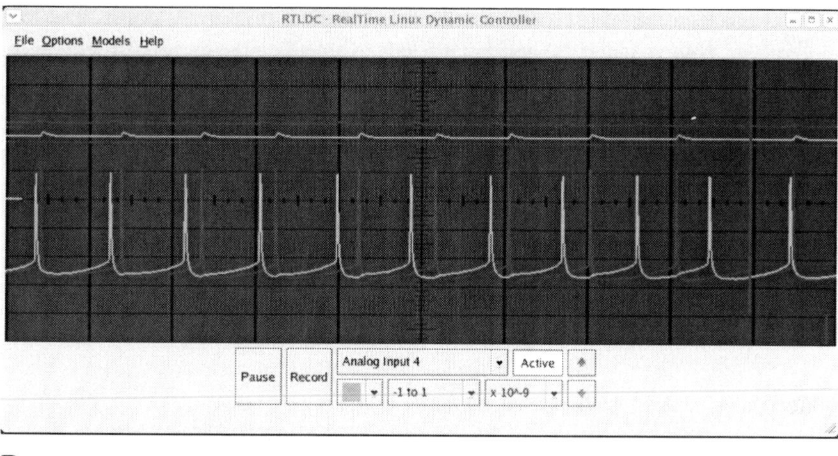

B.

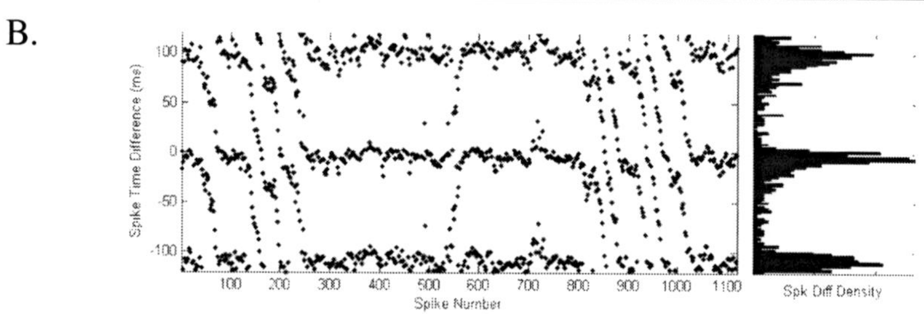

Fig. 3. Results from network 2. (**A**) Screen shot of dynamic clamp running a hybrid network consisting of two patch-clamped neurons whose periodically spiking membrane potentials are depicted. (**B**) Raster plot showing spike time differences in the network. Both cells are rate clamped to give an average interspike interval of 100 ms. Dots depict the firing time of neuron 2 with respect to the firing time of neuron 1. Spikes cluster around 0 and ±100 ms, indicating neuronal synchrony, but show clear signs of phase slipping.

4. Notes

1. The ACSF and artificial neuroplasm listed here are merely examples that we find sufficient for our studies with entorhinal cortical stellate neurons. In addition, you may wish to add synaptic blockers to chemically isolate the desired hybrid network from all other neurons living in the tissue.
2. Why hard real-time OSs?

    a. With a standard OS, computers can become distracted with incoming e-mail or a user moving the mouse. Even if the distraction is for less than 1/1000 of a second, a model synapse being simulated on the computer will not interact with the neuron as a living synapse would and experimental results should then be questioned.
    b. To perform dynamic clamp, the computer must behave on the same time scale as the synaptic channels it is trying to mimic. A hard real-time OS is one in which certain processes are allowed absolute temporal dominance over all others. Hence, when the computer needs to calculate the current flowing through the synapse, it halts all other processes (e.g., ignores the network card, mouse), does the mathematics, and sends the current to the amplifier before allowing the other processes to resume.
    c. Some dynamic clamps overcome the hard real-time burden by running the computations on a second computer or DSP. In this chapter, we use real-time application interface (RTAI) Linux. RTAI allows the dynamic-clamp software to run without interruption when it needs to, is open source, and costs nothing. RTAI can be downloaded with installation instructions from http://www.rtai.org.

3. Multipurpose A/D–D/A boards typically come with a software driver package, but the drivers are rarely written to perform adequately in a hard real-time environment. If you are using RTAI Linux, a free, open-source driver package exists which will interface with most standard multipurpose A/D–D/A boards: Comedi. The RTXI software described in this chapter interfaces exclusively with the Comedi driver package, which can be downloaded with installation instructions from http://www.comedi.org.
4. This chapter assumes that the lowest channel numbers on the A/D–D/A boards are connected to the current-clamp amplifier first. For example, the first neuron in a network occupies AI 0 and AO 0, the second neuron occupies AI 1 and AO 1, and so on.
5. In this chapter, we use the RTXI software package to perform dynamic clamp. It runs on RTAI Linux and requires the Comedi driver package. It can be downloaded from our Web site http://www.rtxi.org.

Acknowledgments

We thank David Christini and Robert Butera for help in development and use of dynamic-clamp techniques. This work was supported by NIH R01 NS34425 (PI: John A. White) and R01 RR020115 (PI: David J. Christini).

References

1. Robinson, H. P., and Kawai, N. (1993) Injection of digitally synthesized synaptic conductance transients to measure the integrative properties of neurons. *J. Neurosci. Methods* 49:157–165.
2. Sharp, A. A., O'Neil, M. B., and Marder, E. (1993) Dynamic clamp: computer-generated conductances in real neurons. *J. Neurophysiol.* 69:992–995.
3. Dorval, A. D., Christini, D. J., and White, J. A. (2001) Real-time Linux dynamic clamp: a fast and flexible way to construct virtual ion channels in living cells. *Ann. Biomed. Eng.* 29:897–907.
4. Butera, R. J., Wilson, C. G., Delnegro, C. A., and Smith, J. C. (2001) A methodology for achieving high-speed rates for artificial conductance injection in electrically excitable biological cells. *IEEE Trans. Biomed. Eng.* 48:1460–1470.
5. Netoff, T. I., Banks, M. I., Dorval, A. D., Acker, C. D., Haas, J. S., Kopell, N., and White, J. A. (2005) Synchronization in hybrid neuronal networks of the hippocampal formation. *J. Neurophysiol.* 93:1197–1208.

16

Cardiac Channelopathies Studied With the Dynamic Action Potential-Clamp Technique

Géza Berecki, Jan G. Zegers, Ronald Wilders, and Antoni C. G. van Ginneken

Summary

The cardiac long QT syndrome (LQTS) is characterized by a delayed repolarization of the ventricular myocytes, resulting in prolongation of the QT interval on the electrocardiogram and increased propensity to cardiac arrhythmias. Congenital LQTS has been linked to mutations in genes encoding ion channel subunits. For a better understanding of LQTS and associated arrhythmias, insight into the nature of ion channel (dys)function is indispensable. Conventionally, voltage-clamp analysis and subsequent mathematical modeling are used to study cardiac channelopathies and to link a certain genetic defect to its cellular phenotype. The recently introduced "dynamic action potential clamp" (dAPC) technique represents an alternative approach, in which a selected native ionic current of the ventricular myocyte can effectively be replaced with wild-type (WT) or mutant current recorded from a human embryonic kidney (HEK)-293 cell that is voltage clamped by the free-running action potential (AP) of the myocyte. Both a computed model of the human ventricular cell and a freshly isolated myocyte can effectively be used in dAPC experiments, resulting in rapid and unambiguous determination of the effect(s) of an ion channel mutation on the ventricular AP. The dAPC technique represents a promising new tool to study various cardiac ion channels and may also prove useful in related fields of research, for example, in neurophysiology.

Key Words: Long QT syndrome; ventricular cell; action potential clamp; repolarization; HEK-293 cell; background current subtraction; *SCN5A*; *HERG*.

From: *Methods in Molecular Biology, vol. 288: Patch-Clamp Methods and Protocols*
Edited by: P. Molnar and J. J. Hickman © Humana Press Inc., Totowa, NJ

1. Introduction

Types 2 and 3 of the congenital long QT syndrome (LQTS) (LQT2 and LQT3, respectively) are linked to mutations in the human ether-a-go-go-related gene (*HERG*) and sodium channel gene (*SCN5A*), respectively. These genes encode the pore-forming subunits of the ion channels carrying the cardiac rapid delayed rectifier current (I_{Kr}) and fast sodium current (I_{Na}), respectively. At present, most effort has been devoted to documenting the effects of identified mutations on the densities and kinetics of various cardiac ion channels upon heterologous expression, and consequences of channelopathies for cardiac function are inferred by testing the functional effects of the experimentally observed changes using mathematical models of cardiac cells ***(1)***. However, despite advances in mathematical modeling, the mechanism by which a given genetic defect leads to the clinically observed electrical disease often remains obscure.

The dynamic action potential clamp (dAPC) technique ***(2,3)***, an extension of the dynamic clamp approach ***(4)***, allows determining the contribution of wild-type (WT) and/or mutated (*HERG* or *SCN5A*) cardiac channels to the cardiac action potential (AP) morphology without making assumptions with regard to (altered) kinetic properties of the studied channel(s). Both a computed model of the human ventricular cell ***(5)*** and a freshly isolated myocyte can effectively be used in dAPC experiments (model cell and real cell mode dAPC, respectively). The first offers an outstanding reproducibility of the results, because the implemented (input) WT or mutant current is the only variable during experimentation. Moreover, it allows "generation" of subendocardial, midmyocardial (M), and subepicardial ventricular cell types by adjusting selected membrane ionic currents in the model cell. Equally, the technically more difficult real cell mode reveals AP waveforms and ion channel kinetics that can be considered close to physiological.

In this chapter, we provide straightforward protocols for setting up the dAPC technique and demonstrate that, during dAPC experiments, the (altered) shape of the AP directly reflects the effect of a mutation. With this methodology, the frequency dependence of the AP durations, the consequence of a pause on AP morphology, as well as the arrhythmogenic nature of LQT-associated mutations can be determined, and special kinetic features of cardiac channels can be revealed. With adequate scaling adjustments and procedures to reduce unwanted and contaminating background currents, this novel technique allows several cardiac ion channel types to be studied. The technique is not limited to the field of cardiac electrophysiology and may also prove useful in, for example, neurophysiology.

2. Materials

2.1. Plasmids

1. pSP64T vector (Invitrogen Europe, Paisley, Scotland, U.K.).
2. pCGI vector ***(6)***. The mammalian expression vector pCGI (provided by Drs David Johns and Edvardo Marbán, Johns Hopkins University, Baltimore, Md., USA.)

2.2. Human Embryonic Kidney-293 Cell (Sub)culture Medium

1. Minimal essential medium (Invitrogen Europe, Paisley, Scotland, UK.) supplemented with 10% fetal bovine serum (FBS) (Invitrogen Europe, Paisley, Scotland, UK.).
2. One percent penicillin/streptomycin (Invitrogen Europe, Paisley, Scotland, UK.) solution (Pen/Strep; 10,000 IU/ml to 10,000 IG/ml), stored in aliquots at −20 °C, and then added to the culture dishes as required (*see* **Note 1**).
3. One percent nonessential amino acids (Invitrogen Europe, Paisley, Scotland, UK.).
4. Store culture medium at +4 °C.

2.3. Human Embryonic Kidney-293 Cell Transfection

1. Phosphate-buffered saline (PBS) without Ca^{2+} and Mg^{2+} (BioWhittaker Europe Verviers, Belgium).
2. Trypsine-Versene (500 mg/l Trypsin [1:250] and 200 mg/l ethylenediaminete-traacetic acid [EDTA]) (Invitrogen Europe, Paisley, Scotland, UK.) (*see* **Note 2**).
3. Lipofectamine 2.0 mg/ml (Invitrogen).

2.4. Isolation of Rabbit Ventricular Myocytes

1. "Normal" Tyrode's solution: 140 m*M* NaCl, 5.4 m*M* KCl, 1.8 m*M* $CaCl_2$, 1.0 m*M* $MgCl_2$, 5.0 m*M* HEPES, and 5.5 m*M* glucose (pH adjusted to 7.4 with NaOH) (*see* **Note 3**).
2. "Ca^{2+}-free" Tyrode's solution: 140 m*M* NaCl, 5.4 m*M* KCl, 0.5 m*M* $MgCl_2$, 1.2 m*M* KH_2PO_4, 5.0 m*M* HEPES, and 5.5 m*M* glucose (pH adjusted to 7.2 with NaOH). The "low-Ca^{2+}" Tyrode's solution was composed of nine parts "Ca^{2+}-free" and one part "normal" Tyrode's solution, yielding a final Ca^{2+} concentration of approximately 180 μ*M*.
3. Myocyte dissociation solution: 150 ml Ca^{2+}-free Tyrode's solution with 10 μ*M* $CaCl_2$ + 22 mg collagenase type B + 7 mg collagenase type P + 30 mg hyalonuridase +15 mg Trypsin inhibitor (all enzymes from Boehringer, Germany) (*see* **Note 4**).

2.5. Solutions and Drugs for Electrophysiology

1. "Normal" Tyrode's solution (*see* **Subheading 2.4.**).
2. Extracellular solution for human embryonic kidney (HEK)-293 cells expressing *HERG* current (I_{HERG}): "modified" Tyrode's solution with 4.5 instead of 5.4 m*M* KCl (*see* **Note 5**).

3. Extracellular solution for HEK-293 cells expressing sodium current (I_{SCN5A}): 140 m*M* NaCl, 10 m*M* CsCl, 2.0 m*M* $CaCl_2$, 1.0 m*M* $MgCl_2$, 5.0 m*M* glucose, 10 m*M* sucrose, and 10 m*M* HEPES (pH 7.4 with NaOH) (*see* **Note 6**).
4. Pipette solution during AP recording from rabbit myocytes: 125 m*M* K-gluconate, 20 m*M* KCl, 1.0 m*M* $MgCl_2$, 5.0 m*M* MgATP, and 10 m*M* HEPES (pH 7.2 with KOH).
5. I_{HERG} pipette solution: 125 m*M* K-gluconate, 20 m*M* KCl, 1.0 m*M* $MgCl_2$, 5.0 m*M* ethyleneglycol-bis-N, N, N′, N′-tetraacetic acid (EGTA), 5.0 m*M* MgATP, and 10 m*M* HEPES (pH 7.2 with KOH).
6. I_{SCN5A} pipette solution: 10 m*M* CsCl, 110 m*M* CsF, 10 m*M* NaF, 11 m*M* EGTA, 1.0 m*M* $CaCl_2$, 1.0 m*M* $MgCl_2$, 2.0 m*M* Na_2ATP, and 10 m*M* HEPES (pH 7.3 with CsOH).
7. E-4031 (Eisai Pharmaceutical Co., Japan) is dissolved in distilled water as a 5.0-m*M* stock solution and used in final concentration of 5.0 μ*M*.
8. Tetrodotoxin (TTX) (Alomone Labs, Israel) is dissolved in distilled water as a 5.0 m*M* stock solution and used in final concentration of 50 μ*M*.

2.6. Dynamic Action Potential Clamp

1. Pentium-4 PC with a 16-bit data aquisition (DAQ) DAQ-board with analog in and out, for example, National Instruments PCI-6052E data acquisition board (*see* **Note 7**).
2. Real-time Linux (RTL) operating system.
3. Software that implements the dAPC.
4. Mathematical model of an isolated cardiac cell, for example, the human ventricular cell model developed by Drs Priebe & Beuckelmann *(5)*.

3. Methods

3.1. Plasmid Construction

1. Clone the coding sequence (cDNA) of ion channel subunit(s) into an appropriate vector (e.g., pSP64T). Subclone the ion channel cDNA into the pCGI vector for bicistronic expression of the channel protein and green fluorescent protein (GFP) reporter in HEK cells (*see* **Note 8**).
2. For co-expression studies, use an additional bicistronic vector: pCCI (cyano-fluorescent protein reporter, CFP instead of GFP).

3.2. HEK-293 Cell (Sub)culture

1. HEK-293 cells are passaged when approaching 80% confluence (check on an inverted microscope). Remove all medium from the culture dish (Falcon, canted neck 50 ml, Polystyrene). Wash the cells with 5 ml PBS (2×). Remove PBS. Add 0.5 ml Trypsine-Versene to the 50-ml culture dish. After 1–2 min incubation, round-shaped cells detach from the bottom of the dish.

2. Add 4.5 ml fresh culture medium, and resuspend the medium 1× with the 5-ml pipette. Add 0.2 ml of this suspension to a new 50-ml cell culture dish and then add 4.8 ml cell culture medium. Place the cell culture dish back in the incubator (5% CO_2, 37 °C, and 8% relative humidity).

3.3. HEK-293 Cell Transfection

1. Grow cells to 50% confluence.
2. Place in a 15-ml centrifuge tube: 0.5–2 μg DNA (plasmid with the cDNA of the ion channel), 20 μl Lipofectamine, and 0.5 ml FBS-free culture medium. Mix gently and incubate for 40 min.
3. Remove the culture medium from the culture dish and wash the cells (2×) with 2 ml FBS-free culture medium. Add the content of the centrifuge tube and 2 ml FBS-free culture medium to the cells. Incubate the dish in the incubator for 6 h.
4. Remove transfection medium and add 5 ml culture medium (with FBS) to the dish. Incubate the cells overnight.
5. Wash the transfected cells with 5 ml PBS (2×). Remove PBS. Add 0.5 ml trypsine-versene to the dish. After 1–2 min of incubation (round-shaped cells detach from the bottom of the dish), add 4.5 ml fresh culture medium. Gently resuspend the medium with a 5-ml pipette to detach all cells from the bottom of the dish.
6. Place 10–20 μl of the above suspension in the recording chamber mounted on the stage of an inverted microscope (Nikon Diaphot, Japan) (*see* **Note 9**); fill the chamber with the extracellular solution (*see* **Note 10**).

3.4. Isolation of Rabbit Ventricular Myocytes and AP Recordings

The experimental protocol should comply with the "Guide for the Care and use of Laboratory Animals" published by the US National Institutes of Health (NIH publication 85-23, revised 1985).

1. Apply heparin (1000 IU IV) and then kill New Zealand White rabbits by injection of pentobarbital (60 mg/ml, 5 ml, IV).
2. Isolate single rabbit ventricular myocytes using an enzymatic dissociation procedure, modified from Tytgat *(7)*. Remove the heart rapidly and mount it on a Langendorff perfusion apparatus; perfuse it retrogradely through the aorta with "normal" Tyrode's solution for 10 min, followed by an additional 10 min of perfusion with "Ca^{2+}-free" Tyrode's solution. Switch perfusion to myocyte dissociation solution for 15 min. During the isolation procedure, the enzymatic solution is recirculated; all solutions are saturated with 100% O_2, and the temperature is maintained at 37 °C.
3. Subsequently, cut the ventricles into pieces; collect pieces in a beaker with myocyte dissociation solution; shake the beaker in a bath at 60 and 100 rpm for 10 and 5 min, respectively; remove supernatant and add fresh myocyte dissociation solution with 1% fatty acid-free bovine serum albumin (Sigma-Aldrich); and shake the beaker at 120 rpm for 10 min. Collect supernatant with isolated myocytes and further digest

the remaining pieces of tissue (120 rpm for 10 min); collect myocyte suspension as in the previous step. Combine both supernatants and allow myocytes to settle for 5–10 min.

4. Remove supernatant and add "low-Ca^{2+}" Tyrode's solution to the cells.
5. After 10 min, replace approximately 75% of the "low-Ca^{2+}" Tyrode's solution by "normal" Tyrode's solution to obtain stepwise increases of the calcium concentration. Repeat this procedure four times at intervals of 10 min. Store the cells at room temperature (20–22 °C) and use them within 8 h.
6. Place 10–20 μl of myocyte suspension in the recording chamber mounted on the stage of an inverted microscope (Nikon Diaphot); fill the chamber with the extracellular (Tyrode's) solution (*see* **Note 11**). Select single, rod-shaped myocytes with smooth surfaces and clear cross-striations for electrophysiological measurements.
7. Superfuse myocytes with Tyrode's solution at 36 °C. Correct for liquid junction potential (*see* **Subheading 3.5.**). Record APs in current clamp mode, using whole-cell configuration of the patch-clamp technique. Elicit the APs by 2-ms current pulses (1.5× threshold) at 0.2, 1, 2, 3, 4 and 5 Hz.
8. Determine AP parameters, that is, overshoot, maximum diastolic potential (MDP), action potential amplitude (APA), maximum upstroke velocity (V_{max}), and AP duration at 50, 80, and 90% repolarization (APD_{50}, APD_{80}, and APD_{90}, respectively) as described in Berecki et al. *(2)*.

3.5. Electrophysiology and Data Acquisition

1. Record membrane potentials and currents at 36 ± 0.5 °C, with Axopatch 200B amplifiers (Axon Instruments, at Molecular Devices, USA) using conventional whole-cell patch-clamp technique *(8)*.
2. Pull electrodes from borosilicate glass with a glass fiber inside the lumen, using conventional one-stage or two-stage pullers. When backfilled with the appropriate pipette solution, electrodes should have resistances of 1–3 MΩ.
3. For I_{Na} recordings, backfilled electrodes should exhibit resistances of 1–1.3 MΩ. The resulting series resistance (R_s) values are typically 2–2.5 MΩ and can be compensated 90–95%. To further minimize possible voltage errors, small HEK cells of 8–12 pF cell capacitance (C_m), expressing peak I_{SCN5A} amplitudes less than 10 nA should be selected (*see* **Note 12**).
4. Adjust the potential between the pipette and bath solution (liquid junction potential) to zero before a high-resistance seal between pipette and cell is formed. All potentials should be corrected for the change in liquid junction potential, which occurs when contact with the cells is made *(10)*.
5. When recording I_{Na} and potentials during dAPC experiments, use low-pass filtering with cutoff frequency of 5 or 10 kHz and digitizing at 20 kHz (*see* **Note 13**). In the experiments using TTX, I_{SCN5A} can be determined as the current blocked by 50 μ*M* TTX (*see* **Note 14**).
6. Accomplish voltage control, data acquisition, and analysis with custom software.

3.6. dAPC Technique

The approach is based on the concept that an isolated (cardiac) cell can be electrically coupled to either another isolated cardiac cell or to a model analog that mimics the electrical properties of the cardiac myocyte. As diagrammed in **Fig. 1**, a single cardiac ventricular cell and a transfected HEK-293 cell can be electrically coupled by means of an electrical circuit. The ventricular cell (with I_{Kr} blocked) is in "current clamp" mode on one patch-clamp setup, whereas the HEK-293 cell is in "voltage-clamp" mode on another setup. The command potential for the HEK-293 cell is the membrane potential (V_m) of the ventricular cell ("action potential clamp"), and the current input applied to the ventricular cell is the I_{HERG} recorded from the transfected HEK-293 cell, a connection resulting in dAPC condition. During dAPC experiments, we define "model cell" and "real cell" modes.

3.6.1. Setting Up the Hardware and the Software for Real-Time Acquisition

The dAPC software has two parts:

1. User program, DynaClamp, which allows the experimenter apply parameter settings for the dAPC experiment.
2. Real-time module that repeatedly performs the following sequence with a fixed time step:

   a. Converts analog to digital (ADC) the HEK-293 cell current.
   b. Calculates V_m from the Priebe & Beuckelmann cell model with the input HEK-293 cell current (*see* **Note 15**).
   c. Outputs the digital to analog converted (DAC) V_m.
   d. Waits for the next time step.

This sequence has to be executed with an equidistant and predictable time interval, and therefore, it is time critical. This is guaranteed by RTL *(11)*, the operating system for DynaClamp. The time critical part of the program is written as an installable real-time module for RTL. With this approach, we achieve 40-μs time steps. For ADC and DAC, we used the freely available package Comedilib written by David Schleef. Results calculated by the real-time module are stored in shared memory accessible from both the user program and the real-time module. We use the shared memory manager written by Tomek Motylewski. To allow floating point calculations in the real-time module, we use the math library rt-math module by Jan Albiez.

1. Get your favorite Linux distribution and install the Linux operating system (*see* **Note 16**).

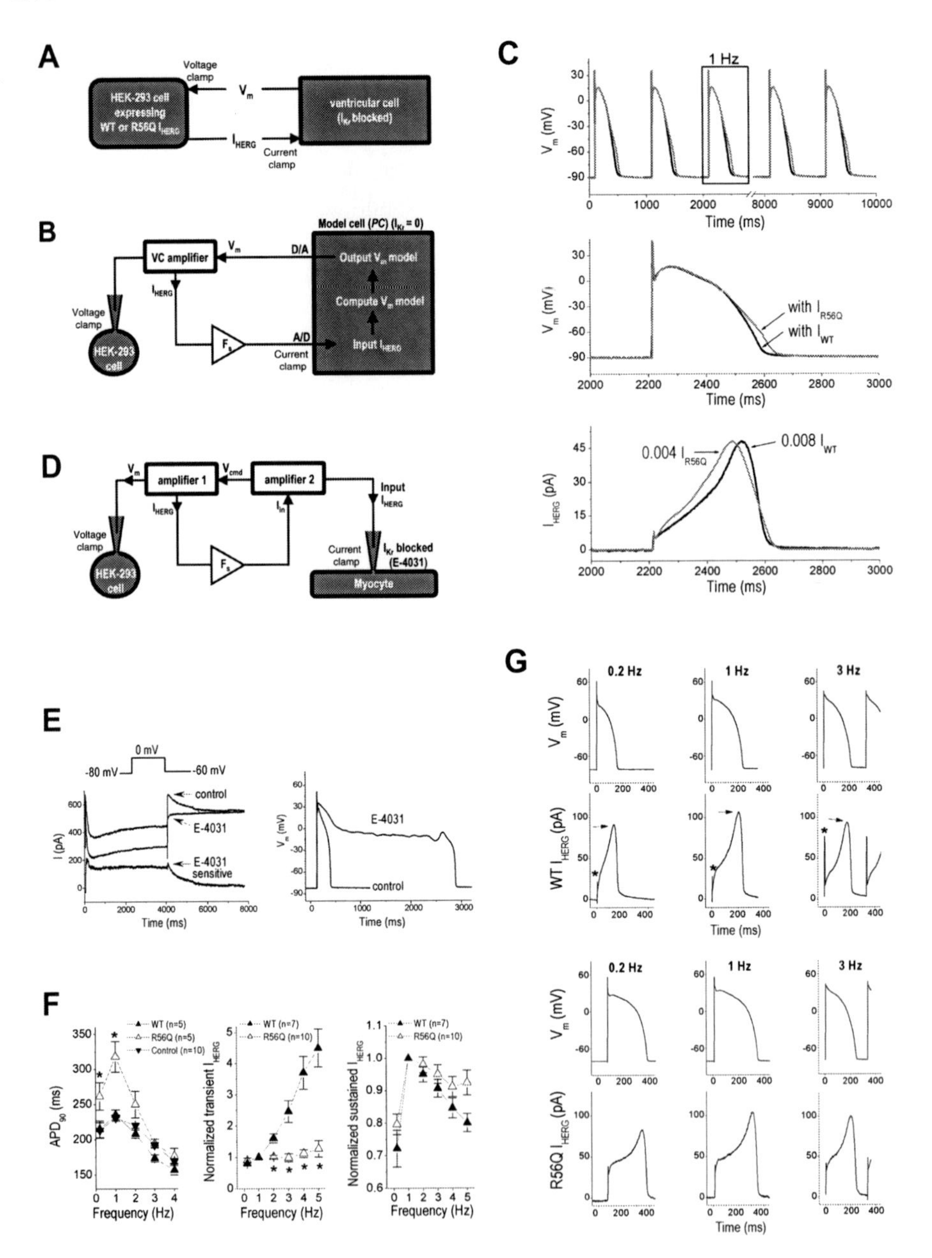

Fig. 1. Use of the dynamic action potential clamp (dAPC) technique to effectively replace native I_{Kr} of a ventricular cell with I_{HERG} from a human embryonic kidney (HEK)-293 cell. (**A**) Overall experimental design. (**B**) Model cell mode. (**C**) Representative action potential (APs) with wild-type (WT) I_{HERG} (black line) or R56Q I_{HERG}

2. Get RTL from http://www.rtlinuxfree.com/and install.
3. Get Comedi from http://www.comedi.org and install.
4. Get the kernel/user space memory driver from http://sourceforge.net/projects/mbuff and install.
5. Get the math library rt-math module from http://mca2.sourceforge.net/and install.
6. Write your own dAPC program or get our sources from http://www.physiol.med.uu.nl/dynaclamp/
7. Connect your VC amplifier to the A/D board in the computer:
   a. Connect the current output of the VC amplifier to A/D input 0.
   b. Connect D/A output 0 to the command input of your VC amplifier.
8. Insert the RTL loadable modules.
9. Insert the mbuff.o and rt_math.o loadable modules.
10. Insert your own real-time loadable module.
11. Run the user part of the dynamic AP program.

3.6.2. Model Cell Mode

In model cell mode (*see* **Fig. 1B**), the membrane potential of a model cell is computed, based on the equations of the underlying mathematical model, and applied as voltage-clamp command to clamp the membrane potential of a HEK-293 cell that expresses the ion current of interest. This current is then recorded and, after appropriate scaling, applied to the model cell as an additional current input. If the corresponding "native" current of the model cell is reduced or set to zero, this current is either partly or fully replaced with the current expressed in the HEK-293 cell. The repetitive process of sending out a command potential for the HEK-293 cell (D/A conversion), reading in the

Fig. 1. (gray line) at 1 Hz (model cell mode with model cell $I_{Kr} = 0$). Boxed APs from top panel (middle) and associated I_{HERG} (bottom) on an expanded time scale (scaling factor, F_s, values indicated). (**D**) Real cell mode. (**E**) Block of I_{Kr} with 5 μM E-4031 in a rabbit ventricular myocyte (inset, pulse protocol). (**F**) Frequency dependence of the myocyte's APD_{90} prolongation (left) and transient (middle) and sustained (right) I_{HERG} amplitudes (real cell mode); asterisks indicate significant difference for R56Q versus WT. (**G**) APs from a myocyte and associated WT (top) or R56Q I_{HERG} (bottom) at different frequencies (real cell mode). The myocyte was successively coupled to HEK-293 cells transfected with WT or R56Q *HERG* channels. Note the different I_{HERG} waveforms (asterisk, transient I_{HERG}; arrow, sustained I_{HERG}) and frequency-dependent AP prolongation with R56Q (adapted from [2] with the permission of the Biophysical Society, http://biophysj.org/).

current elicited in the HEK-293 cell (A/D conversion), and then computing the new membrane potential of the model cell requires a high update rate.

1. The small update time step required for minimization of the phase shift between the ionic current of the HEK-293 cell and the membrane potential of the model cell can be obtained by selecting appropriate hardware and careful design of the dynamic clamp software.
2. Use a data acquisition board that can handle single A/D samples as well as single D/A samples in a time-efficient manner and with reproducible time steps (microsecond resolution). As dAPC is a real-time process, A/D-input and D/A-output buffers are not useful.
3. For numerical accuracy of the cell model, the update time step should be kept less than 80 μs (cf. Fig. 2 of Joyner et al. [12].
4. Select an efficient numerical integration method for the mathematical cell model. A simple Euler-type integration scheme with a fixed time step—identical to the update time step—is the most appropriate. Further improvement can be obtained using the dedicated algorithm by Rush and Larsen ***(13)***.
5. Use lookup tables for (parts of) equations that are functions of membrane potential. Thus, the numerical cost of equations for rate constants can be drastically reduced.
6. With the exception of intracellular calcium, fluctuations in ionic concentrations are negligible during the time of a dAPC run (typically 10 s). Therefore, computational time can be further reduced by setting all ionic concentrations of the model except intracellular calcium concentration to constant values.
7. Always use "stable-start" values, obtained during 1 Hz stimulation, as initial values for model variables rather than steady-state values ("0 Hz" values).
8. Avoid producing inflexible and unreadable computer code while optimizing the model. With current processor speed and the above optimizations, the mathematical model can be updated in a few microseconds. In contrast with early real-time implementations ***(14)***, the model computations are no longer a major limiting factor for achieving small time steps.

3.6.2.1 Implementing a Low Density Current With Delicate Kinetics (I_{HERG})

1. Set I_{Kr} of the model cell to zero.
2. First determine maximal I_{HERG} amplitude in the HEK-293 cell in voltage-clamp configuration using 4-s depolarizing voltage steps to −10, 0, and 10 mV from a holding potential of −80 mV. Measure I_{HERG} amplitudes at the end of 4-s pulses. Use the largest outward current value to estimate the scaling factor (F_s) for the I_{HERG} input to the PB model cell (*see* **Fig. 1B**). In our standard protocol, WT as well as R56Q I_{HERG} amplitude are scaled to 47.6 pA (equivalent to the original I_{Kr} amplitude in the PB model).

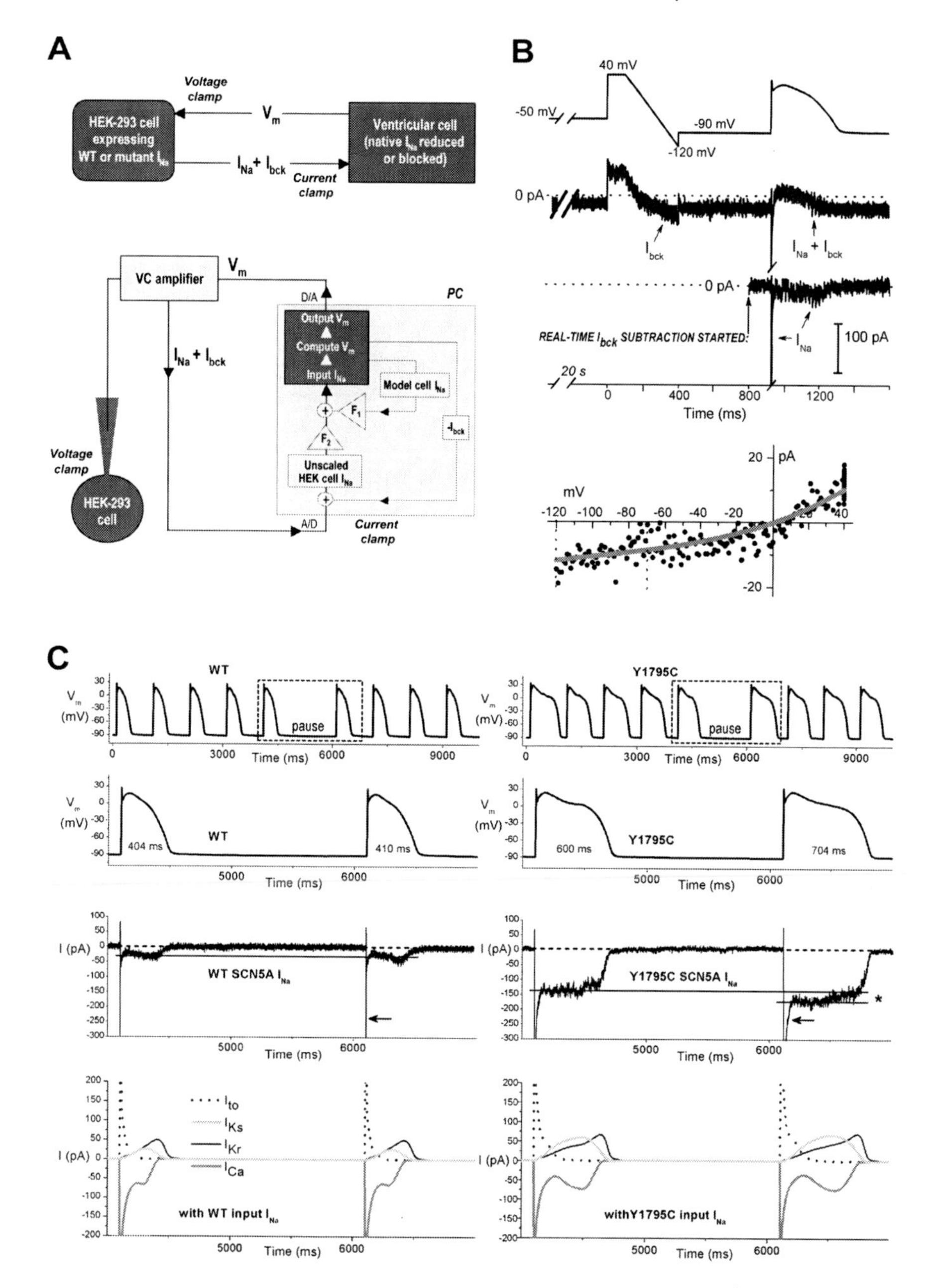

Fig. 2. Design of the "model cell mode" dynamic action potential clamp (dAPC) experiment with I_{Na}. **(A)** I_{SCN5A} from the human embryonic kidney (HEK) cell is continuously applied to the ventricular cell as an external current input, partly or entirely

3. After appropriate scaling, establish dAPC configuration between the model cell and the HEK-293 cell. Apply a series of 2-ms, 4-nA, 1-Hz supra-threshold stimuli to the computer model cell for 10 s. Store the recorded I_{HERG} and computed PB model variables (V_m and ionic currents) and settings of the DynaClamp program on disk for offline analysis (*see* **Note 17**). The time-dependent changes in V_m of the ventricular model cell are derived from WT and/or mutant I_{HERG} input and the model equations. The combination of the cell model and WT I_{HERG} will then result in a "normal" AP. Using the same method for HEK-293 cells with mutant channels will reveal an AP that represents the ventricular AP of the patient from which the mutant was derived.

3.6.2.2 Principles and Considerations for Scaling I_{HERG}

The scaling factor (F_s) will vary between cells. This is unavoidable, because expression levels of *HERG* channels in the HEK-293 cells are also variable. Scaling both WT and R56Q I_{HERG} amplitudes to a value equivalent to the original I_{Kr} amplitude in the PB model (47.6 pA) is based on the following:

Fig. 2. replacing I_{Na} in the ventricular cell (top). After real-time digital subtraction of HEK cell I_{bck} (*see also* **B**), I_{SCN5A} is scaled by factor F_2 (bottom); model cell I_{Na} density is reduced to 40% of the original value (scaling factor F_1). The I_{Na} of the PB model cell incorporates a "heterozygous" I_{Na} composed of reduced model cell I_{Na} and scaled HEK cell (input) I_{SCN5A} (+). (**B**) Real-time I_{bck} subtraction from a WT *SCN5A* cDNA-transfected HEK-293 cell. From top to bottom: step-ramp voltage protocol and the first action potential (AP) elicited in a subepicardial model cell; I_{SCN5A} in the presence of I_{bck}; and I_{SCN5A} after I_{bck} removal: note the transient WT late I_{SCN5A} during AP repolarization (dotted line shows zero current level). In these experiments, the *I-V* relationship of I_{bck} was fit with the $I_{bck} = P \cdot (-0.38 + 0.52 \cdot V_m + 3.69 \cdot 10^{-3} \cdot {V_m}^2 + 1.29 \cdot 10^{-5} \cdot {V_m}^3)$, scaled **Eq. 1** (scaling factor, $P = 0.39$) (*see* **Subheading 3.6.2.4.**). (**C**) dAPC with wild type (WT) (left) or Y1795C I_{SCN5A} (right). From top to bottom: subepicardial cell APs elicited at 1 Hz before and after a 2-s pause. APs with mutant I_{SCN5A} are prolonged compared with WT. Boxed APs from **A** and associated I_{SCN5A} on expanded time scale. Note that late I_{SCN5A} increase after the pause is more pronounced with the mutant (asterisk). Dotted line shows zero current level; peak I_{SCN5A} is off scale; note the slower time course of peak I_{SCN5A} inactivation of the mutant compared with the WT (arrows). Relationships between the APs and selected membrane current components of the human ventricular model cell, showing the changes in the time course of transient outward K^+ current (I_{to}), slowly and rapidly activating components of the delayed rectifier K^+ current (I_{Ks} and I_{Kr}, respectively) and L-type Ca^{2+} current (I_{Ca}), along with changes in mutant I_{SCN5A} (*see* **Note 20**) (adapted from [*3*]).

1. In HEK-293 cells, the transiently expressed WT and R56Q I_{HERG} densities are not significantly different (*see* **Note 18**).
2. By implementing WT I_{HERG}, the shape (and amplitude) of the model cell I_{Kr} can genuinely be reproduced.

3.6.2.3 Implementing a Large Density Current With Fast Kinetics (I_{SCN5A})

1. Replace (partly or completely) I_{Na} of the model cell with transiently expressed *SCN5A* I_{Na}. In the original PB cell model, peak amplitude of the native I_{Na} during the AP upstroke is approximately 58.3 nA (380 A/F at 1 Hz) (*see* **Note 19**). Reducing the ventricular cell's original (peak) I_{Na} density by 34 or 60% results in proportionate model cell AP upstroke velocity decreases. At the same time, these gradual I_{Na}-density reductions do not have APD-shortening effects at physiologically relevant stimulation rates.
2. In all dAPC experiments, use 40% of model cell I_{Na} density (152 A/F at 1 Hz) combined with the implemented HEK cell (input) I_{SCN5A}, the peak amplitude of which is up scaled to approximately 20 nA (approximately 130 A/F). The resulting (combined) I_{Na}s provide realistic rates of membrane depolarizations and approach the heterozygous I_{Na} expression in carriers of LQT3 mutations.
3. When scaling input I_{SCN5A}, first determine the maximal I_{SCN5A} amplitude in voltage-clamped HEK cells during 30-ms depolarizing voltage steps to voltages ranging from −60 to 10 mV (from a holding potential, $V_{hold} = -90$ mV; 10 mV steps). Here, graded activation of the I_{SCN5A} during voltage steps also serves as an index for reliable voltage control. Then use the largest I_{SCN5A} amplitude value to estimate the scaling factor (F_2) for current input to the PB model cell (*see* **Fig. 2A**). In order to minimize the endogenous HEK cell "background" currents (I_{bck}s) as well as noise from several sources including the operational amplifier and associated circuitry, F_2 values should be kept less than 3 and experiments requiring $F_2 > 3$ rejected.
4. To further improve recording conditions, a real-time I_{bck} subtraction procedure should be implemented (*see* **Subheading 3.6.2.4.**).
5. Having completed the mentioned preparative steps, establish dAPC configuration between the ventricular cell and the HEK cell by using DynaClamp. For this, apply a series of 2-ms, 4-nA supra-threshold stimuli to the PB cell model at a fixed rate for 10 s. At each stimulus, the V_m of the HEK cell follows that of the model cell. Supra-threshold stimuli thus elicit I_{SCN5A} in the HEK cell, which—along with 40% of original (cell model) I_{Na}—provide the inward current needed for the upstroke of the AP.
6. Apply various stimulation rates. Optionally, a 2-s pause can be implemented during stimulation, starting at 5 s after onset of pacing.
7. The combination of ionic currents of the model cell and WT I_{SCN5A} of the HEK cell reveals APs that are considered as "normal," representing healthy individuals.

HEK cells containing mutant channels result in APs that are considered to be characteristic for cells from the ventricular tissue of a patient from which the mutant was derived.

3.6.2.4 Real-Time Digital Subtraction of HEK Cell I_{bck}

Contrary to dAPCing with I_{HERG}, where the implemented current is downscaled thus making I_{bck} negligible, in these experiments, I_{SCN5A} needs upscaling. This procedure would increase I_{bck}, so that it would interfere with the late I_{SCN5A} during the AP plateau. To adequately record and implement I_{SCN5A} in dAPC experiments, I_{bck} removal is essential (*see* **Notes 6** and **21**).

1. Record currents in the voltage-clamp configuration of the patch-clamp technique from HEK-293 cells. Use square wave as well as various ramp and AP-shaped protocols (a depolarizing step followed by a 0.43 V/s repolarizing ramp, "step-ramp") at 1-s inter-ramp intervals to characterize I_{bck} in empty and transfected HEK cells (*see* **Fig. 2B**). Ideally, at −90 mV, in most HEK cells, I_{bck} amplitude is less than 40 pA. The averaged I_{bck} traces during the ramp, plotted against V_m, result in slightly outwardly rectifying I_{bck}-V relationships, similar to that obtained with step protocols, and are not influenced by lengthening (to 2 s) or shortening (to 0.2 s) of the inter-ramp intervals.
2. Describe (fit) the average current-voltage (I_{bck}-V) relationship with the appropriately scaled third-order polynomial equation:

$$I_{bck} = -0.38 + 0.52 \cdot V_m + 3.69 \cdot 10^{-3} \cdot V_m^2 + 1.29 \cdot 10^{-5} \cdot V_m^3, \quad (1)$$

   where I_{bck} is in pA and V_m is in mV (*see* **Fig. 2B**).
3. Implement a step-ramp waveform preceded by a 20-s voltage step to −50 or −60 mV in dAPC experiments. The long depolarizing holding potential (V_{hold}) before the step-ramp serves to inactivate I_{SCN5A}, whereas the step-ramp allows defining I_{bck}-V relationships in (transfected) individual HEK cells. Store the ramp-evoked I_{bck}-V relationship in a lookup table. Before establishing dAPC configuration, fit **Eq. 1** to the data points between −120 and −70 mV, using a least square fitting procedure and a scaling factor P as the only variable. Thus, the I_{bck} amplitude values for the whole voltage range are based on the scaled I_{bck}-V of nontransfected HEK cells (*see* **Note 22**). To assess the fit, generate offline results in a separate file.

In summary, I_{bck} should be characterized before DynaClamp establishes dAPC condition and then entirely subtracted in real time from the input current without affecting I_{SCN5A}, including late components.

3.6.3. Real Cell Mode

In real cell mode, a rabbit left ventricular myocyte is used instead of the model cell (*see* **Fig. 1D**). I_{HERG} is recorded from the HEK-293 cell with amplifier 1, which is in voltage-clamp mode, scaled and applied as external current input (I_{in}) to amplifier 2, which is in current clamp mode. The V_m of the myocyte (with I_{Kr} blocked pharmacologically), shaped by the input I_{HERG}, is applied as voltage-clamp command (V_{cmd}) to amplifier 1, thus establishing dAPC.

The use of rabbit ventricular myocytes in dAPC is appropriate because (1) the currents underlying the AP of these cells are comparable with those in the human ventricular cell and (2) the isolation of rabbit ventricular cells is a standard procedure in many laboratories.

1. Define F_s as follows: measure I_{HERG} amplitude in the HEK cell (as described in **Subheading 3.6.2.1.**) and, simultaneously, estimate I_{Kr} density in the rabbit cell (as the E-4031 sensitive current) (*see* **Fig. 1E**).
2. Elicit APs in the myocyte at 1 Hz in the presence of E-4031 and then establish coupling between the myocyte and the HEK-293 cell, implementing scaled I_{HERG} (*see* **Note 23**). A proper F_s value would result in I_{HERG} density comparable to that of the I_{Kr} density in the myocyte and in a typical APD_{90} value of approximately 230 ms at 1 Hz, characteristic for these cells (*see* **Notes 24** and **25**).
3. Apply various stimulation rates.
4. The V_m of the ventricular cell and I_{HERG} of HEK-293 cell can be displayed online, thus providing instant information on the dAPC (*see* **Note 26**).

4. Notes

1. Use sterile technique: all handling in laminar flow hood with sterile disposable pipettes.
2. Small aliquots of this solution can be stored at −20 °C.
3. The pH of solutions should be corrected for the physiological (working) temperature.
4. The yield of myocytes may depend on the batch of enzymes used and on Ca^{2+} concentration. It may prove necessary to adjust the amount of enzymes and/or Ca^{2+} concentration, which should not exceed 50 μ*M*.
5. To obtain a better match between the K^+ equilibrium potential (E_K) of the experimental solutions and the model cell's MDP of −90.7 mV, a modified, 4.5 m*M* KCl, Tyrode's solution should be used (resulting in a E_K of −92.5 mV) during dAPC with I_{HERG}.
6. A significant HEK-293 cell background current (I_{bck}) reduction can be obtained by adjusting the osmolarity of the extracellular solution with sucrose (to 310 mOsm, slightly hypertonic with respect to the 290 mOsm pipette solution) using a semi-micro osmometer (Knauer, Germany). Cesium ions block various K^+ conductances in the HEK-293 cell.
7. A fast 16-bit data acquisition board is required to cover the wide dynamic range of the cardiac sodium current. For other currents, a moderately fast 12-bit board may be sufficient.

8. The mutated insert and ligation regions should be verified by automated sequencing.
9. Ionic currents are recorded within 1–2 days from HEK-293 cells exhibiting cyano and/or green fluorescence in the whole-cell configuration of the patch-clamp technique.
10. The composition of this solution should be adequate for the expressed ion channel protein (*see* **Subheading 2.**).
11. Allow myocytes to settle for 5 min before superfusion with "normal" Tyrode's solution is started.
12. With the given R_s and C_m values, the true membrane potential can be established within about 25 μs after the start of the step command without overshoot or ringing *(9)*, fast enough to voltage clamp rapid ionic currents. The estimated residual (uncompensated) R_s of less than 200 kΩ can theoretically cause a less than 2 mV voltage error in the membrane potential relative to the command potential in the presence of 10 nA amplitude I_{SCN5A}.
13. During dAPC experiments with I_{HERG}, membrane currents are low-pass filtered (cutoff frequency 2 kHz) and digitized at 5 kHz.
14. To quantify late I_{SCN5A}, the largest inward current amplitude is measured during a sustained step depolarization or during phase 2 (plateau) or 3 (rapid repolarization) of the AP.
15. The HEK-293 cell (input) current is scaled during dAPCing (*see* **Subheadings 3.6.2.** and **3.6.3.**). Because I_{bck} depends on the HEK-293 cell V_m in real cell mode, the V_m of the myocyte has to be A/D converted to calculate the correction for the background current. We used A/D input 1 to read the V_m of the real cell. This means an extra A/D conversion in the periodically executed program sequence.
16. Generally, installation guides are available in README and INSTALL files. For help with installation search Internet or visit http://www.physiol.med.uu.nl/dynaclamp/.
17. It is sufficient to store the A/D samples and the parameter settings. With these data, the experiment can be "replayed" using an "offline" version of the DynaClamp user program.
18. In both model cell and real cell modes, we assume that the defects associated with the R56Q mutation are limited to altered gating. Accordingly, we scaled WT and mutant input I_{HERG} to similar magnitudes. However, with other I_{HERG} mutants, the scaling procedure should only be decided after having proper information on basic electrophysiological characteristics of the expressed current. Nevertheless, different expression levels (resulting in different current densities) for WT and mutant I_{HERG} can easily be adopted during the scaling procedure.
19. The PB model of the human cardiac AP does not include a late component exhibited by WT and mutant Na channels. Description of currents in the PB model is based on results from voltage-clamp experiments, and the description of most membrane

current components, but not the sodium current, is based on quantitative data from human ventricular cells.

20. This efficient I_{bck}-subtraction method makes it possible to record I_{SCN5A} traces similar to what can be obtained in experiments where TTX is used to isolate I_{SCN5A}.
21. This is done to avoid any possible interference of late I_{SCN5A} for cases when I_{SCN5A} inactivation during the 20-s V_{hold} of −60 or −50 mV is incomplete.
22. Ca^{2+} loading of the myocytes exhibiting long APs in the presence of E-4031 (as in **Fig. 1E**) is likely. However, when scaled I_{HERG} is implemented, the AP shortens to its initial value (exhibiting similar APD as before the addition of E-4031).
23. As recourse, F_s value may need readjustment to obtain the APD_{90} of approximately 230 ms with the WT I_{HERG} input.
24. DynaClamp allows scaling of the input current to any desired magnitude and subtraction of artifacts (e.g., I_{bck}) before I_{HERG} is applied to the ventricular cell. I_{bck} subtraction, however, is not necessary as I_{HERG} downscaling already reduces endogenous currents to negligible levels.
25. Although much effort was expended on tuning for optimum noise and speed, our attempts to establish real cell mode dAPC with I_{Na} (*SCN5A* cDNA-transfected HEK cell coupled to a freshly isolated rabbit left ventricular myocyte) were not successful because of instabilities and oscillatory distortions of the upscaled and I_{bck}-subtracted input I_{Na}. The shortcomings were probably related to imperfect C_m and/or R_s compensation when coupling two voltage-clamp amplifiers and/or to a delay during the dAPC time steps, occurring even when using an Alembic VE-2 amplifier (Alembic Instruments, Canada) with state estimator R_s compensation. A major problem in connecting to a real ventricular cell is that fast activating and large amplitude (approximately 40–60 nA when up scaled) input I_{Na} might cause inhomogeneities of the membrane potential (V_m) of the real cell. Unfortunately, this cannot be improved with an expression system offering larger currents, because the larger the current, the more difficult to obtain reliable voltage control. A way-out would be to combine "conventional APC" to generate the upstroke and thereafter switching to dAPC with a slower, more stable feedback to study the I_{Na} during the plateau phase.
26. As in most cardiac cell models, autonomic regulation is not implemented in the PB ventricular cell model. Such regulation would not only cause frequency-dependent changes but also have additional effects on various ionic currents.

Acknowledgments

Our dAPC studies were supported by Netherlands Heart Foundation grant 2001B155.

References

1. Wehrens, X.H., Vos, M.A., Doevendans, P.A., and Wellens, H.J. (2002) Novel insights in the congenital long QT syndrome. *Ann Intern Med.* **137**, 981–992.
2. Berecki, G., Zegers, J.G., Verkerk, A.O., Bhuiyan, Z.A., de Jonge, B., Veldkamp, M.W., Wilders, R., van Ginneken, A.C. (2005) HERG channel (dys)function revealed by dynamic action potential clamp technique. *Biophys J.* **88**, 566–578.
3. Berecki, G., Zegers, J.G., Bhuiyan, Z.A., Verkerk, A.O., Wilders, R., and van Ginneken, A.C. (2006) Long-QT syndrome-related sodium channel mutations probed by the dynamic action potential clamp technique. *J Physiol.* **570**, 237–250.
4. Wilders, R. (2006) Dynamic clamp: a powerful tool in cardiac electrophysiology. *J Physiol.* **576**, 349–359.
5. Priebe, L. and Beuckelmann, D.J. (1998) Simulation study of cellular electric properties in heart failure. *Circ Res.* **82**, 1206–1223.
6. Johns, D.C., Nuss, H.B., and Marbán. E. (1997) Suppression of neuronal and cardiac transient outward currents by viral gene transfer of dominant-negative Kv4.2 constructs. *J Biol Chem.* **272**, 31598–31603.
7. Tytgat, J. (1994) How to isolate cardiac myocytes. *Cardiovasc Res.* 28, 280–283.
8. Hamill, O.P., Marty, A., Neher, E., Sakmann, B., and Sigworth, F.J. (1981) Improved patch-clamp techniques for high-resolution current recording from cells and cell-free membrane patches. *Pflügers Arch.* **391**, 85–100.
9. Sigworth, F.J. (1983) Electronic design of the patch clamp, in *Single-Channel Recording* (Sakmann, B. and Neher, E., eds.), Plenum Press, New York, pp. 3–35.
10. Barry, P.H. and Lynch, J.W. (1991) Liquid junction potentials and small cell effects in patch clamp analysis. *J Membr Biol.* **121**, 101–117.
11. Barabanov, M. and Yodaiken, V. (1997) Introducing real-time Linux. *Linux J.* **34**, 19–23.
12. Joyner, R.W., Wang, Y.-G., Wilders, R., Golod, D.A., Wagner, M.B., Kumar, R., and Goolsby, W. N. (2000) A spontaneously active focus drives a model atrial sheet more easily than a model ventricular sheet. *Am J Physiol Heart Circ Physiol.* **279**, H752–H763.
13. Rush, S. and Larsen, H. (1978) A practical algorithm for solving dynamic membrane equations. *IEEE Trans Biomed Eng.* **25**, 389–392.
14. Wilders, R., Verheijck, E.E., Kumar, R., Goolsby, W.N., van Ginneken, A.C.G., Joyner, R.W., and Jongsma, H. J. (1996) Model clamp and its application to synchronization of rabbit sinoatrial node cells. *Am J Physiol.* **271**, H2168–H2182.

III

Biophysics

17

Principles of Single-Channel Kinetic Analysis

Feng Qin

Summary

Single-channel recording provides molecular insights that are nearly unattainable from macroscopic measurements. Analysis of the data, however, has proven to be a difficult challenge. Early approach relies on the half-amplitude threshold detection to idealize the data into dwell-times, followed by fitting of the duration histograms to resolve the kinetics. More recent analysis exploits explicit modeling of both the channel and noise statistics to improve the idealization accuracy. The dwell-time fitting has also evolved into direct fitting of the dwell-time sequences using the full maximum likelihood approach while taking account of the effects of missed events. Finally, hidden Markov modeling provides a new paradigm in which both the amplitudes and kinetics can be analyzed simultaneously without the need of idealization. The progress in theory, along with the advance in computing power and the development of user-friendly software, has made single-channel analysis, once a specialty task, now readily accessible to a broader community of scientists.

Key Words: Ion channel, single channel, single molecule, gating, kinetics, dwell time, idealization, duration histogram, maximum likelihood, global fitting, missed events, segmental k-means, hidden Markov modeling.

1. Introduction

Patch-clamp recording is a primary tool for studying structures and functions of ion channels. It is among the few techniques that allow proteins to be detected at the single-molecule level. The measurement contains much information on the heterogeneity of channel conformations and their dynamic properties. The analysis of single-channel recordings thus provides a unique means for deciphering

From: *Methods in Molecular Biology, vol. 288: Patch-Clamp Methods and Protocols*
Edited by: P. Molnar and J. J. Hickman © Humana Press Inc., Totowa, NJ

the molecular mechanisms underlying channel functions. The dissection of the information requires statistical interpretation of the data. Here, we review the basic principles for analysis of single-channel measurements. The theory is also applicable to modeling of other single-molecule measurements.

2. Markov Model for Single-Channel Kinetics

A common feature of single-molecule measurements is that the activity occurs at discrete steps in a seemingly random manner. In the case of ion channels, the current alternates between "on" and "off" like a random telegraph signal. This type of molecular kinetics can be described by a Markov model *(1)*. The molecule is assumed to exist in a finite number of discrete states, and the transitions between states are governed by the first-order rate constants. The additional assumption that the rate constants are independent of time defines the kinetics as a time-homogeneous Markov process.

The first-order assumption of the transition rates defines the system to be stochastic in nature. It says that within an infinitesimal small time interval Δt, the transition of the system from one state to another follows a probability

$$\Pr[s(t+\Delta t)=j|s(t)=i]=k_{ij}\Delta t, \quad (1)$$

where $s(t)$ designates the state of the channel at time t. As we will see in the next section, this simple rule completely determines how the system evolves in time.

The Markov formalism of the single-molecule kinetics is essentially an extension of the classical chemical reaction mechanism, where a state corresponds to an energetically stable conformation of the molecule and the rate constants are determined by the activation energy or the height of the energy barrier separating the different conformations. Other schemes, such as diffusion models and fractal models, have also been proposed, but they have not proven useful in general *(2–5)*. The Markov scheme remains the most useful and powerful description for molecular kinetics.

The problem of single-channel analysis is to solve the inverse problem, that is, to identify an appropriate model to explain the experimental data. A major challenge to the problem is the limited observability of the system offered by the patch-clamp technique. That is, multiple conformational states may exists with the same conductance. Transitions between these states are not directly observable. Instead, they can only be inferred from the distinct statistical distributions of their durations. This aggregation of multiple states into a single observation class may cause inherent ambiguity in resolving their detailed kinetics *(6)*.

3. Properties of Markov Model

Two probabilities of a Markov model are particularly useful for modeling patch–clamp data. One is the transition probability among all possible states, and the other is the transition probability between a subset of states. Both probabilities can be derived directly from the first-order assumption of the Markov model (*see* **Box 1**). The transition probability between any two states is determined by

$$\mathbf{P}(t) = \exp(\mathbf{Q}t), \tag{2}$$

Box 1. Derivation of transition probability
Consider a two-state system with a starting probability $\pi = (\pi_1, \pi_2)$. The

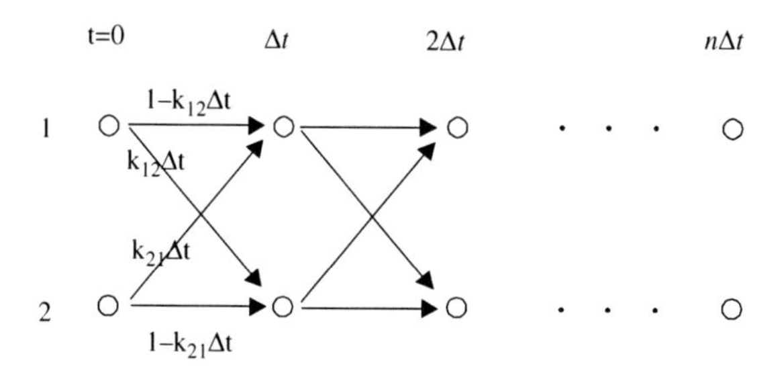

problem is to calculate the probability of the two states at an arbitrary time t. As the Markov assumptions only define transitions in an infinitesimal time, one can divide the interval $[0, t]$ into n infinitesimal ones, each with a duration Δt. The probability that the system is in state 1 after Δt is equal to $\pi_1 \cdot (1 - k_{12}\Delta t) + \pi_2 \cdot k_{21}\Delta t$, which is, the probability to remain in state 1 plus the probability to leave state 2 for state 1. Similarly, the probability being in state 2 after Δt is $\pi_1 \cdot k_{12}\Delta t + \pi_2 \cdot (1 - k_{21}\Delta t)$. In the matrix form, they can be written as

$$(\pi_1, \pi_2) \cdot \begin{pmatrix} 1 - k_{12}\Delta t & k_{12} \\ k_{21} & 1 - k_{21}\Delta t \end{pmatrix}.$$

Repeating the procedure for subsequent Δts leads to the probability at time $t = n\Delta t$:

$$(\pi_1, \pi_2) \cdot \begin{pmatrix} 1 - k_{12}\Delta t & k_{12} \\ k_{21} & 1 - k_{21}\Delta t \end{pmatrix}^n .$$

Recall the relationship $(1 + t/n)^n \rightarrow e^t$ as $n \rightarrow \infty$. The probability at time t can be reduced to $\pi \cdot \exp(\mathbf{Q}t)$, where $\mathbf{Q}$ is the matrix of rate constants defined in the text. The matrix exponential $\exp(\mathbf{Q}t)$ corresponds to the transition probability between states for time t.

where $\mathbf{P}(t)$ is a matrix whose (i,j)th element represents the probability that the channel is in state j at time t given that it starts from state i at the beginning, and $\mathbf{Q}$ is the rate constant matrix whose (i,j)th element, k_{ij}, corresponds to the transition rate from state i to state j (7). The diagonal elements of the matrix are defined by $k_{ii} = -\sum_{j \neq i} k_{ij}$, which is, the total rate of the probability flux leaving state i.

The transition probability within a subset of states defines the probability that the channel stays in the subset of states for a period of time without leaving it during the time. Our interest will be on the states of specific conductance. For simplicity of discussion, we will consider only binary channels. In this case, the transition probability is determined by

$$\begin{aligned} \mathbf{P}_c(t) &= \exp(\mathbf{Q}_{cc}t) \\ \mathbf{P}_o(t) &= \exp(\mathbf{Q}_{oo}t), \end{aligned} \tag{3}$$

where $\mathbf{P}_c(t)$ is a matrix whose (i,j)th element defines the probability that the channel starts from the ith closed state, stays closed for duration t, and ends at the jth closed state at time t, and $\mathbf{Q}_{cc}$ is the submatrix of $\mathbf{Q}$ containing the rate constants for the transitions between the closed states *(7)*. $\mathbf{P}_o(t)$ and $\mathbf{Q}_{oo}$ are defined similarly.

The matrix exponentials in the above equations can be formulated more explicitly as scalar functions of t. For example, by making use of spectral expansions, the $\mathbf{Q}$ matrix can be decomposed into

$$\mathbf{Q} = \lambda_1 \mathbf{A}_1 + \cdots \lambda_N \mathbf{A}_N,$$

where N is the number of states, λ_is are the eigenvalue of $\mathbf{Q}$, and $\mathbf{A}_i$s are the matrices multiplied from the corresponding left and right eigenvectors. The matrix exponential can be represented accordingly by

$$\mathbf{P}(t) = e^{\lambda_1 t}\mathbf{A}_1 + \ldots e^{\lambda_N t}\mathbf{A}_N. \tag{4}$$

The equation indicates that each element of the transition probability matrix is a sum of exponentials with as many components as the number of states. The spectral expansion generally requires the matrix to be diagonable. Fortunately, the rate constant matrix of a physical process usually satisfies the condition given the microscopic reversibility *(8)*.

4. Ensemble Currents

The probabilities described in **Eqs. 2 and 3**. form the basis for computing the probabilities of various measurable quantities. The occupancy probability of a channel, for example, determines the ensemble current recorded from a macro-patch or a whole cell. It can be formulated from the transition probability matrix $\mathbf{P}(t)$ as

$$\boldsymbol{p}(t) = \boldsymbol{\pi} \cdot \exp(\mathbf{Q}t), \tag{5}$$

where $\boldsymbol{\pi}$ is the row vector of the initial probability of each state and $\boldsymbol{p}(t)$ is the vector of the occupancy probability at time t. In other words, the occupancy probability of the channel at time t is its initial probability multiplied by the transition probability over the time t.

From the state occupancy probability, the current of an ensemble of n channels can be predicted by

$$i(t) = n \cdot \boldsymbol{p}(t) \cdot \boldsymbol{I}^{T}, \tag{6}$$

where $\boldsymbol{I} = (\boldsymbol{i}_1, \cdots, \boldsymbol{i}_N)$ is the vector of the unitary current of each state of the channel.

Equation 6 is the basis for fitting the time course of whole-cell recordings. The starting probability $\boldsymbol{\pi}$ corresponds to the equilibrium probability of the channel at the holding condition, which is determined by

$$\boldsymbol{\pi} \cdot \mathbf{Q}_h = 0, \tag{7}$$

where $\mathbf{Q}_h$ is the rate constant matrix at the holding condition. As the transition probability matrix has its elements in the exponential forms, the ensemble current $i(t)$ follows sums of exponentials too. The time constants of the current correspond to the reciprocals of the eigenvalues of the rate constant matrix $\mathbf{Q}$ at the test condition.

When the number of channels is limited, the random openings and closures of the individual channels produce measurable gating noise. At any time t, the variance of the current of a single channel is given by

$$\sum_{k=1}^{N} p_k(t) \cdot [i_k - \bar{i}(t)]^2,$$

where i_k is the unitary current of the kth state and $\bar{i}(t) = \boldsymbol{p}(t) \cdot \boldsymbol{I}^{\tau}$ is the mean current of the channel at time t. The variance of an ensemble of n channels is then equal to

$$\sigma^2(t) = \frac{n \cdot \boldsymbol{p}(t) \cdot \boldsymbol{I}_2^{\tau} - \bar{i}^2(t)}{n}, \qquad (8)$$

where $I_2 = \left(\boldsymbol{i}_1^2, \cdots, \boldsymbol{i}_N^2\right)$. When the channels are either all closed or all open, the variance vanishes. Therefore, if the channels are held at the resting state and then fully stimulated, the variance of the ensemble current will first increase with time and then decrease. The variance contains additional information on the kinetics of the channel and can be combined with the ensemble mean measurement to separate the unitary current and the number of channels in the patch.

In the special case where and the channel has two states (one closed and one open), **Eq. 8** is reduced to the familiar parabolic form

$$\sigma^2 = \frac{i \cdot \langle I \rangle - \langle I \rangle^2}{n}, \qquad (9)$$

where i is the unitary current of the opening and $< I >$ is the mean measurement. The variance has two zero-crossing points at $< I >= 0$ and $< I >= i \cdot n$ and a peak at $i \cdot n/2$. The slope of the curve at the first zero-crossing point determines the unitary current, whereas the peak position can be used in conjunction with the unitary current to estimate the number of channels.

In theory, the ensemble current contains information about the kinetics of a channel. In practice, the ensemble averaging reduces the heterogeneity of the kinetics of individual channels. The number of exponentials that is necessary to fit the time course of an ensemble recording is generally far less than the number of the kinetic components present in single-channel recordings. Thus, the ensemble measurements have a limited resolution on determining the kinetic details of a channel.

5. Dwell-Time Distributions (1D)

Single-channel recording reports directly the individual openings and closures of a channel. The duration of a dwell time at each conductance is a stochastic variable. It consists of three events: the channel enters the conductance at the beginning of the dwell time, remains in the conductance for duration t, and eventually exits the conductance after time t. The probability of observing a dwell time for duration t is thus the product of the probability of the three events, leading to

$$
\begin{aligned}
f_c(t) &= \boldsymbol{\pi}_c \mathbf{P}_{cc}(t)\mathbf{Q}_{co}\boldsymbol{1} \\
f_o(t) &= \boldsymbol{\pi}_o \mathbf{P}_{oo}(t)\mathbf{Q}_{oc}\boldsymbol{1},
\end{aligned}
\tag{10}
$$

where $\boldsymbol{\pi}$s collect the entry probability of the states in the corresponding conductance, $\mathbf{P}(t)$s are the transition probability matrix as previously defined, $\mathbf{Q}_{co}$ is the off-diagonal submatrix of $\mathbf{Q}$ corresponding to the transitions from closed states to open states, $\mathbf{Q}_{oc}$ from open states to closed states, and the $\boldsymbol{1}$ denotes the column vector of ones of appropriate length. The multiplication by the unit vector at the end accounts for the uncertainty of the destination state the channel is leaving for. Strictly speaking, $f(t)$s are probability distributions because $\mathbf{Q}_{co}$ or $\mathbf{Q}_{oc}$ represents the rates rather than the probabilities of the transitions. The latter involves the product of a rate with an infinitesimal time Δt during which the transition occurs.

The entry probability of the states in a conductance can be determined from the equilibrium probability of the channel as *(1)*

$$
\begin{aligned}
\boldsymbol{\pi}_c &= \frac{\boldsymbol{w}_o \mathbf{Q}_{oc}}{\boldsymbol{w}_o \mathbf{Q}_{oc}\boldsymbol{1}} \\
\boldsymbol{\pi}_o &= \frac{\boldsymbol{w}_c \mathbf{Q}_{co}}{\boldsymbol{w}_c \mathbf{Q}_{co}\boldsymbol{1}},
\end{aligned}
\tag{11}
$$

where $\boldsymbol{w}$s are the vectors of the equilibrium probabilities of the states at the corresponding conductance class. The denominators are simply the normalization factors so that the total probability is equal to 1. Alternatively, the entry probability can be determined by *(9)*

$$
\begin{aligned}
\boldsymbol{\pi}_c &= \boldsymbol{\pi}_o \mathbf{Q}_{oo}^{-1}\mathbf{Q}_{oc} \\
\boldsymbol{\pi}_o &= \boldsymbol{\pi}_c \mathbf{Q}_{cc}^{-1}\mathbf{Q}_{co},
\end{aligned}
\tag{12}
$$

where the inverse matrices on the right-hand sides can be considered as the integral of the corresponding transition probability matrix over time. Thus, the product in the right-hand side can be interpreted as the probability entering the conductance of interest irrespective of the duration of its previous stay in the other conductance. The probablities in **Eq. 12** define a set of homogeneous equations and can be solved in combination with the probability totality constraint using matrix singular value decompositions.

The formalisms in **Eq. 10** predict what the dwell-time distributions of a channel should be at each conductance if its gating follows a Markov process. By representing the matrix exponential with spectral expansions, the distributions may be put in the form of sums of exponentials:

$$\begin{aligned} f_c(t) &= \sum_{i=1}^{N_c} (\boldsymbol{\pi}_c \mathbf{A}_i^{(c)} \mathbf{Q}_{co} \boldsymbol{I}) e^{\lambda_i t} \\ f_o(s) &= \sum_{j=1}^{N_o} (\boldsymbol{\pi}_o \mathbf{A}_j^{(o)} \mathbf{Q}_{oc} \boldsymbol{I}) e^{\mu_j s}, \end{aligned} \tag{13}$$

where N_c is the number of closed states, N_o is the number of open states, λ_is are the eigenvalues of $\mathbf{Q}_{cc}$, μ_is are the eigenvalues of $\mathbf{Q}_{oo}$, and $\mathbf{A}_i$s are the spectral expansions of the submatrices. The products in the parentheses are the coefficients of the exponentials. The equations allow one to analyze the experimental dwell-time histograms in linear combinations of exponentials. The information available from the analysis includes the number of open and closed states of the channel and the time constants and the populations of the individual components.

Direct fitting of the dwell-time distributions with sums of exponentials has some limitations. First, the time constants and the coefficients of the exponentials are complicated functions of rate constants. Although they give rise to information on the components of the dwell times, their relation to the model parameters is difficult to interpret. Second, the data obtained under different experimental conditions cannot be combined together for analysis. A better approach is to fit the histograms with a model explicitly *(10)*. In this approach, the time constants and the coefficients of the exponentials are evaluated from the model parameters and the experimental variables. The rate constants of the model are directly optimized to best fit the histograms.

The dwell-time distribution functions in **Eq. 13** have a limited number of degrees of freedom. This places an upper limit on the complexity of a model that may be resolved by histogram fitting. The total number of free parameters involved in the two distributions is $(2N_c - 1) + (2N_o - 1) = 2(N - 1)$. This is exactly the complexity of a linear sequential model or a model with linear branches. Even so, not all models with this complexity can be resolved by histogram fitting. One example is illustrated in Scheme I, which is a model for the Ca^{2+}-activated K^+ channel ***(11)***. The two open states have different lifetimes, one 0.35 ms and the other 3 ms. Regardless of whether the short open state is connected to C_2 or C_3, the resultant model produces identical dwell-time distributions for both openings and closures ***(10,12)***. In the other words, the histogram fitting cannot uniquely determine such a branched model even though its complexity matches the number of degrees of freedom in the distribution functions.

$$\begin{array}{ccccc} C_1 & \underset{k_{21}}{\overset{k_{12}}{\rightleftharpoons}} & C_2 & \underset{k_{32}}{\overset{k_{23}}{\rightleftharpoons}} & C_3 \\ & & k_{42} \updownarrow k_{24} & & k_{53} \updownarrow k_{35} \\ & & O_4 & & O_5 \end{array}$$

(Scheme I)

6. Dwell-Time Distributions (2D)

For models containing branched openings, correlations arise between specific open and closed states. This correlation information is necessary to discriminate the connections between these states. Unfortunately, such information is not preserved in the binned 1D dwell-time histograms that treat the openings and closures independently. In order to exploit the information, a high-order dwell-time distribution has to be used. For binary channels, it has been shown that the 2D distributions of the successive opening and closure pairs contain adequate information on the correlations between open and closed states ***(8)***. Thus, fitting the 2D dwell-time distributions provides a complete approach to make use of all information of the data ***(10,12)***.

Similar to the derivation of the 1D dwell-time distributions, the 2D dwell-time distributions can be obtained as

$$\begin{aligned} f_{co}(t,s) &= \boldsymbol{\pi}_c \mathbf{P}_{cc}(t)\mathbf{Q}_{co}\mathbf{P}_{oo}(s)\mathbf{Q}_{oc}\boldsymbol{1} \\ f_{oc}(s,t) &= \boldsymbol{\pi}_o \mathbf{P}_{oo}(s)\mathbf{Q}_{oc}\mathbf{P}_{cc}(t)\mathbf{Q}_{co}\boldsymbol{1}, \end{aligned} \tag{14}$$

where $f_{co}(t, s)$ represents the probability distribution of observing a closed dwell time t followed by an open dwell time s and $f_{oc}(s, t)$ represents the distribution of an open duration s succeeded by a closed duration t. According to the equations, the distributions are determined by the product of the probability of entering the preceding conductance, the probability of the transitions between the states within the conductance, the rates of the transitions to the following conductance, the probability of the transitions within that conductance, and the rates leaving the conductance.

Substituting the matrix exponentials using spectral expansions, one can expand the 2D dwell-time distributions into

$$f_{co}(t, s) = \sum_{i=1}^{N_c}\sum_{j=1}^{N_o}(\boldsymbol{\pi}_c\mathbf{A}_i^{(c)}\mathbf{Q}_{co}\mathbf{A}_j^{(o)}\mathbf{Q}_{oc}\boldsymbol{1})e^{\lambda_i t+\mu_j s}$$
$$f_{oc}(s, t) = \sum_{i=1}^{N_c}\sum_{j=1}^{N_o}(\boldsymbol{\pi}_o\mathbf{A}_j^{(o)}\mathbf{Q}_{oc}\mathbf{A}_i^{(c)}\mathbf{Q}_{co}\boldsymbol{1})e^{\lambda_i t+\mu_j s} \quad (15)$$

The 2D dwell-time distributions are also combinations of exponential functions. Different from the 1D dwell-time distribution, however, the exponentials are now 2D, with one time constant for closures and the other for openings. These time constants are the same as those of the 1D dwell-time distributions. The coefficient of each 2D component in the distributions measures the volume occupied by the exponential. It is indicative of the strength of the coupling between the closed and open dwell times associated with the component.

Figure 1 illustrates the 2D dwell-time distribution predicted by the model in Scheme I ***(13)***. When a channel contains a single gateway state for opening, the 2D dwell-time distribution is simply the product of the two 1D dwell-time distributions. It has the same appearance on all sections cut at different values at each axis. On the contrary, the open dwell-time distribution adjacent to the long closures in **Fig. 1** exhibits double humps, whereas those adjacent to the short closures show a single sharp peak. Such dependences of the dwell-time distributions on the adjacent dwell times are indicative of the existence of multiple opening pathways of a channel. The analysis of the 2D dwell-time histograms can therefore reveal, for example, whether a channel is able to open when it is partially liganded. As expected, the fitting of the 2D histograms is able to discriminate the models of **Scheme I** that gave the same best fitting of the 1D histograms ***(12)***. Nevertheless, the false solution, which corresponds to the model with the short opening connected to C_2, remains on the surface of the likelihood as a local maxima ***(10)***.

The 2D dwell-time distributions contain a total of $2N_cN_o$ degrees of freedom ***(9)***. This implies that the 2D histogram fitting can only resolve models with up to N_cN_o connections between states. Because the 2D dwell-time distributions contain all the information of the data, this is also the limit of any other approach under a single experimental condition at the equilibrium. The limitation can be ascribed to the inability to observe the transitions between states of the same conductances. It can be shown that this number is equal to the degrees of the

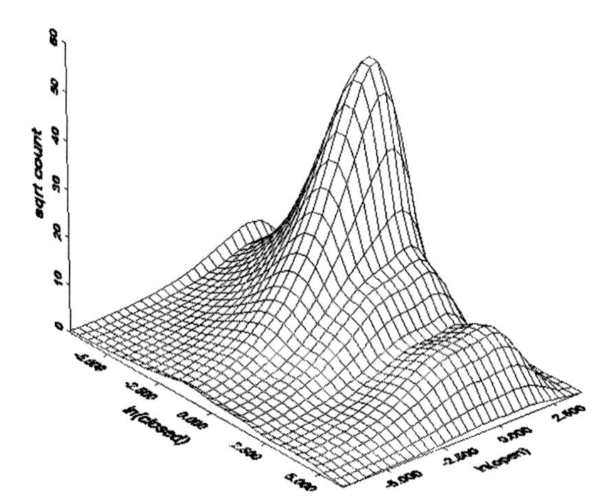

Fig. 1. The 2D dwell-time distribution predicted from **Scheme I** for Ca^{2+}-activated potassium channels. The existence of coupling between closed and open dwell times is evident from the different appearances of the 1D open dwell-time distributions in adjacency to different closed durations.

freedom of the 1D distributions if and only if $N_c = 1$ or $N_o = 1$. In the other words, the 1D dwell-time distributions contain all the information of the data only when the channel has a single open or closed state. Accordingly, the 1D histogram fitting is applicable only to models with a single open or closed state.

7. Maximum Likelihood Fitting

A problem inherent to the dwell-time histogram fitting is that the approach is applicable only to data obtained at equilibrium. The construction of 2D dwell-time histograms also requires a large number of events. An alternative approach is to fit the dwell-time sequence directly using the maximum likelihood estimation (***14–17***). The approach is applicable to both stationary and nonstationary data and allows for multiple conductance levels. The maximum likelihood estimates also have the theoretical advantages of being asymptotically unbiased with the minimum variance and approximate normal distributions as the sample size increases. These desirable properties make the maximum likelihood estimation the most robust parameter estimation technique in many applications.

Consider a dwell-time sequence $t = \{t_1 t_2 \cdots t_L\}$ with the corresponding conductance sequence $a = \{a_1 a_2 \cdots a_L\}$, where a_i is either closed or open. The probability density of the entire dwell-time sequence can be obtained by multiplying the entry probability, the transition probability within a dwell time, and the transition rates between dwell times. This leads to

$$f(t,a) = \boldsymbol{\pi}_{a_1} \left\{ \prod_{i=1}^{L} \left[P_{a_j a_i}(t_i) Q_{a_j a_{j+1}} \right] \right\} \boldsymbol{1}, \tag{16}$$

where a_{L+1} denotes the states not in a_L. When viewed as a function of the model, the probability distribution is also called the likelihood of the model, which is often designated as $L(\theta)$, where θ represents the model parameters. The likelihood value provides a natural measure on the goodness of a fit. The maximum likelihood estimation is then to seek the maxima of the likelihood function that defines the most likely estimates of the parameters.

The maximum likelihood concept can be illustrated using a simple two-state model. The likelihood function for a dwell-time sequence can be written as

$$L(k_{co}, k_{oc}) = \exp(-k_{oc}t_1) \cdot k_{oc} \cdot \exp(-k_{co}t_2) \cdot k_{co} \cdots \exp(-k_{oc}t_L) \cdot k_{oc},$$

where the first and the last dwell times are assumed to be open and k_{co} and k_{oc} are the opening and closing rate constants, respectively. Suppose the data are generated by a channel with comparable lifetimes on the closed and open states. Then, for a model with very different k_{co} and k_{oc}, its likelihood will be small. This occurs because of the mismatch between the model rate constants and the observed dwell-time durations in the exponents. The likelihood can be maximized by calculating the logarithm of the likelihood and setting its derivatives to zero. The maximum occurs at

$$k_{co}^{-1} = \frac{2\sum t_{2i}}{L}$$

$$k_{oc}^{-1} = \frac{2\sum t_{2i-1}}{L}$$

which are the same as the mean estimates of durations of the observed closed and open dwell times, respectively.

For more complicated models, the likelihood function has to be calculated numerically. One approach is to evaluate it recursively by taking advantage of the matrix product structure of the likelihood function *(18)*. Starting from the entry probability $\alpha_0^{\mathsf{T}} = \pi_{a_1}$, one then calculates

$$\alpha_k^{\mathsf{T}} = \alpha_{k-1}^{\mathsf{T}} \mathbf{P}_{a_k}(t_k) \mathbf{Q}_{a_k a_{k+1}}, \tag{17}$$

for $k = 1, 2, \cdots, L$. As the procedure progresses through each dwell time of the sequence, the probabilities of all partial dwell-time sequences, $\{t_1 \cdots t_k\}$, $k =$

$1 \cdots L$, are computed. The likelihood for the entire sequence is simply the sum of the components of the final probability vector, that is, $L(\theta) = \alpha_L^{\prime} \cdot \mathbf{1}$.

One potential problem with the previously described procedure is numeric overflow. For long data sequence or poor starting values of parameters, the likelihood value could become extremely small and exceed the machine precision range. The problem can be alleviated by adaptively scaling αs into proper numeric ranges and calculating the log likelihood instead of the likelihood itself. The procedure has an overall computational complexity on the order of N^2L, where N is the number of states and L is the number of dwell times. For models of moderate sizes, the procedure is generally fast, taking on the order of seconds. This efficiency renders the histogram-fitting techniques largely obsolete. The latter become advantageous only when the full likelihood approach becomes too slow, in which case the binning of the dwell times can significantly speed up the calculations.

8. Missed Events Correction

The theory described above assumes a perfect signal from a Markov model. In practice, the recording apparatus has a limited time response that causes some short events to inevitably go undetected. The missed events could introduce large errors in the detection of dwell-time durations. For example, if a channel contains long openings interrupted by brief closures, omissions of the short closures will prolong the apparent openings. Without correction for the effects of the missed events, the results from the dwell-time analysis will be inaccurate.

Because of their stochastic occurrences, the missed events cannot be recovered deterministically. Instead, they can only be taken into account statistically by correcting the probabilistic distribution for the apparent dwell times. The missed events are usually characterized by a fixed dead time, so that all events longer than the dead time are detected, and all shorter ones are missed. According to this criterion, an apparent dwell-time t may result from any dwell-time sequence $\{\tau_i, i = 1 \cdots n\}$ that satisfies the following conditions:

1. The first dwell time has the same conductance as the apparent dwell time, and its duration is $\tau_1 \geq t_d$.
2. The departures to the other conductance are shorter than the dead time, that is, $\tau_{2i} \leq t_d$ for all i.
3. The total duration of all dwell times of the sequence is equal to the duration of the apparent dwell time, that is, $\sum \tau_i = t$.

The transition probability matrix of the apparent dwell time is then the sum of the probability of all such dwell-time sequences. The final solution can be written analytically in the Fourier domain as *(19)*

$$
\begin{aligned}
{}^{e}P_{c}(t) &= -\frac{1}{2\pi} e^{\mathbf{Q}_{cc} t_d} \int_{-\infty}^{\infty} e^{-j\omega(t-t_d)} \left[j\omega \boldsymbol{I} + \mathbf{Q}_{cc} - \mathbf{Q}_{co} \left(\boldsymbol{I} - e^{(j\omega \boldsymbol{I} + \mathbf{Q}_{oo}) t_d} \right) (j\omega \boldsymbol{I} + \mathbf{Q}_{oo})^{-1} \mathbf{Q}_{oc} \right]^{-1} d\omega \\
{}^{e}P_{o}(t) &= -\frac{1}{2\pi} e^{\mathbf{Q}_{cc} t_d} \int_{-\infty}^{\infty} e^{\,j\omega(t-t_d)} \left[j\omega \boldsymbol{I} + \mathbf{Q}_{cc} - \mathbf{Q}_{co} \left(\boldsymbol{I} - e^{(j\omega \boldsymbol{I} + \mathbf{Q}_{oo}) t_d} \right) (j\omega \boldsymbol{I} + \mathbf{Q}_{oo})^{-1} \mathbf{Q}_{oc} \right]^{-1} d\omega
\end{aligned}
\tag{18}
$$

These are the equivalents of the transition probability matrices defined in **Eq. 3** in the absence of missed events. Substituting them into the various dwell-time distributions results in corrections of these distributions for the effects of missed events.

The exact solutions in **Eq. 18** are difficult to evaluate because of the appearance of the inversions of matrix functions in the integrand. Various approximate solutions have been proposed *(16,19–22)*. One simple solution is the first-order approximation:

$$
\begin{aligned}
{}^{e}P_{c}(t) &= \exp({}^{e}Q_{cc} t) \\
{}^{e}P_{o}(t) &= \exp({}^{e}Q_{oo} t),
\end{aligned}
\tag{19}
$$

where ${}^{e}Q_{cc}$ and ${}^{e}Q_{oo}$ are defined by

$$
\begin{aligned}
{}^{e}Q_{cc} &= \mathbf{Q}_{cc} - \mathbf{Q}_{co} \left(\boldsymbol{I} - e^{\mathbf{Q}_{oo} t_d} \right) \mathbf{Q}_{oo}^{-1} \mathbf{Q}_{oc} + t_d \mathbf{Q}_{co} \left(\boldsymbol{I} - e^{\mathbf{Q}_{oo} t_d} \right) \mathbf{Q}_{oo}^{-1} \mathbf{Q}_{oc} \\
{}^{e}Q_{oo} &= \mathbf{Q}_{oo} - \mathbf{Q}_{oc} \left(\boldsymbol{I} - e^{\mathbf{Q}_{cc} t_d} \right) \mathbf{Q}_{cc}^{-1} \mathbf{Q}_{co} + t_d \mathbf{Q}_{oc} \left(\boldsymbol{I} - e^{\mathbf{Q}_{cc} t_d} \right) \mathbf{Q}_{cc}^{-1} \mathbf{Q}_{co}
\end{aligned}
\tag{20}
$$

The corrected transition probability matrix has the same form as in the ideal case. The only change is the replacement of the original **Q** matrix with the corrected one given in **Eq. 20**. The first-order approximation is obtained under the assumption that the total missed event durations are negligible compared to the apparent duration. There are other solutions that are less sensitive to the assumption *(21,22)*. It has also been shown that the exact solutions could be evaluated piecewisely by dividing an apparent dwell time into multiples of the dead time *(22)*. The advantage of the first-order approximation is the simplicity of its evaluations and the availability of its analytical derivatives to rate constants, which are useful for the optimization of model parameters. It can also be easily generalized to channels with multiple conductance levels *(16)*.

9. Practical Issues of Analysis

The theory described **above** explains the basic principles of analysis of single-channel currents. There are many issues concerning the use of the theory in practice. Some typical ones include, for example, how to optimize the likelihood, how to allow for simultaneous fitting of data from multiple conditions, how to impose constraints on rate constants, and how to test whether a model is indeed optimal.

9.1. Global Fitting

The rate constants of a biological channel are often scaled by experimental conditions. To fit data across multiple conditions, it is necessary to represent the rate constants in terms of some intrinsic parameters that do not vary with the conditions. For voltage-gated channels, the rates typically depend on voltage in an exponential manner. The same is true for force in mechanosensitive channels. For ligand-gated channels, the rates change linearly with ligand concentration. A unified representation of these dependencies may be written as

$$q_{ij}=C_{ij}\exp(\mu_{ij}+\nu_{ij}V), \tag{21}$$

where C_{ij} is the concentration of the drug that the rate k_{ij} is sensitive to and V is the membrane potential or other global variables such as force. Each rate constant is represented by two variables: μ_{ij} and ν_{ij}. For the rates that are independent of concentration, one can set $C_{ij}=1$. Similarly, for the rates that are independent of voltage, $\nu_{ij}=0$. The parameters μ_{ij} and ν_{ij} do not vary with the experimental conditions. They can be chosen as the intrinsic parameters of the channel for fitting.

The parameterization of a rate constant in terms of μ_{ij} and ν_{ij} automatically restricts the rate constant to be nonnegative. It also has the advantage to facilitate the handling of the detailed balance constraints. These constraints are highly nonlinear in q_{ij} but become linear with μ_{ij} and ν_{ij}. As described **below**, the linear constraints can be taken into account explicitly during optimization.

9.2. Constraints

Typical constraints imposed on rate constants include holding some rate constants at fixed values, linear scaling by other rate constants, and detailed balance conditions. All these constraints can be represented as linear equations in terms of variables μ_{ij} and ν_{ij}, leading to

$$\Gamma\boldsymbol{\theta}=\xi, \tag{22}$$

where $\boldsymbol{\theta} = (\cdots, \mu_{ij}, \nu_{ij}, \cdots)^{\tau}$ is the vector of variables, $\boldsymbol{\Gamma}$ is the coefficient matrix, and $\boldsymbol{\xi}$ is the constant vector. If the imposed constraints are mutually independent, $\boldsymbol{\Gamma}$ has a full rank.

The inclusion of constraints into optimization leads to constrained optimization, which is generally difficult to solve. For linear equality constraints, however, it is possible to eliminate them by reducing the degree of freedom of the problem. That is, one can formulate the linearly constrained variables into linear combinations of a set of unconstrained variables,

$$\theta = \mathbf{A} \cdot \mathbf{x} + \mathbf{b}, \tag{23}$$

where $\mathbf{A}$ is a constant matrix and $\mathbf{b}$ is a constant vector, both of which can be readily determined from the matrix $\boldsymbol{\Gamma}$ and the vector $\boldsymbol{\xi}$ in **Eq. 22** using standard matrix decomposition techniques ***(16)***.

For any value of $\mathbf{x}$, the $\boldsymbol{\theta}$ given by **Eq. 23** automatically satisfies the constraints in **Eq. 22**. Thus, one can choose $\mathbf{x}$ as the free variable to optimize. By doing so, the optimization subject to the constraints in **Eq. 22** is reduced to an equivalent unconstrained optimization problem, which can be solved by the conventional optimization techniques.

9.3. Derivatives

For efficient optimization and accurate estimation of model parameters, it is necessary to have gradient information of the objective functions. There are optimization algorithms that require only function evaluations, such as the downhill simplex method and the Powell's direction set method. These derivative-free approaches are usually not efficient for complex functions, such as the likelihood of dwell-time sequences. They are also limited to applications with fewer parameters. Analytical derivatives have been derived for the 1D and 2D dwell-time distributions ***(10)*** and the full likelihood function ***(16,18)***. One central step of the derivation is the calculation of the derivatives of a matrix exponential that follows ***(15)***

$$\frac{\partial}{\partial\theta}\mathbf{P}_{cc}(t) = \sum_{i=1}^{n_c}\sum_{j=1}^{n_c}\left(\mathbf{A}_i^{(c)}\frac{\partial\mathbf{Q}_{cc}}{\partial\theta}\mathbf{A}_j^{(c)}\right)\gamma(\lambda_i, \lambda_j; t), \tag{24}$$

where γ is a scalar function determined by the eigenvalues. The derivatives of the dwell-time distributions can then be determined by applying **Eq. 23** to the transition probability matrices of the individual dwell times using the chain rule. In practice, the calculation of the derivatives of an objective function involves the following steps: the derivatives of the parameters $\boldsymbol{\theta}$ with respect to the free variables $\mathbf{x}$, the derivatives of the $\mathbf{Q}$ matrix to the parameters $\boldsymbol{\theta}$, the derivatives of the corrected $\mathbf{Q}^e$ matrix to the elements of $\mathbf{Q}$, and the derivatives of the dwell-time distributions or the likelihood function with respect to the elements of $\mathbf{Q}^e$. The final derivatives of the objective function with respect to the free variables $\mathbf{x}$ are determined by combining all intermediate derivatives with the chain rule.

9.4. Optimization

There are three types of gradient-based approaches for unconstrained optimization: the steepest descent method, the conjugate directions method, and the variable metric method *(23)*. Among them, the variable metric method generally performs well for fitting dwell-time distributions or the likelihood function. The steepest descent method has a poor performance on the convergence near the optimal solution. The conjugate directions method requires an exact line search at every iteration, which generally takes many function evaluations. The variable metric method, on the other hand, has a quadratic convergence around the optimal point. It also exploits approximate line search for step size determination. The search direction resulting from the parabola interpolation is often sufficiently accurate so that only a few functional evaluations are needed for each line search. Thus, the method is particularly suitable to functions whose evaluation requires extensive computing. The method does not need the exact Hessian matrix of the objective function. Instead, it builds up the curvature information from successive parabola interpolations of the objective function along with its first-order gradients.

9.5. Error Estimates

The uncertainties on the fitted parameters can be estimated from the curvature of the likelihood function at its maximum:

$$\mathrm{Cov}(\mathbf{x}) \approx -\mathrm{H}^{-1}(\mathbf{x}), \tag{25}$$

where Cov is the covariance matrix of the parameters and $\mathbf{H}$ is the Hessian matrix of the second derivatives. The exact Hessian matrix could be calculated numerically but involves intensive computing. A faster, although less accurate, solution is the approximate inverse Hessian matrix generated by the variable metric optimizer. For a quadratic function of n variables, the matrices generated by the method converge to the correct inverse Hessian matrix after n steps.

For a nonquadratic function, this no longer holds. However, in the vicinity of a maximum point, the likelihood function may be well approximated by a quadratic function. Thus, it can still be expected to provide information on the local Hessian at the point.

The error estimates on the free variables $\mathbf{x}$ can be converted to the error estimates of the rate constants. The two are related by *(24)*:

$$\mathrm{Var}(q_{ij}) \approx \sum_{k} \left(\frac{\partial q_{ij}}{\partial x_k} \right)^2 \cdot \mathrm{Var}(x_k), \tag{26}$$

where Var denotes variance and $\mathrm{Var}(x_k)$ corresponds to the diagonal elements of the covariance matrix obtained in **Eq. 25**.

9.6. Multiple Channels

Often, there will be more than one channel in the recording. If the constituent channels are statistically independent, the data of multiple channels will still be a Markov process. Suppose the patch contains n channels and the ith individual channel has n_i kinetic states. The patch at any time is then fully described by a vector:

$$\mathbf{s}_t = \left(\mathbf{s}_t^{(1)}, \mathbf{s}_t^{(2)}, \cdots, \mathbf{s}_t^{(n)} \right), \tag{27}$$

where $s_t^{(i)}$ specifies which state the ith channel is in at time t *(14)*. The state space of $\mathbf{s}_t$ has a dimension of $n_1 n_2 \cdots n_n$. The transition rate of $\mathbf{s}_t$ from state $(i_1, \ldots, i_l, \ldots i_n)$ to state $(i_1, \ldots, j_l, \ldots i_n)$ is equal to the rate from i_l to j_l of the lth channel. In other words, the data of multiple channels can be considered as if generated from a single channel in which the states are defined as vectors and the transition rates are constrained to be those of the constituent channels. The theory developed above can still be applied to these data by operating on the induced Markov model for $\mathbf{s}_t$ and by imposing appropriate constraints on its rate constants. All constraints are of the linear scaling type and are therefore accommodated by the linear equality constraints.

The vector state defined in **Eq. 27** is applicable to both identical and nonidentical channels. For identical channels, there exists a more efficient definition by exploiting the indistinguishability of the constituent channels, namely,

$$\boldsymbol{n}_t = \left(\boldsymbol{n}_t^{(1)}, \boldsymbol{n}_t^{(2)}, \cdots \boldsymbol{n}_t^{(k)} \right), \tag{28}$$

where $n_t^{(i)}$ represents the number of channels in state i at time t *(14)*. The state space of n_t comprises all possible compositions of n into k parts. So the dimension of the state space is given by the binomial coefficient $\binom{n+k-1}{n}$. The rate constant for the transition from state $(m_1 \cdots m_i \cdots m_j \cdots m_k)$ to state $(m_1 \cdots m_i - 1 \cdots m_j + 1 \cdots m_k)$ is equal to $m_i q_{ij}$, where q_{ij}s are the rate constant of the constituent channel.

The composition vectors $\boldsymbol{n}_t$ for given n and k may be generated automatically by a standard combinatorial analysis algorithm *(25)*. In general, $\binom{n+k-1}{n}$ is much less than k^n. So the state space of $\boldsymbol{n}_t$ is much smaller than that of $\boldsymbol{s}_t$. Considering that the evaluation of a dwell-time distribution has a complexity at least quadratic on the number of states, a considerable reduction in both processing time and memory space may be expected by using $\boldsymbol{n}_t$ for modeling multiple identical channels.

9.7. Optimality

For maximum likelihood fitting, regardless of the model used, the fitting always produces a set of parameters. The question is whether the resultant model has the right features of the data. There are two possible scenarios: the model is overly complex or the model is inadequate. When a model is overly complex, the resultant fit often contains rate constants that are either very small or very large. In either case, it signals that the data do not contain sufficient information on the corresponding transitions. On the other hand, when a model is invalid or too simplistic, the predicted distributions may become significantly deviated from the experimental histograms. In this case, the superposition of the fitted dwell-time distributions with the histograms provides a visual clue on whether the fit is inadequate. However, it should be cautious that a good fit of the histograms does not imply the correctness of the model as the 1D dwell-time histograms do not contain all information of the data. A more sensitive test is to calculate the maximum attainable likelihood, that is, the likelihood that would be obtained when the true model is used. This true maximum likelihood can be obtained by fitting data with a fully connected and uncoupled model as illustrated in **Scheme II**.

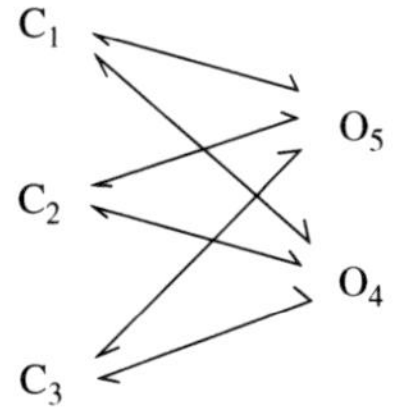

C_1
O_5
C_2
O_4
C_3

(Scheme II)

The model has $2N_cN_o$ parameters that coincide with the maximum number of degrees of freedom that can be resolved from the data for the given number of states *(8)*. Because the open and closed states are uncoupled, the model is fully identifiable *(6)*. If the likelihood value of the test model is significantly less than that of this uncoupled model, it would suggest that the test model does not have adequate capacity to account for the data. For fitting multiple data sets over different conditions, one may calculate the maximum log likelihood with the uncoupled model for each condition. The maximum log likelihood over the entire data sets can be obtained as the sum of the maximum log likelihood over the individual data sets.

10. Hidden Markov Model

Traditionally, single-channel currents are analyzed by first idealizing the records into dwell times, followed by fitting the dwell times with appropriate models. The idealization is typically performed by threshold detection. The technique relies on heavy filtering, which results in band-limiting distortions. In some extreme cases, for example, when the current is too small or the kinetics too fast, a reliable idealization of the data may become impossible. For these problems, the hidden Markov modeling provides an alternative means for analysis of data. It can be used for both idealization *(26)* and direct estimation of model parameters (*27–32*). Furthermore, it does not have the missed events problem.

A hidden Markov model refers to a Markov process whose states are not directly observable, such as a Markov process embedded in noise *(33)*. When the noise is high, the observations cannot be associated with states unambiguously. Instead, they can only be characterized probabilistically by

$$b_i(y_t) = \Pr(y_t | s_t = i),$$

which is the probability of observing a sample y_t given that the underlying Markov process s_t is at state i. In the case of white noise, this observation probability is determined by the Gaussian distribution:

$$b_i(t) = \frac{1}{\sqrt{2\pi}\sigma_\mu} \exp\left(-\frac{(y_t - I_\mu)^2}{2\sigma_\mu^2} \right) \tag{29}$$

where μ designates the conductance class that state i belongs to, σ_μ^2 is noise variance at the corresponding conductance level, and I_μ is the unitary current.

Because of the presence of noise, the data need to be modeled in discrete samples instead of continuous dwell times. For a sampled Markov process, the transitions between states can be more conveniently characterized by the transition probabilities between adjacent samples that are collectively represented by a matrix $\mathbf{A} = [a_{ij}]$, where a_{ij} defines the probability of the channel being in state j at the next sampling clock given that it was in state i at the previous sampling clock. The transition probability matrix $\mathbf{A}$ is related to the infinitesimal rate constant matrix $\mathbf{Q}$ by

$$\mathbf{A} = \exp(\mathbf{Q}\Delta t), \tag{30}$$

where Δt is the sampling interval. The parameters of a hidden Markov model thus consists of the transition probabilities a_{ij}s, the current amplitudes I_μs, and the noise variances σ_μ^2, which are collectively designated by $\lambda = \{\mathbf{A}, \boldsymbol{I}_\mu, \sigma_\mu^2\}$.

Patch-clamp recording contains noise from a variety of sources. In general, the noise is not white. Furthermore, the data is filtered. The filtering introduces correlations in both signal and noise. The standard hidden Markov model can be extended to accommodate these distortions. For the illustration of concepts, we will restrict ourselves to the case of an ideal Markov signal in white noise. For generalization to filtered signal or correlated noise *see* **refs.** ***28–30,32***.

11. Restoration of Single-Channel Currents

One application of hidden Markov modeling for single-channel analysis is the detection of dwell times. In particular, it allows one to consider an idealization that is optimal in some statistical sense. One such criterion is to seek a state sequence, $\mathbf{s} = s_1 s_2 \ldots s_T$, to maximize the a posteriori probability ***(34)***,

$$\Pr(s_1 s_2 \ldots s_T;\, y_1 y_2 \ldots y_T;\, \lambda\} = \max, \tag{31}$$

where $\boldsymbol{y} = y_1 y_2 \ldots y_T$ is the sequence of the observation samples and λ designates the underlying hidden Markov model. The solution satisfying the criterion represents the most likely idealization of the observations for the given model. Accordingly, the probability can be viewed as the likelihood of the idealization.

By Bayes laws, the probability in **Eq. 31** can be cast into the probability of the state sequence itself, multiplied by the probability of observing the samples given the state sequence, that is,

$$\Pr(s_1 s_2 \ldots s_T;\, y_1 y_2 \ldots y_T;\, \lambda\} = \Pr(s_1 s_2 \ldots s_T;\, \lambda\} \cdot \Pr(y_1 y_2 \ldots y_T | s_1 s_2 \ldots s_T;\, \lambda\}.$$

The probability of the state sequence is determined by the initial probability of the states π_is, multiplied by the transition probabilities through the subsequent samples

$$\Pr(s_1 s_2 \ldots s_T; \lambda\} = \Pr(s_1) \cdot \Pr(s_2|s_1) \ldots \Pr(s_T|s_{T-1}) = \Pi_{s_1} \cdot a_{s_1 s_2} a_{s_2 s_3} \cdots a_{s_{T-1} s_T} \cdot$$

The probability of the observed samples given the state sequence is simply the product of the observation probability of individual samples

$$\Pr(y_1 y_2 \ldots y_T|s_1 s_2 \ldots s_T) = \Pr(y_1|s_1) \ldots \Pr(y_T|s_T) = b_{s_1}(y_1) b_{s_2}(y_2) \ldots b_{s_T}(y_T).$$

The likelihood function in **Eq. 31** can be represented by

$$\Pr(s, y; \lambda) = \Pi_{s_1} a_{s_1 s_2} \cdots a_{s_{T-1} s_T} \times b_{s_1}(y_1) b_{s_2}(y_2) \ldots b_{s_T}(y_T).$$

The problem of the idealization is then to choose among all possible choices a state sequence **s** and a set of model parameters λ so that the probability is maximal.

A straightforward solution to the problem is to enumerate all state sequences and determine the one that gives the maximal probability. Unfortunately, the strategy is unrealistic. Even in the simplest case with two states, there are 2^T sequences for T samples. For a small number of 100 samples, this will result in approximately 10^{30} state sequences. A simple enumeration of all the sequences, even without calculating their probability, would take greater than 10^{23} years on a computer operating at 1 GHz.

The exhaustive search fails because it evaluates each state at each sample point repetitively for different state sequences. An approach that can avoid the problem is the Viterbi algorithm based on dynamic programming *(35)*. The algorithm is recursive and proceeds as follows. Let $\phi_1(i) = \pi_i b_i(y_1)$ for $1 \leq i \leq N$. Then the following recursion for $2 \leq t \leq T$ and $1 \leq j \leq N$:

$$\phi_t(j) = \max_{1 \leq i \leq N} \{\phi_{t-1}(i) a_{ij} b_j(y_t)\} \tag{32a}$$

and

$$\psi_t(j) = i*, \tag{32b}$$

where i^* is a choice of an index i that maximizes $\phi_t(i)$. Upon termination, the likelihood is given by $\Pr(s, y; \lambda) = \max_{1 \leq i \leq N} \phi_t(i)$. The most likely state sequence can be recovered from ψ as follows. Let $s_T = i^*$, which maximizes $\phi_T(i)$. Then for $T \geq t \geq 2$, $s_{t-1} = \psi_t(s_t)$.

The basic idea of the Viterbi algorithm is schematically illustrated in **Fig. 2**. It performs the idealization through time successively. At each time t, it keeps track of the optimal state sequences (pathways) leading to all possible states at that point. Then, an optimal sequence up to the next time $t+1$ is constructed by examining all existing N sequences up to time t in combination with an appropriate transition from time t to $t+1$. Because the probabilities of the state sequences up to time t are remembered, the construction of the new extended sequences requires only N^2 computations, as implied by **Eq. 32a**. The idealization of the entire data set then takes on the order of N^2T operations, which is quadratic on the number of states and linear on the number of samples, as opposed to the exponential dependence required by an exhaustive search.

The result of the Viterbi detection is optimal relative to the model used. In practice, the model is unknown priori to analysis. As a result, the model parameters need to be estimated. Given an idealization, the estimation can be done empirically. To estimate the current amplitudes and noise variances, one can classify the samples into clusters according to their conductance. The

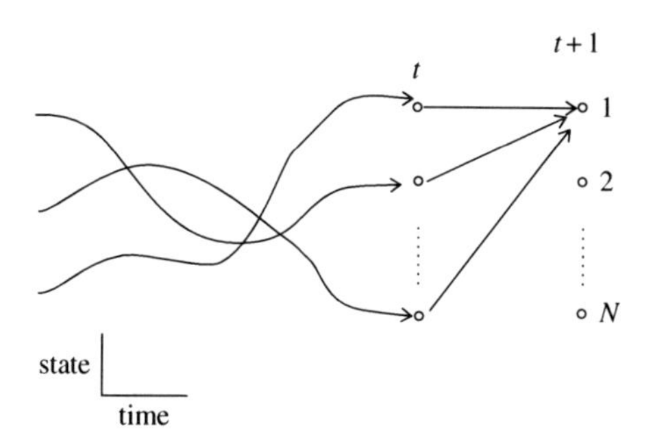

Fig. 2. Illustration of the Viterbi algorithm. The optimal sequence leading to a given state at time $t+1$ can be constructed from the sequences up to time t combined with a single transition from the previous ending state at time t to the given state at time $t+1$.

current amplitudes and noise variances can then be estimated as the means and variances of the samples within each cluster, respectively,

$$\hat{I}_\mu = \frac{\sum_{s_t \in C_\mu} y_t}{\sum_{s_t \in C_\mu} 1} \tag{33a}$$

$$\hat{\sigma}_\mu^2 = \frac{\sum_{S_t \in C_\mu} (y_t - \hat{I}_\mu)^2}{\sum_{S_t \in C_\mu} 1}, \tag{33b}$$

where $\mathcal{C}_\mu$ denotes the states of the μth conductance, so the denominator corresponds to the number of samples that are idealized into $\mathcal{C}_\mu$. Similarly, the transition probability can be estimated by counting the number of transitions occurring from each state:

$$\hat{a}_{ij} = \frac{\boldsymbol{n}(i, j)}{\boldsymbol{n}(i)}, \tag{33c}$$

where $\boldsymbol{n}(i)$ is the number of occurrences of state i over the entire sample sequence and $\boldsymbol{n}(i, j)$ is the number of occurrences that state j is an immediate successor of state i.

Ideally, the new estimates of the model parameters should agree with those that initiate the idealization. When the model is unknown, however, they may not be equal, in which case the estimates can be used to upgrade the model. This leads to an iterative loop, where an initial model, λ_0, is chosen, and the Viterbi algorithm is used to find an optimal idealization, from which the model parameters are reestimated. The iteration continues until it converges, for example, when the difference of the parameter values in two consecutive iterations becomes less than a preset small tolerance. This is the essence of the segmental k-means method ***(34)***.

Experimental testing indicates that the approach has a high tolerance for noise ***(25)***. Reliable idealization can be achieved at a noise-to-signal ratio as high as 2:1. It also provides good estimates on amplitude parameters including the unitary currents and the variances of noise. The approach, however, shows a poor performance on kinetics parameters. When there are multiple states with the same conductance, it generally cannot resolve the transitions between these states. This occurs because the estimation is based on only a single state sequence. A reliable estimation of kinetics requires the use of the full likelihood approach that examines all possible state sequences.

The insensitivity of the approach to kinetic parameters, on the other hand, provides ease for selection of models. Data generated by complex models can often be idealized accurately with a two-state model in which each state represents a conductance level. More states become necessary when the channel contains dwell times that are orders of magnitude different in durations. Under such conditions, introducing a new aggregated state can improve the detection of fast transitions. In practice, an adequate model can be obtained retrospectively. The data can be first idealized with a two-state model. Then the resultant dwell-time distributions can be explored for additional components. Once the number of components is determined, a fully connected and uncoupled model (**Scheme II**), which attains the maximal complexity for the given number of components, can be used for full idealization.

12. Direct Estimation of Kinetics From Single-Channel Recordings

The single-channel analysis based on idealized dwell times has the advantage of being computationally fast. This occurs because there are fewer dwell times than samples. The hidden Markov modeling, on the other hand, provides a general paradigm allowing estimation of kinetics directly from single-channel recordings. The concept is similar to maximum likelihood fitting of dwell-time sequences. That is, a model is determined to maximize the probability of the observations

$$L(\lambda) = \Pr(y_1 y_2 \ldots y_T | \lambda), \tag{34}$$

where λ represents the underlying hidden Markov model and $\boldsymbol{y} = y_1 y_2 \ldots y_T$, the observed samples. In the absence of noise, the observations reduce to dwell times. So the maximum likelihood analysis of dwell times can be considered as a special case of hidden Markov modeling.

12.1. Likelihood Calculation

As with most applications of maximum likelihood estimation, the core of the problem is how to evaluate the likelihood function and how to search for its maximum *(36)*. Conceptually, the probability of the observations is determined by the sum of the a posteriori probability in **Eq. 31** over all state sequences. This can be equivalently written in the matrix form as

$$L(\lambda) = \pi^{\tau}\mathbf{B}_1 \cdot \mathbf{A}\mathbf{B}_2 \cdots \mathbf{A}\mathbf{B}_T \cdot 1, \tag{35}$$

where $\pi = (\pi_1, \cdots, \pi_N)^{\tau}$ is the starting probability, $\mathbf{A} = [a_{ij}]_{N\times N}$ is the transition probability matrix, and $\mathbf{B}_t = \mathrm{diag}[b_1(y_t), \cdots, b_N(y_t)]$ is a diagonal matrix of the observation distributions for each state at time t. The multiplication of the matrices through samples effectively sums the probability over all possible combinations of states through the entire sequence of samples. Intuitively, the equation says that the likelihood is equal to the initial probability of entering the states, multiplied by the probability of observing the first sample, and then multiplied by the probability to make a transition to the next sample, followed by the probability of observing the second sample, and so on.

The matrix product form of the likelihood function suggests that the likelihood can be computed recursively. This leads to the so-called forward–backward procedure *(33)*. Let $\alpha_t(i) = \Pr(Y_1 \cdots Y_t, s_t = i),\ 1 \le i \le N,\ 1 \le t \le T$ and $\beta_t(j) = \Pr(Y_{t+1} \cdots Y_T,\ s_t = j),\ 1 \le j \le N,\ 1 \le t \le T$ denote the forward and backward variables, respectively, where $\alpha_t(i)$ is the joint probability of the partial observation sequence up to time t assuming the channel is in state i at time t and $\beta_t(j)$ is the joint probability of the complement observations from the last sample back to time t assuming the channel is in state j at time t. Essentially, they are the elements of the forward and backward partial products of **Eq. 35** up to time t, respectively. Thus, the forward and backward variables can be formulated recursively as

$$\alpha_{t+1}(j) = \sum_{i=1}^{N} \alpha_t(i) a_{ij} b_j(y_{t+1}), \quad j = 1 \ldots N, \quad t = 1 \ldots T$$

$$\beta_t(i) = \sum_{j=1}^{N} a_{ij} \beta_{t+1}(j) b_j(y_{t+1}), \quad i = 1 \ldots N, \quad t = T \ldots 1, \qquad (36)$$

where the initial conditions $\alpha_0(i) = \pi_i$ and $\beta_T(j) = 1$. These equations imply that the forward and backward variables can be calculated recursively.

From the forward and backward variables, the likelihood function is given by

$$\Pr(Y) = \sum_{i=1}^{N} \alpha_t(i) \beta_t(i). \qquad (37)$$

The equality holds for every $t = 1 \ldots T$. In particular, taking $t = T$, one obtains the likelihood as the sum of the last forward variables over all states. In other words, the forward variables alone suffice for evaluating the likelihood. But as we will see later, the backward variables are needed for reestimation of parameters or calculation of the derivatives of the likelihood function.

12.2. Baum–Welch Reestimation

The standard hidden Markov modeling relies on Baum's reestimation to estimate model parameters *(36)*. The algorithm is a precursor of the more general expectation-maximization (EM) approach for maximum likelihood estimation, which states that the maximum likelihood estimates of the model parameters can be constructed iteratively from their expected values based on their current estimates ***(37)***. For a hidden Markov model, there are two fundamental expected probabilities, namely,

$$\xi_t(i, j) = \Pr(s_t = i, s_{t+1} = j|\lambda^{(n)}, y)$$

and

$$\gamma_t(i) = \Pr(s_t = i, |\lambda^{(n)}, y),$$

where $\gamma_t(i)$ is the probability of being in state i at time t given the current estimate of the model $\lambda^{(n)}$ and the observations $\boldsymbol{y}$, and $\xi_t(i, j)$ is the probability in state i at time t and state j at time $t+1$, also conditioned on $\lambda^{(n)}$ and $\boldsymbol{y}$. They can be calculated from the forward and backward probabilities by

$$\xi_t(i, j) = \frac{\alpha_t(i)a_{ij}b_i(t)\beta_t(j)}{\sum_i \alpha_t(i)\beta_t(i)}$$

$$\gamma_t(i) = \frac{\alpha_t(i)\beta_t(i)}{\sum_i \alpha_t(i)\beta_t(i)} \tag{38}$$

where the numerator of $\xi_t(i, j)$ is $\Pr(s_t = i, s_{t+1} = j, y)$, the numerator of $\gamma_t(i)$ is $\Pr(s_t = i, y)$, and the denominator is $\Pr(y)$. The divisions thus give the desired probabilities conditioned on the observations.

The quantities $\xi_t(i, j)$ and $\gamma_t(i)$ can be used to construct the expected values of the model parameters. For example, the summation of $\xi_t(i, j)$ over time t leads to the expected number of transitions from state i to state j, and the summation of $\gamma_t(i)$ over t gives the expected occupancy of state i. The ratio of the two then gives the expected value of the transition probability a_{ij},

$$\hat{a}_{ij} = \frac{\sum_{t=1}^{T-1} \xi_t(i, j)}{\sum_{t=1}^{T-1} \gamma_t(j)} \tag{39a}$$

The initial probability, the unitary current, and the noise variance can be similarly represented by

$$\hat{\pi}_i = \gamma_1(i) \tag{39b}$$

$$\hat{I}_\mu = \frac{\sum_{t=1}^{T-1} \sum_{i \in C_\mu} \gamma_t(i) y_t}{\sum_{t=1}^{T-1} \sum_{i \in C_\mu} \gamma_t(i)} \tag{39c}$$

$$\hat{\sigma}_\mu^2 = \frac{\sum_{t=1}^{T-1} \sum_{i \in C_\mu} \gamma_t(i)(y_t - I_\mu)^2}{\sum_{t=1}^{T-1} \sum_{i \in C_\mu} \gamma_t(i)}, \tag{39d}$$

where C_μ denotes the states of the μth conductance.

These expected values of the model parameters provide a reestimate of the underlying model. According to Baum's theory, the new estimate always results in an improved likelihood value. Repetition of the reestimation eventually leads to the (local) maximization of the likelihood function.

The reestimates of the parameters given in **Eq. 39** should be distinguished from those in **Eq. 32**. The reestimation in the segmental k-means method is made from the most likely state sequence. The information from a single state sequence is generally inadequate to resolve the transitions between aggregated states. The true maximum likelihood estimation overcomes the problem by gleaning the information from all state sequences.

Figure 3 illustrates the performance of hidden Markov modeling on noise levels and channel kinetics *(31)*. For a fixed data length, increasing the noise level reduces the curvature of the likelihood surface. The position of the maximum, however, remains unchanged. In other words, the noise level does not affect the mean values of the estimates but only their confidence limits. Furthermore, the reduced curvature of the likelihood surface at high noise levels can be restored by increasing the data length. This suggests that the algorithm has no intrinsic limit on the noise level provided a sufficient amount of data. Contrary to the effect of noise, a hard limit appears to exist for kinetic parameters. The likelihood surface is well defined for $k\Delta t < 1$. But when $k\Delta t$ increases up to 2, the upper half of the function becomes virtually flat. The fastest rates that can be extracted reliably are about as fast as the sampling rate. Beyond that, the algorithm can only tell whether a rate is too slow to fit the data but cannot discriminate its upper limit because all rates faster than the sampling rate give equal likelihood values.

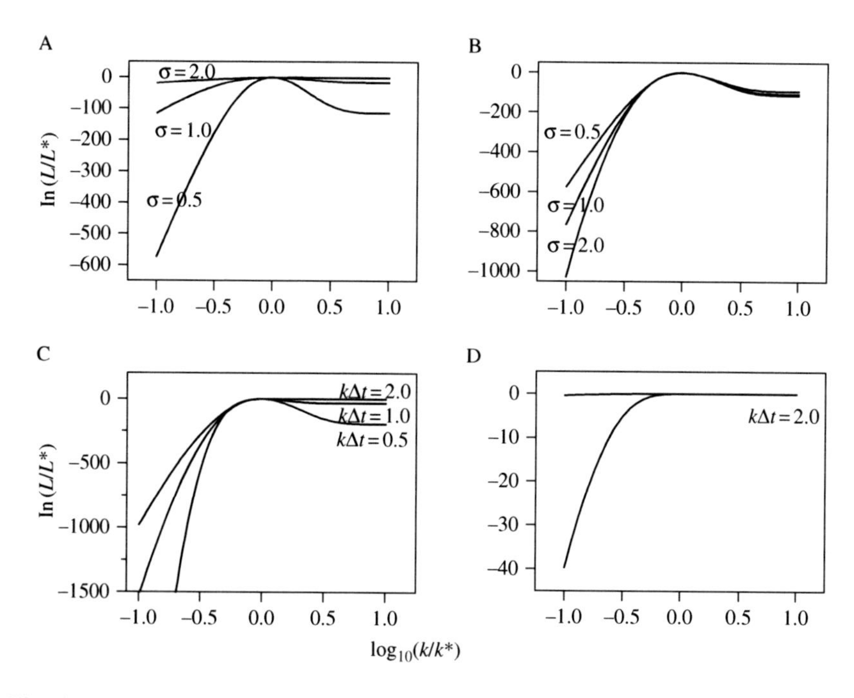

Fig. 3. Dependence of the log likelihood surface on noise and kinetics. **(A)** For a fixed data length, the likelihood surface becomes flatter as the data get noisier. The algorithm reaches the limit at a standard deviation of about 2 pA, corresponding to a single-to-noise ratio (SNR) $= 0.5$. **(B)** When the noise increases, the data length is also increased so that the log likelihood function maintains about the same curvature at the maxima as in the low noise case. **(C)** As the sampling interval increases, the likelihood surface becomes increasingly asymmetric, where the lower half ($k \leq k^*$) still maintains a good curvature, but the upper half ($k \geq k^*$) becomes flat. The maxima of the likelihood can be reliably identified until $k\Delta t = 2$. **(D)** Two log likelihood functions at $k\Delta t = 2$ with the data length different by a factor of 100. Increasing the data length in this case improves the curvature of the lower half ($k \leq k^*$) of log likelihood function but has little effect on the flatness of the upper half ($k \geq k^*$). Data were simulated from a two-state model with both rates equal to 100,000/s with the unitary currents at 0 and 1 pA, respectively. For the noise test, the sampling duration was fixed at $\Delta t = 5\,\mu s$, corresponding to $k \cdot \Delta t = 0.5$. For the kinetics test, a small amount of noise with $\sigma = 0.1$ pA was added. The curves correspond to the cross sections of the log likelihood surface cut along the direction where the two rates are equal.

12.3. Direct Optimization of Rate Constants

Baum's theory reestimates transition probabilities rather than rate constants of a model. This incurs limitations on the handling of global fitting, constraints on rate constants, specification of model topology, and so on. One solution to the problem is to convert **A** into **Q** at each iteration and applying the constraints to the resultant **Q** *(28)*. A more straightforward approach is to treat the maximum likelihood estimation as an optimization problem and resort to optimization techniques to optimize the rate constants directly *(31)*.

The derivative of the likelihood function with respect to the model parameters can be calculated analytically. Applying the chain rule to **Eq. 35**, one obtains the derivative of the likelihood function to any variable x:

$$\frac{\partial L}{\partial x} = \frac{\partial \pi' \mathbf{B}_1}{\partial x}\beta_1 + \cdots + \sum_t \alpha_t^{\mathsf{T}} \frac{\partial \mathbf{AB}_{t+1}}{\partial x}\beta_{t+1}.$$

Letting x be the initial probabilities, the transition probabilities, the current amplitudes, and the noise standard deviations respectively, then

$$\frac{\partial L}{\partial \pi_i} = \beta_1(i) b_1(1) \tag{40a}$$

$$\frac{\partial L}{\partial a_{ij}} = \sum_t \alpha_t(i)\beta_{t+1}(j) b_j(t+1) \tag{40b}$$

$$\frac{\partial L}{\partial I_k} = \sum_t \sum_{\kappa(i)=k} \alpha_t(i)\beta_t(i)\sigma_k^{-2}(Y_t - I_k) \tag{40c}$$

$$\frac{\partial L}{\partial \sigma_k^2} = \sum_t \sum_{\kappa(i)=k} \alpha_t(i)\beta_t(i)\left[-\frac{1}{2\sigma_k^2} + \frac{(Y_t - I_k)^2}{2\sigma_k^4}\right], \tag{40d}$$

where $\kappa(i)$ denotes the conductance class of state i. The derivatives of the transition probabilities with respect to the rate constants can be obtained from **Eq. 24** by setting t to the sampling duration. When the starting probability is chosen as the equilibrium probability at the holding condition, it is also a function of the rate constants. Its derivatives to the rate variables follow

$$\frac{\partial \pi^{\mathsf{T}}}{\partial x}\mathbf{Q}_h = \pi^{\mathsf{T}}\frac{\partial \mathbf{Q}_h}{\partial x}, \tag{41}$$

which can be solved as linear system equations. **Equations 40** and **41**, along with the derivatives of the transition probabilities to the rate constants, constitute the basis for the calculation of the derivatives of the likelihood with respect to the kinetic constants.

Figure 4 compares the convergence of Baum's reestimation and direct optimization of the likelihood function by the variable metric method ***(31)***. The two algorithms proceeded basically in the same direction toward the maximum. Baum's algorithm tends to converge rapidly in the beginning but becomes slow when it gets close to the maximum. This is particularly evident when the likelihood surface is relatively flat in the case of either low signal-to-noise ratio or aggregation of states. The optimization approach, on the other hand, shows little dependence on either the noise level or the aggregation of states. The convergence near the maximum is always approximately quadratic. When the likelihood surface is well defined, however, Baum's reestimation can sometimes outperform the optimization approach. The algorithm is thus well suited for estimation of amplitude parameters, where the likelihood function usually attains relatively large curvatures.

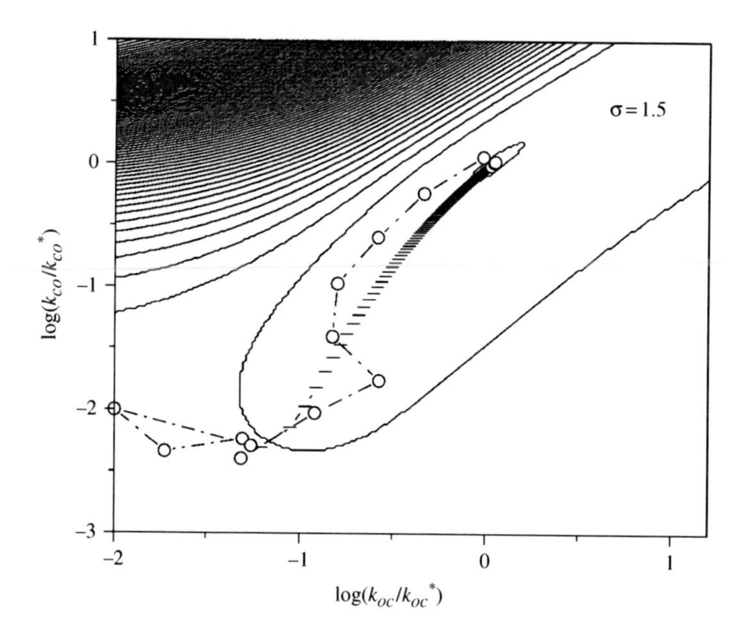

Fig. 4. Comparison of the convergence behavior between Baum's reestimation and the optimization method at a relatively high noise level $\sigma/i = 1.5$. The circles represent the optimization approach, and the bars correspond to Baum's reestimation. Each symbol represents one iteration.

13. Application Notes

A software suite (QuB) implementing the algorithms described above with user-friendly graphic interface on the 32-bit Windows has been developed (www.qub.buffalo.edu). A different implementation with a pClamp-like data visualization interface is also available (qinlab.med.buffalo.edu). Besides tools for data analysis, the software also contains modules for data acquisition and supports some cost-effective cards such as those from National Instrument. Examples and tutorials on the practical applications of the algorithms and the usage of the programs are part of the package.

The study of single channel analysis has an extensive history, and numerous techniques and tools have been developed over the past years. For advanced kinetics analysis, another free program for the maximum-likelihood dwell-time fitting is HJCFIT *(38)*, which employs the asymptotic approximation for missed event correction. Practical fitting of the 2D dwell-time distributions was first described by Magleby and his coworkers *(12)*. The commercial TAC software also integrates hidden Markov modeling for single-channel analysis. The method is based on Baum-Welch's reestimation as developed by the Sigworth lab (*28–30*). One significant difference between QuB and these other programs is the use of analytically calculated derivatives for optimization of the likelihood function, a feature that is important for fitting complex models and large datasets.

References

1. Colquhoun D, Hawkes AG (1977) Relaxation and fluctions of membrane currents that flow through drug-operated channels. *Phil Trans R Soc Lond B* 199:231–262.
2. McManus OB, Weiss DS, Spivak CE, Blatz AL, Magleby KL (1988) Fractal models are inadequate for the kinetics of four different ion channels. *Biophys J* 54:859–870.
3. McManus OB, Spivak CE, Blatz AL, Weiss DS, Magleby KL (1989) Fractal models, Markov models, and channel kinetics. *Biophys J* 55:383–385.
4. Korn SJ, Horn R (1988) Statistical discremination of fractal and Markov models of single channel gating, *Biophys J* 54:871–877.
5. Sansom MSP, Ball FG, Kerry CJ, McGee R, Ramsey RL, Usherwood PNR (1989) Markov, fractal, diffusion, and related models of ion channel gating. *Biophys J* 56:1229–1243.
6. Kienker P (1989) Equivalence of aggregated Markov models of ion-channel gating. *Proc. R Soc Lond B* 236:269–309.
7. Colquhoun D, Hawkes AG (1981) On the Stochastic Properties of Single Ion Channels. *Proceedings of the Royal Society of London Series B-Biological Sciences*, 211:205–235.

8. Fredkin DR, Montal M, Rice JA (1985) *Identification of Aggregated Markovian Models: Application to the Nicotinic Acetylcholine Receptor*, pp. 269–289. *Proc of the Berkeley Conference in Honor of Jerzy Neymann and Jack Kiefer,* Belmont, CA: Wadsworth.
9. Labarca P, Rice JA, Fredkin DR, Montal M (1985) *Kinetic Analysis of Channel Gating: Application to the Cholinergic Receptor Channel and the Chloride Channel From Torpedo California*, pp. 469–478 *Proc of the Berkeley Conference in Honor of Jerzy Neymann and Jack Kiefer,* Belmont, CA: Wadsworth.
10. Qin F, Li L (2004) Model-based fitting of single-channel dwell-time distributions. *Biophysical J* 86:1657–1671.
11. Magleby KL, Pallotta BS (1983) *Calcium Dependence of Open and Shut Interval Distributions From Calcium-Activated Potassium Channels in Cultured Rat Muscle*, J Physiol. 344: pp. 585–604.
12. Magleby KL, Weiss DS (1990) Identifying kinetic gating mechanisms for ion channels by using two-dimensional distributions of simulated dwell times. *Proc R Soc Lond B Biol Sci* 241:220–228.
13. Magleby KL and Song L (1992) Dependency plots suggest the kinetic structure of ion channels. *Proc R Soc Lond B Biol Sci* 249:133–142.
14. Horn R, Lange K (1983) Estimating kinetic constants from single channel data. *Biophys. J* 43:207–223.
15. Ball FG, Sansom MSP (1989) Ion-channel gating mechanisms: model identification and parameter estimation from single channel recordings. *Proc R Soc Lond B* 236:385–416.
16. Qin F, Auerbach A, and Sachs F (1996) Estimating single channel kinetic parameters from idealized patch-clamp data containing missed events. *Biophys J* 70: 264–280.
17. Colquhoun D, Hawkes AG, and Srodzinski K (1996) Joint distributions of apparent open and shut times of single-ion channels and maximum likelihood fitting of mechanisms. *Philosophical Transactions of the Royal Society of London Series A-Mathematical Physical and Engineering Sciences,* 354:2555–2590.
18. Qin F, Auerbach A, Sachs F (1997) Maximum likelihood estimation of aggregated Markov processes. *Proc R Soc Lond [Biol]* 264:375–383.
19. Roux B, Sauve R (1985) A general solution to the time interval omission problem applied to single channel analysis. *Biophys J* 48:149–158.
20. Blatz AL, Magleby KL (1986) Correcting single channel data for missed events. *Biophys J* 49:967–980.
21. Crouzy SC, Sigworth FJ (1990) Yet another approach to the dwell-time omission problem of single-channel analysis. *Biophys J* 58: 731–743.
22. Hawkes AG, Jalali A, Colquhoun D (1992) Asymptotic distributions of apparent open times and shut times in a single channel record allowing for the omission of brief events. *Phil Trans R Soc Lond B* 337:383–404.

23. Press WH, Teukolsky SA, Vetterling WT, Flannery BP (1992) *Numerical Recipes in C*. Cambridge: Cambridge University Press.
24. Kendall MG, Stuart A (1977) *The Advanced Theory of Statistics*. London: Griffin.
25. Nijenhuis A, Wilf HS (1978) *Combinatorial Algorithms*. New York: Academic Press Inc.
26. Qin F (2004) Restoration of single-channel currents using the segmental k-means method based on hidden Markov modeling. *Biophysical J* 86:1488–1501.
27. Fredkin DR, Rice JA (1992) Maximum likelihood estimation and identification directly from single-channel recordings. *Proc R Soc Lond* 239:125–132.
28. Venkataramanan L, Walsh JL, Kuc R, and Sigworth FJ (1998) Identification of hidden Markov models for ion channel currents - Part I: Colored background noise. *IEEE Trans. Signal Processing*, 46:1901–1915.
29. Venkataramanan L, Kuc R, and Sigworth FJ (1998) Identification of hidden Markov models for ion channel currents - Part II: State-dependent excess noise. *IEEE Trans. Signal Processing*, 46:1916–1929.
30. Venkataramanan L, Kuc R, and Sigworth FJ (2000) Identification of hidden Markov models for ion channel currents - Part III: Bandlimited, sampled data. *IEEE Trans. Signal Processing*, 48:376–385.
31. Qin F, Auerbach A, and Sachs F (2000) A direct optimization approach to hidden Markov modeling for single channel kinetics. *Biophys J* 79:1915–1927.
32. Qin F, Auerbach A, Sachs F (2000) Hidden Markov modeling for single channel kinetics with filtering and correlated noise. *Biophys J* 79:1928–1944.
33. Rabiner LR (1989) A tutorial on hidden Markov models and selected applications in speech recognition. *Proc. IEEE*, 77:257–286.
34. Rabiner LR, Wilpon JG, and Juang BH (1986) A segmental k-means training procedure for connected word recognition. *AT & T Tech. J* 65:21–31.
35. Forney GD (1973) The Viterbi algorithm. *Proc IEEE* 61:268–278.
36. Baum LE, Petrie T, Soules G, and Weiss N (1970) A maximization technique occuring in the statistical analysis of probabilistic functions of Markov chains. *Ann Math Stat* 41: 164–171.
37. Dempster AP, Laird NM, and Rubin DB (1977) Maximum likelihood from incomplete data via the EM algorithm. *J Roy Stat* 39:1–38.
38. Colquhoun D, Hatton CJ, and Hawkes AJ. 2003. The quality of maximum likelihood estimates of ion channel rate constants. *J Physiol* 547:699–728.

18

Use of *Xenopus* Oocytes to Measure Ionic Selectivity of Pore-Forming Peptides and Ion Channels

Thierry Cens and Pierre Charnet

Summary

The *Xenopus laevis* oocyte is a widely used system for heterologous expression of exogenous ion channel proteins (***1***, ***2***). Among other advantages, these easy to obtain, mechanically and electrically stable, large-sized cells enable multiple types of electrophysiological recordings: two-electrode voltage-clamp, single-cell attached or cell-free patch-clamp, and macropatch recordings. The size of an oocyte (1 mm in diameter) also allows the use of additional electrodes (1–3) for injection of diverse materials (Ca^{2+} chelators, peptides, chemicals, antibodies, proteic-partners, and so on) before or during the course of the electrophysiological experiment. We have successfully used this system to analyze the biophysical properties of pore-forming peptides. Simple perfusion of these peptides induced the formation of channels in the oocyte plasma membrane; these channels can then be studied and characterized in diverse ionic conditions. The ease of the perfusion and the stability of the voltage-clamped oocyte make it a powerful tool for such analyses. Compared with artificial bilayers, oocytes offer a real animal plasma membrane where biophysical properties and toxicity can be studied in the same environment.

Key Words: Voltage clamp; reversal potential; calcium channels; anomalous mole fraction.

1. Introduction

The electrophysiological analysis of naturally occurring as well as synthetic channel-forming peptides has proved useful to identify, characterize, and understand many aspects of their insertion and their ionotropic, antibacterial, or toxic activities. The basic steps of this approach rely on the incorporation of the peptide in a membrane system that can be voltage clamped in known ionic

From: *Methods in Molecular Biology, vol. 288: Patch-Clamp Methods and Protocols*
Edited by: P. Molnar and J. J. Hickman © Humana Press Inc., Totowa, NJ

conditions for a fine description (in terms of ionic and electric environment) of the conditions of insertion, of the channel conductance and selectivity, and, possibly, of the regulation of these properties. In these approaches, the determination of the ionic selectivity, by identifying the nature and the direction of the membrane currents induced by insertion of the peptide in the natural membrane, provides important information on their toxicity and, coupled with mutagenesis, helps to identify the residues involved in channel formation and permeation. All these experiments can be performed in *Xenopus* oocytes with a simple two-electrode, and/or single-channel recording, set up. The ease with which oocytes can be microinjected and their capacity to express foreign RNA provide an additional technical advantage when pore-forming peptides require intracellular protein partners for the formation of the channel. This chapter describes two methods that have been used to analyze ionic selectivity and to determine the multi-ion nature of channel pores by reversal potential and anomalous mole fraction measurements.

2. Materials

2.1. Xenopus laevis *Oocyte Preparation*

1. Adult *X. laevis* females are from Centre National d'Elevage de Lavalette, France.
2. Anesthetic solution: 0.2% MS222 (ethyl 3-aminobenzoate methanesulfonate salt; Sigma, France; cat. no. A5040) is dissolved in tap water.
3. Dissociation solution: collagenase (type IA; Sigma, cat. no. C9891) is dissolved at 1 mg/ml in the following solution: 82.5 m*M* NaCl, 2 m*M* KCl, 1 m*M* $MgCl_2$, 5m*M* HEPES, pH = 7.2 with NaOH (*see* **Note 1**).
4. Washing solution: same as item 3 without collagenase.
5. Survival solution: 96 m*M* NaCl, 2 m*M* KCl, 2 m*M* $CaCl_2$, 1 m*M* $MgCl_2$, 5 m*M* HEPES, 2.5 m*M* pyruvic acid, 0.05 g/l gentamycin, pH = 7.2 with NaOH.

2.2. *Recording Solutions*

2.2.1. Monovalent Solutions

1. Basic recording solution: 100 m*M* NaCl, 5 m*M* HEPES, 2 m*M* $MgCl_2$, pH = 7.2 with NaOH (Na100 solution).
2. For ionic selectivity measurements, NaCl is replaced by an equimolar concentration of KCl, TEA-Cl, LiCl, CsCl, or choline-Cl and pH is adjusted using KOH, TEAOH, LiOH, CsOH, or NaOH, respectively (K100, TEA100, Li100, Cs100, or choline100 solutions). $MgCl_2$ was found to be necessary for the stability of the recordings. Removing $MgCl_2$ is possible but induces the appearance of a leak current that increases with time. For chloride permeability, NaCl is replaced with equimolar Na-acetate, and $MgCl_2$ is removed from both Cl-containing and Cl-free solutions.

2.2.2. Solutions for Anomalous Mole Fraction Effects

1. Bant10: 10 m*M* BaOH, 20 m*M* TEAOH, 50 m*M* N-methyl-D-glucamine (NMDG), 2 m*M* CsOH, 10 m*M* HEPES, pH = 7.2 using methane sulfonic acid.
2. Cant10: 10 m*M* CaOH, 20 m*M* TEAOH, 50 m*M* NMDG, 2 m*M* CsOH, 10 m*M* HEPES, pH = 7.2 using methane sulfonic acid.
3. The solutions of different Ba^{2+} and Ca^{2+} concentrations (respective Ba^{2+}/Ca^{2+} concentrations in m*M* 10/0, 9/1, 8/2, 6/4, 4/6, 2/8, and 0/10) are prepared by appropriate mixing of the two Bant10 and Cant10 solutions.

2.2.3. BAPTA Solution and Injection

1. BAPTA solution has the following composition: 100 m*M* 1,2-bis (2-aminophenoxy)ethane-N, N, N′, N′-tetraacetic acid (BAPTA; Sigma, cat. no. A4926), 10 m*M* HEPES, 10 m*M* CsOH, pH = 7.2.
2. Injection needles: pulled from Clark Electromedical Instrument borosilicate glass capillaries (GC150T-10) using a P-30 Sutter Instrument Company (USA) puller and subsequently broken under the stereomicroscope (×50 magnification) at 5–10 μm. This is done by gently touching the tip of the needle on the bottom of a plastic Petri dish.

2.3. Electrophysiological Measurements

2.3.1. Two-Electrode Voltage Clamp

1. Agar bridges: Clark capillaries (with filament, GC150F10) are bent at approximately 120 °C under flame and subsequently immersed in 60-mm Petri dishes filled with almost boiling agar (high gel strength, Sigma, ref A6924) dissolved at 1% in 3 *M* KCl (5–10 capillaries can be placed per dish). When immersed, these bridges are usually filled naturally (by capillarity) by the hot agar solution (*see* **Note 2**). After cooling, two bridges are "dissected" from the agar and placed in the bath-electrode holder previously filled with 3 *M* KCl.
2. Voltage and current electrodes: pulled from Clark Electromedical Instrument borosilicate glass capillaries (GC150T-10) using a P-97 Sutter Instrument Company puller. They have a resistance of 1–2 MΩ when filled with 3 *M* KCl.
3. Gravity-driven perfusion: The home-made recording chamber (50 μl) is connected to an array of eight reservoirs (50-ml syringes) containing the various solutions. The flow of solution from each syringe can individually be switched on or off by micro-electrovalves (Fisher-Bioblock, France, cat. no. C92901).

2.3.2. Single-Channel Recordings

1. Hyperosmotic solution: 200 m*M* NaCl, 2 m*M* KCl, 2 m*M* $MgCl_2$, 10 m*M* HEPES, pH = 7.2 with NaOH.
2. Depolarizing solution: 100 m*M* KCl, 2 m*M* $MgCl_2$, 10 m*M* HEPES, 10 m*M* ethyleneglycol-bis-N, N, N′, N′-tetraacetic acid (EGTA), pH = 7.4 with KOH.

3. Patch-clamp pipettes, pulled from KG33 borosilicate glasses (Garner Glass, USA), treated with Sylgard™ (*see* **Note 3**) and fire-polished (Narishige microforge MF-830), have a resistance of 10–15 MΩ when filled with Na100.

3. Methods

3.1. Oocyte Preparation

1. Ovaries are surgically removed from *X. laevis* females anesthetized by 15- to 20-min immersion into the anesthetic solution. A small abdominal incision through the skin and the gut wall is sufficient. Skin and gut wall are stitched separately after ovarian lobe removal.
2. Pieces (approximately 10 ml) of dilacerated lobe of ovaries containing the oocytes are placed in a 50-ml Falcon tube and washed three to four times with 40 ml washing solution.
3. The Falcon tube is then filled with 30 ml dissociation solution and placed at 20 °C under gentle agitation using an orbital shaker (120 rpm, Rotamax120 Heidolph). The isolation procedure is usually stopped after 2–3 h, once the follicular cells have detached from the oocytes. This can be estimated by observation of a sample under the stereomicroscope.
4. After extensive washing (three to five times with 40 ml wash solution), a last wash is made with the survival solution.
5. Isolated stage V–VI oocytes are placed in a 100-mm Petri dish filled with the survival medium, and batches of 20–100 oocytes are selected under a stereomicroscope and kept in the survival solution under gentle agitation in 35-mm Petri dishes at 20 °C.
6. The survival solution is changed every day (and dead oocytes discarded) until use. Oocytes may be used directly for peptide perfusion or microinjected with either RNA or cDNA and then left 48–72 h before electrophysiological recordings.

3.2. Estimation of the Junction Potential

The junction potential between the different solutions is estimated by two methods (*see* **Note 4**).

1. The first one uses the tool "junction potential calculator" provided by version 9.0 of the pClamp program (Axon Instruments, USA).
2. The second one simply consists in zeroing the junction potential of both current and voltage electrodes when they are in the bath in the presence of the Na100 solution. We then perfuse the different recording solutions and measure the junction potentials on the voltage or current electrode DC meter of the Geneclamp 500 amplifier (Axon Instruments). Both the methods give differences in junction potentials compared with Na100 solution smaller than 2 mV, except for the solution with acetate ions replacing chloride (junction potential= 5 mV).

3.3. BAPTA Solution and Injection

In some experiments using external divalent cations (Ca^{2+}, Ba^{2+}, and Sr^{2+}), the contaminating endogenous Ca^{2+}-activated Cl^- current must be suppressed. The available inhibitors of this current (e.g., anthracene-9-carboxylic acid [9AC] or niflumic acid) are usually not sufficiently efficient and specific to completely eliminate this conductance, and thus, we find it preferable to inject a Ca^{2+} chelator (BAPTA has our preference for its fast chelating capabilities, *see* **Note 5**) into the oocyte during the course of the experiment.

1. Pull an injection needle as described in **Subheading 2.**
2. Place this needle in the holder of a pneumatic injector connected to vacuum.
3. Backfill this needle by immersing the tip in a drop (1–2 μl) of BAPTA solution placed on the bottom of a Petri dish (this is done under a stereomicroscope at ×30 magnification).
4. Apply vacuum to the needle (this is most conveniently done with a footswitch connected to the pneumatic injector) and follow the backfilling under the stereomicroscope (*see* **Note 6**).
5. Remove the filled needle and place it on an electrode holder fixed on a micromanipulator on the top of the recording chamber and connected to air pressure (1 bar) through a solenoid valve (Parker Hannifin Corp., Fair feild, NJ, USA, serie3, three ways) with time-controlled aperture (*see* **Note 7**; from 10 ms to several seconds).
6. The needle is impaled into the oocyte at the beginning of the recording session, and the injection of BAPTA is made by repetitive application of a positive pressure (100 ms two to five times). The inhibition of the chloride conductance is followed online during injection.
7. The final intra-oocyte BAPTA concentration can be estimated to be around 2–5 m*M* (*see* **Note 8**).

3.4. Electrophysiological Characterization of Pore-Forming Peptides

3.4.1. Reversal Potential Measurements

Macroscopic whole-cell currents are recorded under two-electrode voltage clamp using the GeneClamp 500 amplifier (Axon Instruments). Voltage command, sampling, acquisition, and analysis are done using a Digidata 1200 and the pClamp program (version 9, Axon Instruments). All experiments are performed at room temperature (20–25 °C).

1. An oocyte is placed in the recording chamber filled with the desired solution (usually Na100).
2. Junction potentials (typically less than 3–5 mV) at the two electrodes are first cancelled using the zero button of the amplifier in Na100 solution, and the electrode resistance is checked before introduction into the oocyte.

3. Both the electrodes are impaled into the oocyte, and the resting membrane potential is measured (typically $-30/-50\,mV$). The oocyte is then voltage clamped at $-80\,mV$.
4. A third electrode is impaled to inject BAPTA as described in section 3.3. This step is necessary when contaminations by Ca^{2+}-activated Cl^- currents are suspected.
5. Then, 50 µl tested protein or peptide is applied directly to the bath (with the main perfusion stopped) at the final working concentration (usually 20–50 µ*M*), after proper dilution from a stock solution (500 µ*M* in water) into the desired recording solution. We assume that there is no dilution in the recording chamber that has a volume of around 50 µl.
6. Membrane currents are measured during voltage ramps (from −80 to +20 mV, 450 ms long), applied from the holding potential of −80 mV every 3 s, continuously, before and after addition of the peptide. The change in membrane conductance is followed by the modification in the slope of the current–voltage curve (current recorded during the voltage ramp, with the corresponding voltage axis; *see* **Fig. 1**). Similar recordings can be done at different holding potentials to analyze the voltage dependence of peptide insertion (*see* **Fig. 2**).
7. Washing out of the peptide is done by a gravity-driven perfusion of the chamber with the same solution without protein (*see* **Note 9**).
8. Similar experiments are performed with the different solutions using different oocytes (a new oocyte for each ionic condition).
9. Offline analysis of the changes in the membrane conductance (slope of the current–voltage curve) is made for each membrane potential by a least-squares fit of the current using the pClamp software (version 9.0, Axon Instruments). These slopes can be plotted as a function of time during peptide application (*see* **Fig. 1**).
10. Current reversal potential (*Erev[X]*) is measured. This is the point where the current traces cross each other during peptide application (potential "*Erev*" in **Fig. 1B**). It can be measured as the zero-current potential after digital subtraction of traces

Fig. 1. **(A)** Membrane currents elicited in an oocyte voltage clamped at −100 mV and submitted to 400-ms-long voltage ramps from −80 to +20 mV, before and after addition of 50 µl peptide (in this case, puroindoline at 50 µ*M*). Top, voltage protocol; bottom, membrane currents. Ramps are applied every 15 s, and the 400-ms-long recording traces have been concatenated for iconographic purposes. Note the increase in membrane current following addition of the peptide. **(B)** Superimposed current–voltage curves recorded during the experiments shown in panel **A** before and after the addition of the peptide. *Vm* denotes the potential at which the current is measured, and the shaded area depicts the zone where the linear regression is calculated to measure the slope conductance. *Erev*, the crossing point of all the currents recorded during peptide application, is the current reversal potential of the peptide-induced current *(3)*. *Erev* is the

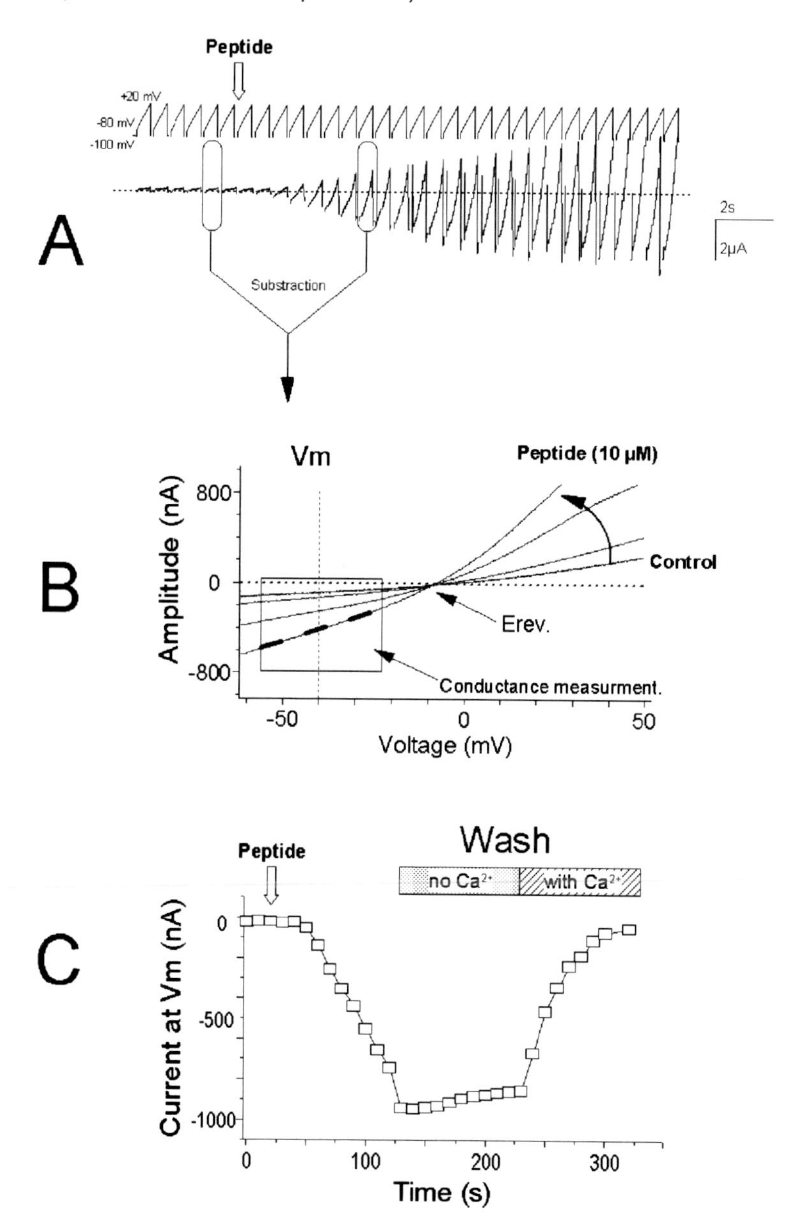

Fig. 1. zero-current potential when traces recorded before and after peptide application are subtracted (*see* **A**). **(C)** Time course of current amplitude recorded at the potential *Vm* before and after application of the peptide and during perfusion of peptide-free solutions without (no Ca^{2+}) and with (with Ca^{2+}) 5 m*M* calcium. Note that reversibility is only obtained with Ca^{2+}-containing peptide-free solution.

recorded before and after peptide/protein application. This potential is measured for each ionic condition (Na100, K100, TEA100, Li100, Cs100, or choline100 solutions) and eventually corrected for the junction potential (see Section 3.2).

11. Ionic permeability ratios of peptide-induced channels (permeability of cation X relative to Na permeability: P_X/P_{Na}) can be deduced for each cation X from the shift in reversal potentials *Erev(X) – Erev(Na)* using the Goldman–Hodgkin–Katz (GHK) equation *(5)*:

$$Erev\,(X) - Erev\,(Na) = \frac{RT}{zF}\mathrm{Ln}\frac{P_X\,[X]_o}{P_{Na}\,[Na]_o}$$

$$\frac{P_X}{P_{Na}} = \frac{[Na]_o}{[X]_o}\mathrm{e}^{[Erev(X)-Erev(Na)](zF/RT)},$$

where $[X]_o$ is the extracellular concentration of ion X and R, T, z, and F have their usual meanings (*RT/F* approximately 25.3 mV at 20 °C).

12. Washout of the peptide as well as effect of regulators of permeability (Ca^{2+}, e.g., *see* **ref. 4**) can be assayed using the perfusion system.

3.4.2. Determination of Single-Channel Conductance

1. The vitelline envelope is removed manually under a stereomicroscope (×50 magnification) using tweezers style no. 5 (cat.no. T4537, Sigma) after placing the oocytes for 10–20 min in the hyperosmotic solution.
2. Carefully place this "denuded" oocyte in the recording chamber filled with the depolarizing solution that nulls the oocyte membrane potential.
3. Patch pipettes are filled with Na100 with the desired concentration of peptide (usually, with the ones we have tested, 5–10 μM was necessary, *see* **Note 10**).
4. They are placed on a holder connected to the 100 G-CV5 headstage of a Geneclamp500 amplifier (Axon Instruments). For stability reasons, the single-channel recordings are performed preferentially on cell-attached patches.
5. After establishing a gigaohm seal (usually > 10GΩ), the patch potential is set at the desired value and recordings are made using Clampex 7 (Axon Instruments) in event-driven mode and digitized by a Digidata1200. The time necessary for insertion of the peptide is function of the sequence, the solution, and the applied voltage. Insertion typically takes several minutes.

Fig. 2. **(A)** Effect of application of a peptide on an oocyte held at −120, −100, or −80 mV and submitted to 400-ms-long voltage ramps from −80 to +20 mV. The change in membrane current is null at −80, small at −100, and large at −120 mV *(4)*. **(B)** Left: averaged slope conductance (calculated as described in **Fig. 1**) recorded

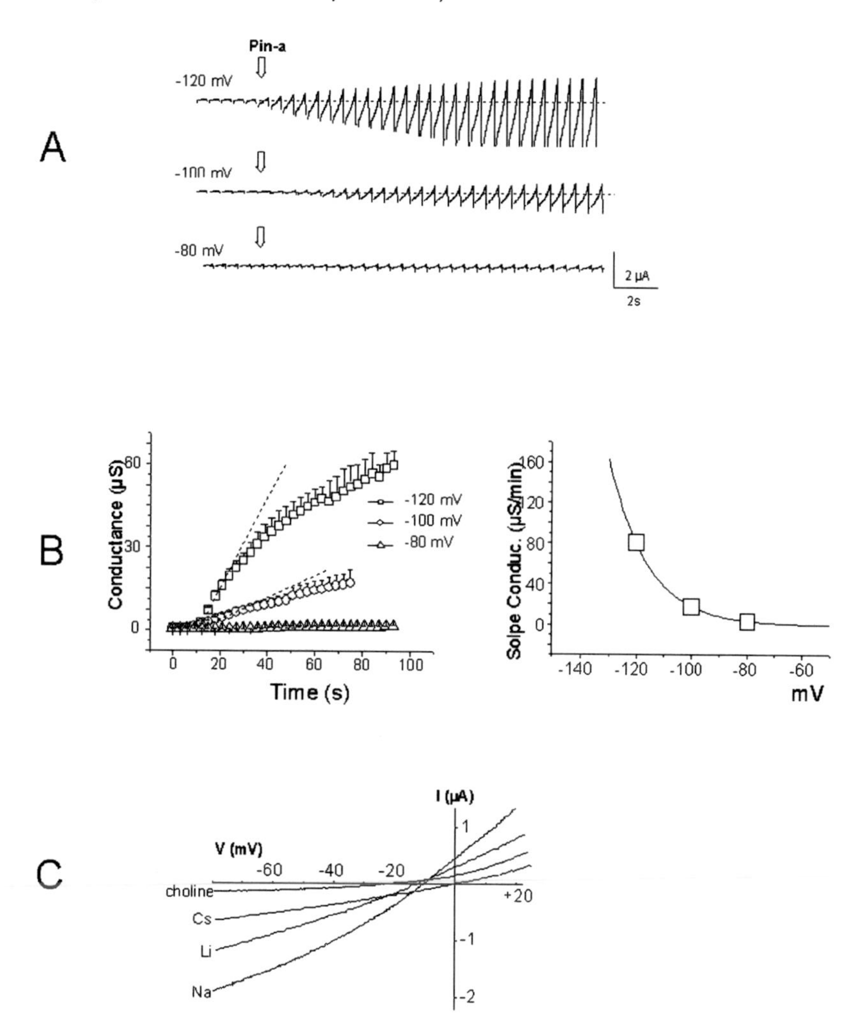

Fig. 2. during application of the peptide at −80, −100, or −120 mV. Right: voltage dependence of the increase in membrane current induced by the peptide. In this case, the maximal variation of conductance (depicted as a dashed line in the left panel) is calculated by linear regression as the maximal slope variation of the conductance for each potential and plotted as a function of the potential. **(C)** Peptide-induced currents recorded in different ionic conditions. Current traces recorded before and after perfusion of the peptide in different ionic solutions are digitally subtracted and plotted against membrane potential (*see* **Subheading 2.** for the composition of the different solutions). The potentials at which the current traces cross the abscissa (zero current) are the reversal potentials of the currents in the different solutions, which are used to calculate the permeability ratios (*see* Section 3.4.1.)

6. Once single-channel events start to appear, the patch potential can be set at different values to construct single-channel current–voltage curves. It is very important to start recording immediately as the continuous insertion of peptides in the membrane will give rise, with time, to an increased probability of having multiple openings *(6)*, complicating the analysis of the results.
7. Amplitude histograms (Clampfit 9.0, Axon Instruments) of these single-channel events recorded at different patch potentials are made and single-channel current amplitudes determined from the fit of such histograms to Gaussian functions. The slope of the linear regression of the single-channel amplitude–voltage curve gives the single-channel conductance.
8. The continuous insertion of the peptide in the membrane patch can make the recordings very unstable after several minutes. As the insertion is often voltage dependent, it can be useful to decrease the patch potential as soon as the peptide starts to form channels. In this case, the patch potential can be set at different value in an episodic mode just during the recording time (episodic stimulation mode of Clampex 9.0).

3.5. Anomalous Mole Fraction for Ca^{2+} Channels

For ionic channels with multiple ion-binding sites in the pore, the concentration-dependent permeability ratios between different ions display the so-called anomalous mole fraction effect. This phenomenon is characterized by a channel conductance of smaller amplitude in the presence of a mixture of two ions than in the presence of a pure solution of either of these ions (at the same total concentration, *see* **Fig. 3A, C** and **(ref. 5)**). In the example described in

Fig. 3. **(A)** Membrane currents elicited in an oocyte voltage clamped at −80 mV and submitted to 400-ms-long voltage ramps from −80 to +80 mV. Currents were recorded from oocytes expressing the $Ca_V1.2$ voltage-gated Ca^{2+} channels (with $\alpha_2-\delta$ and β_2 auxiliary subunits) in solutions of different Ca^{2+} mole fractions (Ba 10 m*M*, Ba/Ca 9/1, 8/2, 6/4, 4/6, 2/8, and Ca 10 m*M*). Current traces are superimposed for iconographic purposes. **(B)** Current–voltage curves recorded in 10 m*M* Ba^{2+} (open squares) or Ca^{2+} (open circles) using 400 ms voltage steps from −80 to 50 mV in 10-mV increments. The superimposed continuous lines are the current–voltage curves recorded with 400-ms-long voltage ramps of the same voltage amplitude. Similar curves are obtained by the two methods. **(C)** Anomalous mole fraction curves obtained from currents recorded in oocytes expressing the $Ca_V1.2$ voltage-gated Ca^{2+} channels (with α_2–δ and β_2 auxiliary subunits). Each point represents the peak of the current–voltage curve obtained in a given Ca^{2+} mole fraction (Ba 10 m*M*, Ba/Ca 9/1, 8/2, 6/4, 4/6, 2/8, and Ca 10 m*M*). The same curves were obtained whether the current–voltage curves were obtained with voltage steps (open circles) or voltage ramps (open squares). The anomalous mole fraction curve for $Ca_V2.1$ (with the same auxiliary subunits) is also shown (open inverted triangle) and displays a specific shape.

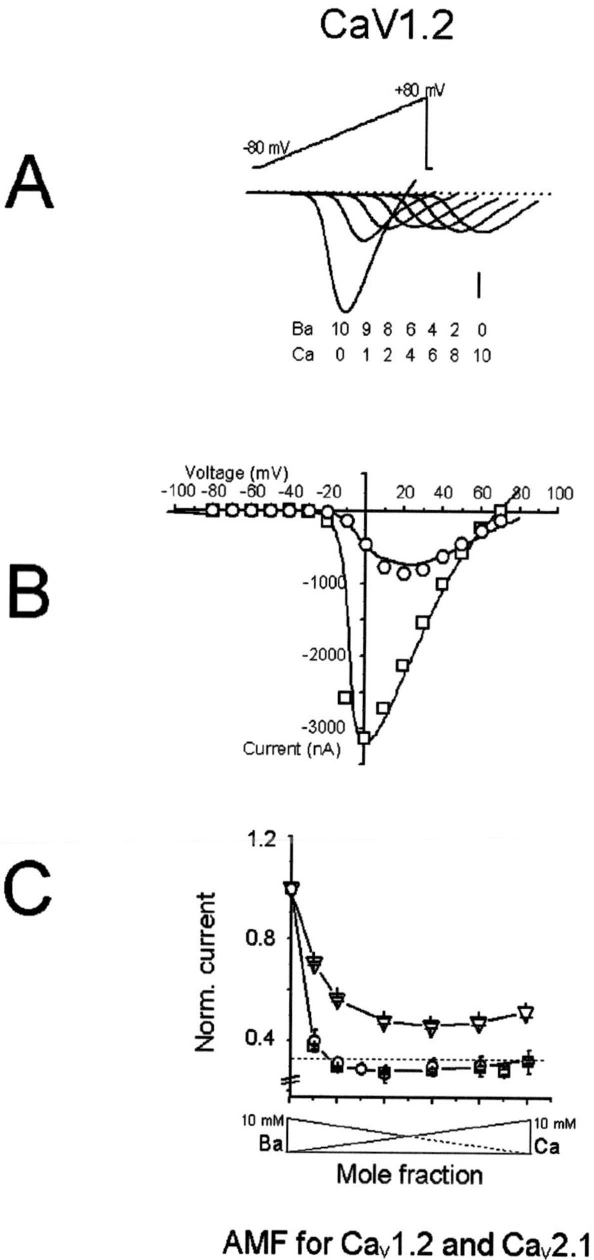
CaV1.2
A
+80 mV
-80 mV
Ba 10 9 8 6 4 2 0
Ca 0 1 2 4 6 8 10
B
Voltage (mV)
-100 -80 -60 -40 -20 20 40 60 80 100
-1000
-2000
-3000
Current (nA)
C
1.2
0.8
0.4
Norm. current
10 mM
Ba
10 mM
Ca
Mole fraction
AMF for $Ca_V1.2$ and $Ca_V2.1$
using ramps or steps

Fig. 3, Ca^{2+} channels display a higher permeability for Ba^{2+} than for Ca^{2+}. But in the presence of a mixture of Ba^{2+} and Ca^{2+}, the measured conductance is even smaller (*see* **Fig. 3A, C**). In these recordings, the conductance measured at the peak of the current–voltage curve or the reversal potential can be used *(5)*. The profile of the anomalous mole fraction curve (*see* **Fig. 3C**) is determined by (1) the number and height of the ion-binding sites, and energy barriers between these sites, present in the pore of the channel and (2) the type and concentration of the ions used. Such experiments can be useful to interpret the effects of mutations on channel pore, which affect channel selectivity or conductance. We describe here our method to measure anomalous mole fraction in *Xenopus* oocytes expressing voltage-gated Ca^{2+} channels (either $Ca_V1.2$ or $Ca_V2.1$ [7]).

1. After isolation, the *Xenopus* oocytes have to be injected with 40 nl mixture of RNA encoding the different Ca^{2+} channel subunits ($Ca_V\alpha$, $Ca_V\beta$, and $Ca_V\alpha2–\delta$ at a concentration of 1 ng/nl each) and left 3–4 days for proper expression of these proteins. Alternatively, 10 nl DNA encoding the same subunits can be injected in the nucleus. In this case, expression vectors with eukaryotic promoters have to be used (e.g., pcDNA3.1 from Invitrogen Cailsbad, CA, USA), and nuclear injection is simply performed by choosing the center of the animal (dark) pole for injection.
2. Put one of these oocytes in the recording chamber and impale both the recording electrodes, for two-electrode voltage clamp, and the BAPTA injection needle.
3. Voltage-clamp the oocyte at −80 mV and perfuse the pure 10 m*M* Ba solution (Bant10).
4. After injection of BAPTA and stabilization of the Ba^{2+} currents observed with repetitive depolarization from −80 to +10 mV (400 ms long), record current–voltage curves using either a voltage-ramp protocol (from −60 to +60 mV in 400 ms, *see* **Note 11**) or a series of conventional voltage steps of increasing amplitude (from −60 to +60 mV in 5 or 10 mV increments). These two methods should give similar results if the speed of the voltage ramp is chosen carefully (*see* **Fig. 3B, C**).
5. Repeat these recordings with the different solutions of increasing Ca^{2+} molar fraction (with a total concentration $[Ba^{2+}]+[Ca^{2+}]$ kept constant at 10 m*M*, that is, 9/1, 8/2, 6/4, 4/6, 2/8, and 0/10, *see* **Subheading 2.** and **Fig. 3C**).
6. Often check the stability of the recordings and the lack of current rundown by switching back to the original solution at the end of the experiments.
7. Plot the peak of the current–voltage curves recorded in the presence of each solution of increasing Ca^{2+} molar fraction.
8. Repeat the same experiment with each Ca^{2+} channel type or mutant.

In a typical experiment (*see* **Fig. 3**), such mole fraction curves are measured for different Ca^{2+} channels ($Ca_V1.2$ and $Ca_V2.1$). In this particular case, the recording of different mole fraction curves with the two types of Ca^{2+} channels suggests that the permeability profiles of these channels are different. Mutation

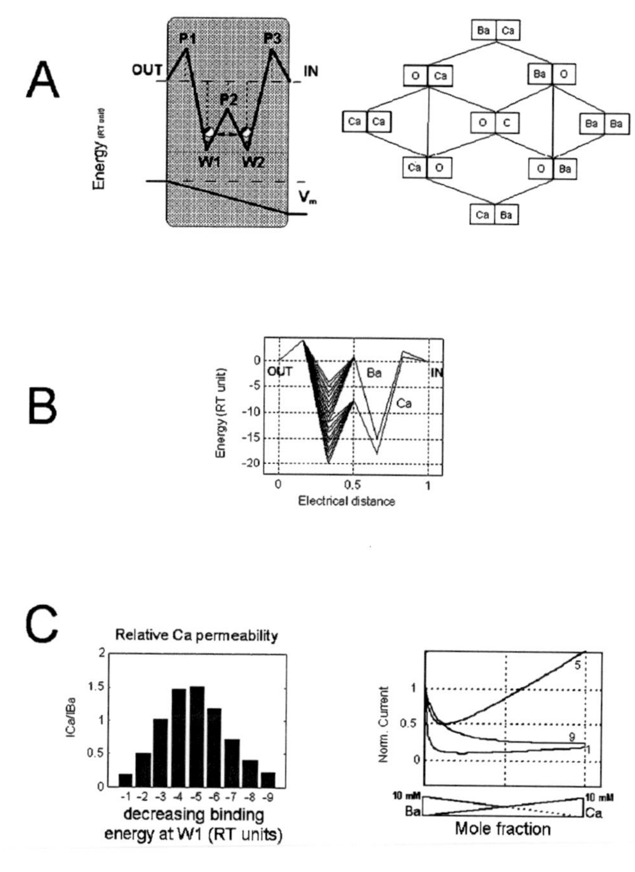

Fig. 4. **(A)** Left: schematic of the free energy profiles for Ca^{2+} and Ba^{2+} of a Ca^{2+} channel with its two Ca^{2+}-binding sites (W1 and W2) and three energy barriers (P1–P3). *Vm* denotes the membrane potential and schematizes the applied electric field. Right: state diagram of the different occupied states of the channel in the presence of two different cations (Ba^{2+} and Ca^{2+}), assuming single filling (only one ion per site). **(B)** Different energy profiles for Ba^{2+} and Ca^{2+} ions used in the following numerical simulation. The binding energy of the extracellular Ca^{2+}/Ba^{2+}-binding site is decreased progressively. **(C)** Numerical simulation of current mole fraction curves of a two binding site Ca^{2+} channel with the energy profiles displayed in panel **B**. Left: relative Ca^{2+} permeability (current at 10 m*M* Ca^{2+} divided by the current at 10 m*M* Ba^{2+}) for the channels with the different profiles. Right: current mole fraction curves for channels with an external site of high (1), intermediate (5), or low (9) affinity for divalent cations. High-binding and low-binding energies gave the same relative permeability, but anomalous mole fraction is only recorded for high energy binding.

of negatively charged amino acids that are suspected to form a divalent cation-binding sites within the pore can lead to more drastic changes with a complete suppression of the anomalous mole fraction (not shown). Using Eyring rate theory and modeling a channel with two Ca^{2+}-binding sites (W1 and W2) separated by three energy barriers (P1–P3) allows to build a state diagram of all possible occupancies of these two sites by Ba^{2+} and/or Ca^{2+} and to assign rate constants to the transitions between these states (*see* **Fig. 4A**). The rate constants depend on the channel-dependent energy profiles of the pore for both Ba^{2+} and Ca^{2+}, on an electrostatic repulsion factor arising when the pore is doubly occupied and on the electric field generated by the membrane potential (*see* **Note 12**). We have numerically solved these differential rate equations and calculated the theoretical mole fraction current for channel with an external cation-binding site of decreasing energy (*see* **Fig. 4B**). It can be seen that this decrease in the binding energy induces a bell-shaped curve of the relative Ca/Ba current, with large Ca current for intermediate binding energies (labeled 5 on **Fig. 4C**) and small Ca currents for either strong (labeled 1) or weak (labeled 9) binding energies on the external site. Interestingly, in these later cases, the mole fraction curve is only "anomalous" (i.e., displaying a minimum) for strong binding energies (trace 1), whereas it decreases monotonically for weak binding energies (trace 9). These data suggest that decreasing the binding energy of an external cation-binding site (by removal of the negative charges) can modify the anomalous mole fraction curve, whereas the relative Ca^{2+} current is only slightly modified.

4. Notes

1. All solutions are prepared with deionized water. After equilibration and pH adjustment, they are filtered at 0.22 μm using disposable Nalgen sterilization filter units. They can be stored at 4 °C for several months.
2. After cooling, these Petri dishes are sealed with Parafilm™ and can be stored at 4 °C for several months.
3. Once pipettes have been pulled, the tip is painted with Sylgard under a stereomicroscope up to 200–500 μm of the aperture and rapidly baked at moderate temperature (for Sylgard polymerization) by insertion into the center of a hot coiled filament. Classical fire polishing is then performed.
4. For a complete description of the liquid junction potential correction, the following reference should be consulted *(8)*.
5. We have also used with success DM-NITROPHEN™ (Caltriochem, Merck KGaA, Germany and derivatives of BAPTA. EGTA and ethylenediaminetetraacetic acid (EDTA) work less efficiently probably because of their slower chelating kinetics

and their inability to chelate Ca^{2+} in the close vicinity of the membrane because of the lack of interaction with phospholipids ***(9)***.

6. Be careful to stop backfilling before any air bubbles enter the needle.
7. When the needle is not connected to the pressure, it should be kept open to air pressure and not closed, to avoid continuous injection by the pressure established during the previous injection.
8. For experiments using oocytes expressing voltage-gated Ca^{2+} channels, the BAPTA solution should be injected in the animal (dark) pole of the oocyte where the translation machinery and the Cl channels are present in higher concentration. Injection in the other pole will significantly delay the inhibition of the Cl conductance. The same injection needle can be used for several oocytes provided that it is not clogged-up.
9. From our personal experience, reversibility of peptide insertion is difficult to obtain. The determination of the reversal potential in different ionic conditions thus requires a new oocyte for each condition. For some peptides, a block of the current can be recorded by perfusing Ca^{2+}-containing (10 m*M*) solution ***(4)***. This block disappears when Ca^{2+} is removed (even in the absence of continuous perfusion of the peptide).
10. The presence of the peptide in the pipette solution can sometimes prevent the seal formation. In these conditions, it can be advantageous to backfill the tip of the pipette with a peptide-free solution.
11. The voltage-ramp protocol has to be adjusted for the type of Ca^{2+} channel used, considering the specific kinetics of inactivation of the channel. Great care should be taken to avoid any contamination of the shape of the current–voltage curves generated using these ramps, because of the inactivation kinetics of the channel. A comparison of the current–voltage curves recorded during voltage ramps and measured at the peak of voltage steps of increasing amplitude should confirm the lack of contamination by inactivation (*see* **Fig. 3B**). It is often preferable to use voltage ramps because the determination of an anomalous mole fraction curve requires the recording of current–voltage curves in different ionic conditions. These recording are obtained relatively rapidly with ramps but can last more than 30–40 min if voltage steps are used, thus increasing the likelihood of rundown of the current. Moreover, the peak of the current–voltage curve should be taken for these curves and not the current at a fixed potential, to take into account the modification in the surface charges induced by the modification of the ionic composition of the extracellular solutions ***(10)***.
12. A detailed description of this numerical simulation is beyond the scope of this article but can be found elsewhere ***(11–13)***.

Acknowledgments

This work was supported by CNRS, INSERM, Association Française contre les Myopathies, Association pour la Recherche contre le Cancer, Fondation pour la Recherche sur le Cerveau, and Fondation Simone et Cino del Duca. The authors thank Dr. I. Lefevre for critical reading of the manuscript, Dr. V. Lullien-Pellerin for Discussion and J-M Donnay for oocyte preparation.

References

1. Snutch, T. P. (1988) The use of Xenopus oocytes to probe synaptic communication. *Trends Neurosci.* **11**, 250–256.
2. Dascal, N. (1987) The use of Xenopus oocytes for the study of ion channels. *CRC Crit. Rev. Biochem.* **22**, 317–387.
3. Pellegrin, P., Menard, C., Mery, J., Lory, P., Charnet, P. & Bennes, R. (1997) Cell cycle dependent toxicity of an amphiphilic synthetic peptide. *FEBS Lett.* **418**, 101–105.
4. Charnet, P., Molle, G., Marion, D., Rousset, M. & Lullien-Pellerin, V. (2003) Puroindolines form ion channels in biological membranes. *Biophys. J.* **84**, 2416–2426.
5. Hille, B. (2001) *Ion Channels of Excitable Membranes.* 3rd edition. Sinauer Associates Inc., Sunderland, MA.
6. Chaloin, L., De, E., Charnet, P., Molle, G. & Heitz, F. (1998) Ionic channels formed by a primary amphipathic peptide containing a signal peptide and a nuclear localization sequence. *Biochim. Biophys. Acta* **1375**, 52–60.
7. Mangoni, M. E., Cens, T., Dalle, C., Nargeot, J. & Charnet, P. (1997) Characterisation of alpha1A Ba2+, Sr2+ and Ca2+ currents recorded with ancillary beta1-4 subunits. *Receptors Channels* **5**, 1–14.
8. Neher, E. (1992) Correction for liquid junction potentials in patch clamp experiments. *Methods Enzymol.* **207**, 123–131.
9. Rousset, M., Cens, T., Vanmau, N. & Charnet, P. (2004) Ca2+-dependent interaction of BAPTA with phospholipids. *FEBS Lett.* **576**, 41–45.
10. Green, W. N. & Andersen, O. S. (1991) Surface charges and ion channel function. *Annu. Rev. Physiol.* **53**, 341–359.
11. Begenisich, T. B. & Cahalan, M. D. (1980) Sodium channel permeation in squid axons. I: reversal potential experiments. *J. Physiol.* **307**, 217–242.
12. Campbell, D. L., Rasmusson, R. L. & Strauss, H. C. (1988) Theoretical study of the voltage and concentration dependence of the anomalous mole fraction effect in single calcium channels. New insights into the characterization of multi-ion channels. *Biophys. J.* **54**, 945–954.
13. McCleskey, E. W. (1999) Calcium channel permeation: a field in flux. *J. Gen. Physiol.* **113**, 765–772.

19

Estimation of Quantal Parameters With Multiple-Probability Fluctuation Analysis

Chiara Saviane and R. Angus Silver

Summary

The functional properties of central synapses are difficult to study because they can be modulated either presynaptically or postsynaptically, each connection has multiple contacts and release at each contact is stochastic. Moreover, studying central synapses with electrophysiology is complicated by the fact that synapses are often remote from the recording site and signals are often difficult to resolve above the noise. This together with the fact that central synapses often have few release sites and have nonuniform quantal parameters makes classical quantal analysis methods difficult to apply. Here, we discuss an alternative approach, multiple-probability fluctuation analysis (MPFA), which can be used to estimate nonuniform quantal parameters from fits of the relationship between the variance and mean amplitude of postsynaptic responses, recorded at different release probabilities. We illustrate the experimental protocols and the analysis procedure that should be followed to perform MPFA and interpret the estimated parameters.

Key Words: Synapse; transmitter release; quantal analysis; fluctuation analysis; MPFA; synaptic plasticity; CV analysis.

1. Introduction

According to the quantal hypothesis of transmitter release (*1,2*), neurotransmitter is packaged in discrete quantities (*quanta*) that are released in an all-or-none fashion. As release occurs at specific sites, synaptic efficacy can be described using three quantal parameters (*2,3*): the number of functional release sites (N), the probability of release (P_R), and the amplitude of the response to a single *quantum* (Q). Changes in one or more of these parameters account for

From: *Methods in Molecular Biology, vol. 288: Patch-Clamp Methods and Protocols*
Edited by: P. Molnar and J. J. Hickman © Humana Press Inc., Totowa, NJ

modifications in synaptic strength. Assuming that each site acts independently, binomial and Poisson statistics can be used to model the stochastic behavior of synaptic transmission, estimate quantal parameters, and identify the site of changes in synaptic efficacy. However, classical quantal analysis methods, based on the analysis of amplitude histograms or failure rate, can be difficult to apply at central synapses because of the low signal-to-noise ratio due to the small quantal size and because of the nonuniformity in P_R and Q ***(4)***. Therefore, new approaches inspired by stationary and nonstationary noise analysis of ion channels ***(5,6)*** have been developed. These include stationary analysis of mean and variance at different release probabilities ***(7,8)***, the analysis of mean, variance, and covariance ***(9,10)*** during repetitive stimulation, and the analysis of higher cumulants ***(11)*** during prolonged release from voltage-clamped presynaptic terminals. Multiple-probability fluctuation analysis (MPFA) ***(8)*** (also known as variance–mean analysis [*7*]) is a method that can be applied to determine uniform or nonuniform quantal parameters when the signal-to-noise ratio is low. MPFA involves fitting the relationship between the variance and mean of the synaptic responses measured under different release probability conditions ***(8,12,13)*** with a binomial or multinomial model.

1.1. Statistical Models of Quantal Transmission

According to a binomial model of neurotransmitter release, the mean peak amplitude of the synaptic response I and the associated variance σ^2 can be expressed as a function of the number of release sites, the release probability, and the quantal size:

$$I = NP_{\mathrm{R}}Q, \tag{1}$$

$$\sigma^2 = NQ^2P_{\mathrm{R}}(1 - P_{\mathrm{R}}), \tag{2}$$

so that a parabolic relationship holds between them (*see* **Fig. 1A**):

$$\sigma^2 = IQ - \frac{I^2}{N}. \tag{3}$$

As several studies have shown that the assumptions required for applying a simple binomial model often do not hold for central synapses, a multinomial model has therefore been developed ***(8,14)*** (*see* **Fig. 1B**) in order to take into account quantal variability both at the single-site level (intrasite variability) and across sites (intersite variability). Intrasite variability arises from fluctuations in the size of quantal events from trial to trial and from asynchrony in their latency

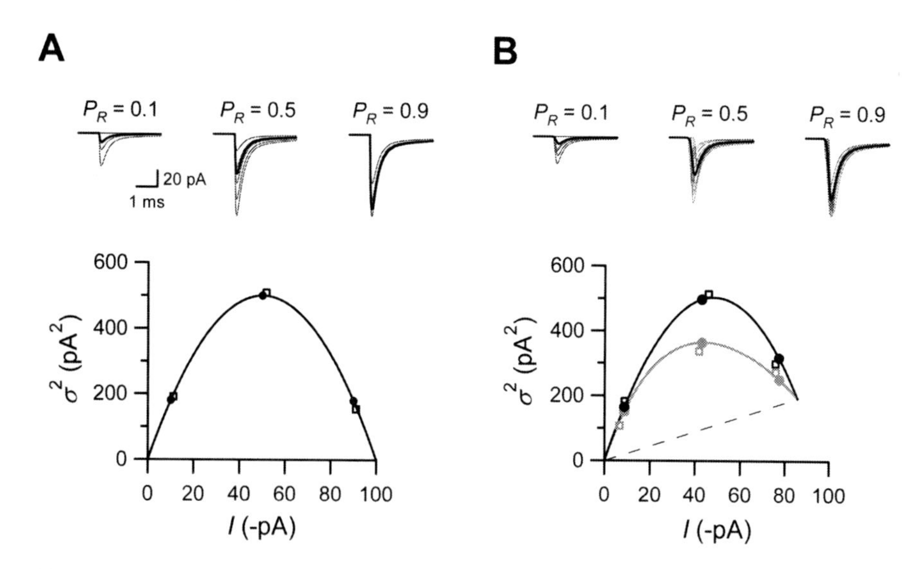

Fig. 1. Binomial and multinomial models of synaptic transmission. **(A)** Monte Carlo simulations of synaptic currents (15 traces superimposed) and mean current (black trace, average from 200 traces) for a synapse with $N = 5$, $Q = -20\,\text{pA}$ at three different release probabilities and associated variance–mean relationship (bottom; *see* **Eq. 3**). The black dots represent the theoretical data points associated with the tested release probabilities (*see* **Eqs. 1** and **2**), whereas the empty squares are values from sets of 200 simulated synaptic currents. **(B)** Same as in panel **A** for synapses with intrasite and intersite quantal variability ($\text{CV}_{QS} = \text{CV}_{QII} = 0.3$ and $\text{CV}_{QL} = 0.2$). The current traces and the black line and markers in the variance–mean plot refer to a synapse with uniform P_R (*see* **Eqs. 4–7**), whereas the grey line and markers in the graph refer to a nonuniform release probability case ($\alpha = 1$; *see* **Eq. 10**). The dashed line shows the variance contributed by intrasite variability. The mean quantal amplitude across sites was set to $-20\,\text{pA}$ that gave $\bar{Q}_\text{p} = -17.2\,\text{pA}$ in the presence of asynchronous release. (Modified from **ref.** ***17*** with permission from Elsevier.)

(13) and can be quantified in terms of the associated coefficients of variation (CV_{QS} and CV_{QL}, respectively). Intersite variability arises from differences in the mean quantal size from site to site, and the CV of this component is defined as CV_{QII}. Both intrasite and intersite variability can be quite easily implemented in a multinomial model, where the mean response and the variance are given by:

$$I = NP_\text{R}\bar{Q}_\text{p} \tag{4}$$

and

$$\sigma^2 = N\bar{Q}_p^2 P_R (1 - P_R)\left(1 + CV_{QII}^2\right) + N\bar{Q}_p^2 P_R CV_{QI}^2, \quad (5)$$

and $\bar{Q}_p$ represents the mean quantal size at the time of the peak of the mean synaptic response and is therefore affected by asynchronous release. CV_{QI} includes all the variability observed at the single-site level, being defined by:

$$CV_{QI}^2 = CV_{QL}^2 + CV_{QS}^2. \quad (6)$$

Of course, the use of constant coefficients of variation assumes that the release latency and the quantal variability are independent of P_R.

Substituting **Eq. 4**, the variance can be expressed in terms of I:

$$\sigma^2 = \left(\bar{Q}_p I - \frac{I^2}{N}\right)\left(1 + CV_{QII}^2\right) + \bar{Q}_p I CV_{QI}^2. \quad (7)$$

According to the model of release used, fitting **Eq. 3** or **7** to the relationship between the variance and the mean of the synaptic responses recorded in different release probability conditions provides estimates of the mean quantal size and N; P_R can then be calculated from the mean response (*see* **Eqs. 1** or **4**).

Several groups have shown that the probability of release can also be nonuniform across release sites at central synapses ***(15,16)*** (*see* **Fig. 1B**). In this case, assuming that there are no correlations between release probability and quantal size across sites ***(8,14)***, the relationship between the variance and the mean response ***(13)*** is given by:

$$\sigma^2 = \left[\bar{Q}_p I - \frac{I^2}{N}\left(1 + CV_P^2\right)\right]\left(1 + CV_{QII}^2\right) + \bar{Q}_p I CV_{QI}^2, \quad (8)$$

where CV_P represents the CV of release probability across sites and changes with the mean release probability across sites at the time of the peak of the response ($\bar{P}_R$). CV_P and how it changes with $\bar{P}_R$ has been modeled ***(13)*** using families of beta functions $B(\alpha, \beta)$, so that

$$CV_P = \sqrt{\frac{1 - \bar{P}_R}{\bar{P}_R + \alpha}}, \quad (9)$$

and the variance–mean relationship depends on only one additional parameter α:

$$\sigma^2 = \left[\bar{Q}_p I - \frac{\bar{Q}_p I^2 (1 + \alpha)}{I + N\bar{Q}_p \alpha}\right]\left(1 + CV_{QII}^2\right) + \bar{Q}_p I CV_{QI}^2. \quad (10)$$

It follows that low α values indicate nonuniform release probability and the model reduces to the uniform P_R case for $\alpha \rightarrow \infty$.

Fitting **Eq. 10** to the variance–mean relationship also gives an estimate of the nonuniformity in P_R. However, α is the least well-constrained parameter in **Eq. 10** and its estimate should only be used as a rough indicator ***(13,17)***. In practice, the variance–mean relationship is rather insensitive to α values larger than 2–3, which can give high CV_P at low $\bar{P}_R$, thus allowing the use of a uniform model ***(18)***.

MPFA can be combined with CV analysis of Excitatory Postsynaptic Current (EPSC) amplitudes to track changes in quantal parameters during repetitive stimulation when P_R is uniform ***(19)***. After estimating quantal parameters for the first pulse in the train using MPFA, P_R can then be estimated for the subsequent synaptic responses in the train from the CV, assuming that N is constant:

$$CV = \frac{\sigma}{I} = \sqrt{\frac{(1 - P_R)\left(1 + CV_{QII}^2\right) + CV_{QI}^2}{NP_R}}; \qquad (11)$$

$\bar{Q}_p$ can then be calculated from the mean response (*see* **Eq. 4**).

1.2. Interpretation of Quantal Parameters

The interpretation of the results from MPFA is straightforward if quantal responses are independent and, therefore, sum linearly: $\bar{Q}_p$ represents the mean elementary postsynaptic response at the time of the peak, P_R is the release probability of a single *quantum* at the time of the peak [which is equivalent to the cumulative probability if the release function has a small tail ***(18)***], and N corresponds to the number of release sites. N will also equal the number of synaptic contacts if a maximum of one *quantum* of transmitter is released at each site (uniquantal release). However, several studies have shown that at some synapses, multiple independent vesicles can be released at the same synaptic contact ***(20–22)***. If quantal responses sum linearly, N will indicate the number of functional release sites, but in this case, it will be larger than the number of synaptic contacts. Interpretation of quantal parameters becomes complicated when *quanta* do not sum linearly because of receptor saturation ***(23)*** which causes the quantal size to change with P_R. In the limiting case when the receptors are saturated by a single vesicle, $\bar{Q}_p$ would be independent of the number of vesicles released per site and N and P_R would represent the number of synaptic contacts at which release occurred and the probability that release occurred at each contact.

Whether multivesicular release occurs, or not, can be investigated by examining the fractional block of the synaptic response in the presence of a rapidly equilibrating competitive antagonist under low and high release probability conditions *(18,20,22,24)* (*see* **Fig. 2**). If the fractional block is reduced at high P_R, this suggests that the synaptic receptors are exposed to higher concentrations of glutamate at higher P_R and therefore that multiquantal release is present (*see* **Note 1**). In contrast, if the fractional block is independent of P_R, this suggests that uniquantal release predominates. However, it should be noted that this result relies on the assumption that the competitive antagonist can equilibrate during the rise of the EPSC.

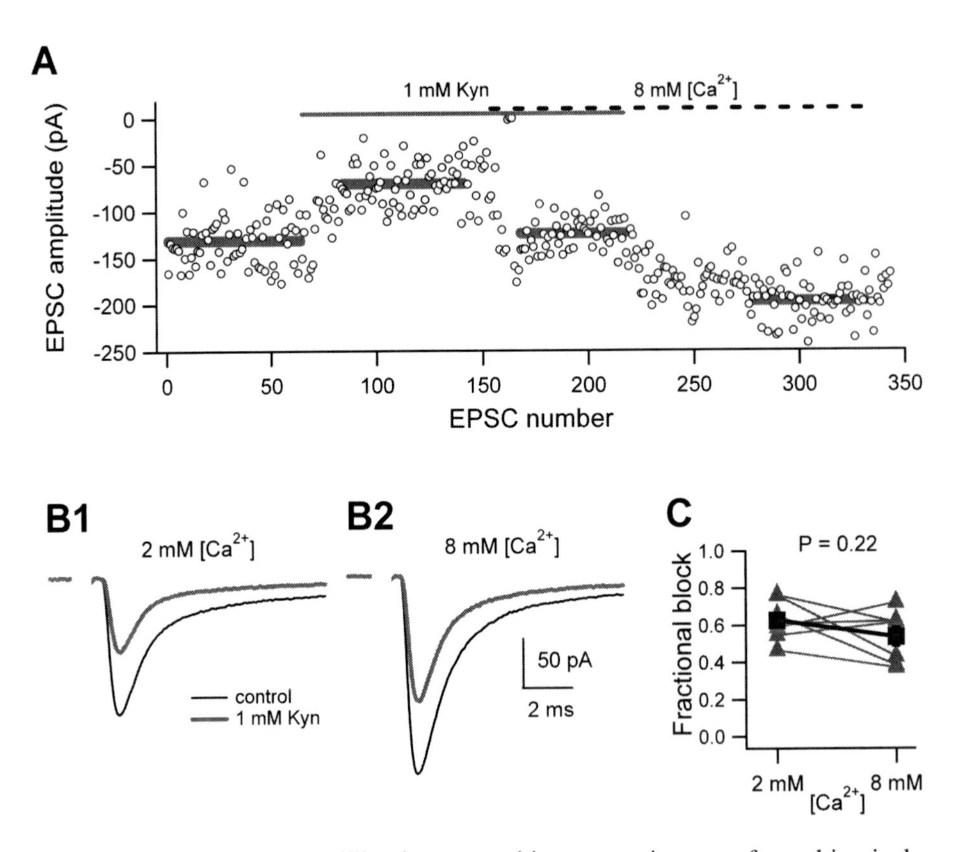

Fig. 2. Use of a rapidly equilibrating competitive antagonist to test for multivesicular release. (**A**) EPSC amplitudes in 2 and 8 m*M* (dashed bar) [Ca^{2+}], in the absence or presence of 1 m*M* kynurenic acid (Kyn; thin grey bar). The thick grey bars indicate stable epochs as assessed with the Spearman rank-order test. (**B**) Mean EPSCs from the four stable epochs in panel **A** (artifact blanked). (**C**) Fractional block of the EPSC by 1 m*M* Kyn in 2 and 8 m*M* [Ca^{2+}] across cells. (Modified from **Fig. 4**, **ref.** *18* with permission from The Society for Neuroscience.)

Here, we describe the experimental protocols and analysis procedures that should be used to estimate and interpret quantal parameters using fluctuations analysis, explaining how to quantify intrasite and intersite variability for a multinomial model of release.

2. Materials

1. Extracellular solution (artificial cerebrospinal fluid [ACSF]) typically used for experiments in acute cerebellar or cortical slices ***(19,24)***: 125 m*M* NaCl, 2.5 m*M* KCl, 26 m*M* $NaHCO_3$, 1.25 m*M* NaH_2PO_4, 25 m*M* glucose, 2 m*M* $CaCl_2$, 1 m*M* $MgCl_2$ (pH 7.4 when bubbled with 95% O_2, 5% CO_2).
2. Different release probability conditions are imposed by changing the [Ca^{2+}] and [Mg^{2+}] in ACSF ***(18,19,24)***: for example, 1 and 5, 1.5 and 2, 2 and 1, 3 and 0.75, 5 and 0.5, 8 and 0 mM, respectively. The osmolarity is kept constant by varying (glucose) between 10 and 25 mM. A phosphate-free solution (with NaCl increased to 126.25 m*M*) is used in high [Ca^{2+}] (8 m*M*) to avoid phosphate precipitation.
3. Specific combinations of drugs might need to be added to ACSF to isolate a particular synaptic component. These will depend on the type of receptors and on the preparation. In the case of glutamatergic connections, blockers of NMDA receptors should be used to prevent Ca^{2+}-mediated plasticity processes ***(18,19,24)***. A rapidly equilibrating competitive antagonist is added to ACSF to test for multivesicular release ***(18,24)*** and interpret the estimated parameters (for example, kynurenic acid or γ-D-glutamylglycine (DGG) for alpha-amino-3-hydroxy-5-methyl-4-iroxazole propionic acid (AMPA) receptors).
4. Intracellular solution for whole-cell recordings, voltage-clamp configuration ***(19)*** (*see* **Note 2**): 110 m*M* $KMeSO_4$, 40 m*M* 4-(2-hydroxyethyl)-1-piperazineethane sulfonic acid (HEPES), 4 m*M* NaCl, 1 m*M* KCl, 1.78 m*M* $CaCl_2$, 5 m*M* ethylene glycol tetraacetic acid (EGTA), 4 m*M* MgATP, 0.3 m*M* MgGTP (pH 7.3; free [Ca^{2+}] approximately 50 n*M* at physiological temperature).
5. Data acquisition and analysis can be performed using different programs, such as Axograph, PClamp (Molecular Devices, USA), or Igor PRO (Wavemetrics, USA), for which the acquisition and analysis packages NClamp and Neuromatic are freely available on http://www.physiol.ucl.ac.uk/research/silver_a/software.

3. Methods

MPFA is best applied at single-synaptic connections as the number of stimulated fibers must be constant from trial to trial. This can be achieved by paired recordings ***(24)*** or, in some preparations, by extracellular stimulation ***(8)***. Quantal parameters are estimated by fitting the variance–mean relationship obtained by recording synaptic responses in different [Ca^{2+}]/[Mg^{2+}]. However, the first step is to verify that it will be possible to interpret the estimated quantal parameters, by testing for multiquantal release. Moreover, if a multinomial model of release is used, the different types of quantal variability have to be estimated.

In the following sections, we will describe the general procedures for data acquisition and analysis and then focus on the different types of experiments.

3.1. Data Acquisition

1. To avoid aliasing and obtain a reliable reconstruction of the signal, the sampling frequency should be set at least four to five times the highest frequency component of interest. For MPFA at the mossy fiber-granule cell (MF-GC) synapse in the cerebellum ***(18,19)***, currents were filtered to 7.1 kHz, with a cascade of two Bessel filters with 10 kHz cutoff frequency, and digitized at 100 kHz.
2. If extracellular stimulation is used across different recording conditions, it is important to know how the stimulation threshold changes under the different conditions. Stimulation threshold should be checked in each condition, and if multiple fiber preparations are used, additional controls, such as measuring the afferent volley, should be carried out.
3. The fastest stimulation frequency that gives time-stable EPSCs in each condition (*see* Section 3.2.3) and the minimum time necessary for solution exchange should be established (for example, we have used 2, 0.5, 0.2 Hz at 1, 2, 8 m*M* [Ca^{2+}], respectively, at the MF-GC synapse [*18,19*]).

3.2. Data Analysis

1. Individual traces are baseline subtracted using a 0.5 to 1.0 ms window (*see* **Fig. 3A**, W_B) just before the stimulus.
2. The stimulus artifact is removed by subtracting a double-exponential fit to the decay of the artifact current.
3. In each recording condition, a statistical test, such as the Spearman rank-order test, should be used to find the largest set of contiguous events that are time stable ***(25)*** (*see* **Fig. 2A**). This can be done by measuring for each event either the peak amplitude or the mean amplitude of the points within a 0.1ms window around the time of the peak of the mean current (*see* point 4 below). Both individual and mean currents can be digitally low-pass filtered using a Gaussian filter to improve the accuracy of the peak detection (for example, at the MF-GC synapse [***26***], we use a cutoff frequency of 1.9–4 kHz). However, all the amplitude measurements for MPFA should be made on the original traces.
4. The synaptic response amplitude and the background noise are measured for each event using two brief measurement windows of equal duration (for example, 0.1 ms; *see* **Fig. 3A**, W_P and W_N, respectively). The former should be centered at the time at which the average waveform peaks and the latter should be "reflected" before the stimulus at a time such that the center of the baseline window is equidistant between the centers of the two measurement windows (*see* **Note 3**). For each event,

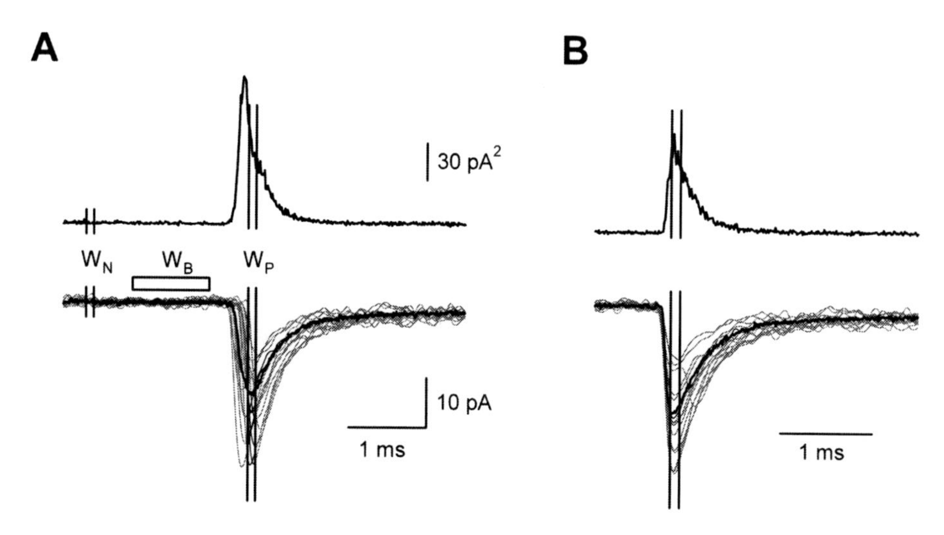

Fig. 3. Measuring quantal variability in low probability conditions. Five hundred synaptic currents were simulated for a multinomial model of release with uniform $P_R = 0.02$. Fifteen successes (bottom) are shown stimulus aligned **(A)** and rise time aligned **(B)**. The black lines represent the mean (bottom) and variance (top) obtained from all successes. Open bar (W_B) in **A** indicates an appropriate window for baseline-correcting individual traces. Vertical lines, marked W_N and W_P, indicate windows for measuring mean and variance of the baseline noise and peak of the EPSC, respectively. Because of the asynchrony in quantal release, the mean of the stimulus-aligned successes is smaller and thus the variability larger.

the synaptic amplitude and prebaseline amplitude are calculated from the mean of the points within the two windows.

5. Once that measurements have been made from all the stable events in a certain condition, the overall mean, sample variance, and the error in the sample variance (*see* **Note 4**) can be calculated from the amplitude measurements in W_P. The synaptic variance can then be corrected for the baseline variance by simply subtracting the variance of the prebaseline window from the variance of W_P.

3.3. Discriminating Between Univesicular and Multivesicular Release

1. Synaptic responses should be recorded in control and in the presence of a rapidly equilibrating competitive antagonist under low and high release probability conditions ***(18,20,22,24)*** (*see* **Fig. 2**).
2. The mean peak amplitude in the different condition is then measured to examine the fractional block (*see* **Fig. 2**). If it is independent of release probability, MPFA can be performed and the interpretation of quantal parameters will be straightforward.

Otherwise, MPFA should be performed in the presence of the antagonist *(27)* to estimate N and P_R. However, this can be complicated if the quantal size becomes very small because it is important to quantify the quantal variance in the presence of antagonist as it may increase as the receptor occupancy is reduced.

3.4. Quantifying Quantal Variability

1. Total quantal variability ($CV_{QT} = \sqrt{CV_{QI}^2 + CV_{QII}^2}$) can be determined from evoked stimulus-aligned successes under low probability conditions when the proportion of multiquantal responses is negligible *(13,18)* (*see* **Fig. 3A**). Ideally, a failure rate of approximately 80–90% and at least 20 successes should be used.
2. After baseline and artifact subtraction, successes and failures in the stable epoch are distinguished using an amplitude criterion set to three times the standard deviation of the noise. This can be done using 0.75 to 1ms measuring windows, with W_P centered slightly later than the time of the peak of the mean response to capture most of the amplitude.
3. CV_{QL} is determined from the difference in variability of stimulus-aligned (*see* **Fig. 3A**) and rise time-aligned successes (*see* **Fig. 3B**). Events are usually aligned on the 10 or 20% rise time point (*see* **Note 5**).
4. CV_{QS} can be estimated from rise time-aligned events recorded at single site (*see* **Note 6**). If the estimation of CV_{QI} is not possible, the best strategy is to use $CV_{QI} = CV_{QII}$, which is likely to introduce little error *(28)*. This value can be compared with the upper estimate of CV_{QI} that can be obtained from the variance remaining when release probability is maximal ($P_R \sim 1$; $\sigma^2 \approx N\bar{Q}_p^2 CV_{QI}^2$).

3.5. MPFA

1. Ideally 100 EPSCs per condition and six to eight different conditions should be used to construct the variance–mean relationship. However, this is feasible only at connections where a relatively high stimulation rate can be applied with no synaptic depression and reasonably long recordings can be performed. Otherwise, the number of traces and/or the number of probability conditions can be reduced. At least three different recording conditions are needed to test for the statistical significance of the fit for a model of release with uniform release probability and four for the nonuniform case. The minimum number of traces depends on P_R (*see* **Note 7**) and the type of analysis to be done. High $[Ca^{2+}]$ should be used last to avoid any plasticity processes.
2. The quantal parameters $\bar{Q}_p$ and N and their uncertainties are estimated by a weighted fit *(17)* of the variance–mean relationship to a binomial (*see* **Eq. 3**) or multinomial (*see* **Eqs. 7** or **10**) model of release where each data point is weighted by the reciprocal of its variance (*see* **Fig. 4A**). P_R is calculated at each $[Ca^{2+}]$ from the mean current and its uncertainty can be calculated by error propagation.

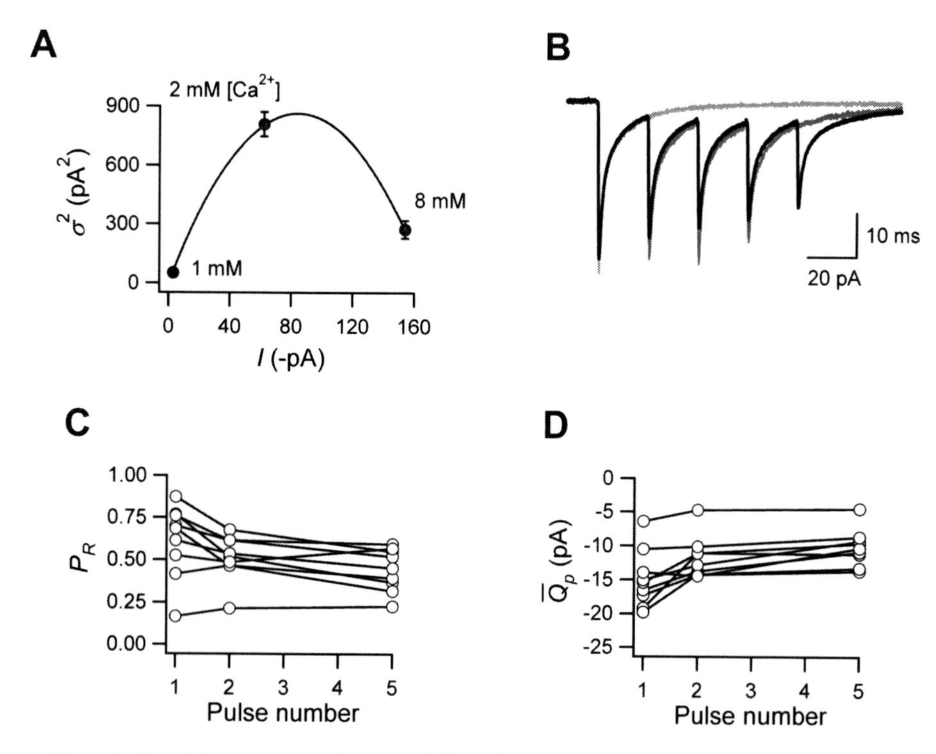

Fig. 4. Combining multiple-probability fluctuation analysis and coefficient of variation analysis to determine quantal parameters during trains. **(A)** Variance–mean plot obtained from EPSCs recorded in different release probability conditions at a mossy fiber-granule cell connection. Errors are calculated according to **Eq. 12**. The line shows the weighted fit according to a multinomial model (*see* **Eq. 7**). **(B)** Mean EPSCs for single pulses, 4-pulse, and 5-pulse 100 Hz trains in 2 m*M* [Ca^{2+}]. Double-exponential fit of the mean single pulse or 4-pulse train was scaled and subtracted from each trace for analyzing the second or the fifth EPSC, respectively. **(C and D)** P_R and $\bar{Q}_p$ during 5-pulse 100 Hz trains for nine connections. (Modified from **ref.** ***19***.)

3. Fits should be accepted when the χ^2 value gives a $P \geq 0.05$. If an accurate estimate of N is to be made, it is important that the P_R at the highest [Ca^{2+}] is greater than 0.60 ***(13)***.

3.6. Quantal Determinants of Short-Term Plasticity

1. Trains of stimulation can be used to look at changes in quantal parameters during short-term plasticity (*see* **Fig. 4B–D**). If long trains are used, stability should be checked on the first EPSC and on the steady-state level at the end of the trains. A lower repetition rate might be needed to avoid depression.

2. When performing CV analysis (using **Eq. 11**), the number of traces recorded should be maximized to reduce the sampling error in the CV estimate. We have recently used ***(19)*** a minimum of approximately 150 traces at the MF-GC synapse in 2 m*M* [Ca^{2+}], but this depends on P_R in the condition under investigation.
3. It is important to identify the time of the peak of the current at each pulse in the train and measure the CV at the right time point. If the current does not go back to baseline within the interpulse interval, it is necessary to subtract the residual component of the preceding EPSC. Trains of different length can be alternated, and the fit to the preceding EPSC scaled and subtracted from each trace ***(19)*** (*see* **Fig. 4B**).

4. Notes

1. The linear summation of *quanta* can also be tested by comparing the estimate of N, obtained from the quantal content when release is maximal (for example, 8 m*M* [Ca^{2+}] when $NP_R \sim N$), in the presence and absence of 1 m*M* kynurenic acid ***(18)***. This can be obtained by dividing the mean currents by their respective quantal sizes measured from the σ^2/I in low probability conditions ***(13)*** (for example, 1.25 m*M* [Ca^{2+}]). If the estimate of N is unchanged in the presence of kynurenic acid, no new release sites are revealed by lowering receptor occupancy, suggesting that *quanta* summate linearly over the entire range of P_R.
2. Synaptic responses should be measured in voltage-clamp configuration to alleviate driving force issues and avoid dependence on voltage-gated channels. However, if synapses are far from the soma and the quality of the voltage-clamp is poor, postsynaptic potentials should be recorded and a correction for the deviation in driving force should be applied ***(24,29)***.
3. This arrangement ensures that the peak measurement and prebaseline measurements pick up similar frequency components of the background noise.
4. The expression used to estimate the error on the sample variance depends on the number of release sites ***(17)***. A general expression for all likely conditions is given by:

$$\mathrm{var}(s^2) = \frac{n}{(n-2)(n-3)}\left[\frac{3(3-2n)(n-1)^2 - n(n-2)(n-3)^2}{(n^2-2n+3)(n-1)^2}m_2^2 + m_4\right] \tag{12}$$

that, for large N, simplifies to:

$$\mathrm{var}(s^2) = \frac{2n}{(n-1)(n-2)(n-3)}\left[\frac{n^2(n-2)(n-3) - 3(3-2n)(n-1)}{(n^2-2n+3)(n-1)}m_2^2 - m_4\right], \tag{13}$$

where n is the number of traces acquired and m_2 and m_4 are sample central moments about the mean and can be calculated as $m_r = (1/n)\sum_{i=1}^{n}(X_i - \bar{X})^r$. In case of large sample size, neglecting terms to the order n^{-1}, m_2 can be used to estimate the variance, whereas proper estimators of its variance are given by:

$$\mathrm{var}(m_2) = \frac{m_4 - m_2^2}{n}, \tag{14}$$

which simplifies for large N to

$$\mathrm{var}(m_2) = \frac{2m_2^2}{n}. \tag{15}$$

As the reliability of individual estimates is higher for the simplified expression, the use of this estimator could be more appropriate if it is known *a priori* that a large number of release sites are present. Moreover, maximizing the number of samples improves the estimators' reliability and is therefore anyway desirable.

5. To minimize the effects of sampling error, CV_{QL} can also be measured by simulating stochastic quantal release at a release site using the mean measured quantal waveform and the mean latency distribution measured under low probability conditions ***(18)***. For each connection, the mean quantal response is obtained by averaging the rise time-aligned successes and by fitting it with an expression ***(30)*** modified from **ref. *31*** to minimize errors due to noise:

$$\begin{aligned}\mathrm{EPSC}(t) = & A_1\left[1-\exp\left(-\frac{t-t_0}{\tau_{\mathrm{rise}}}\right)\right]^n\left[A_2\exp\left(-\frac{t-t_0}{\tau_{\mathrm{decay1}}}\right)\right.\\ & \left.+(1-A_2)\exp\left(-\frac{t-t_0}{\tau_{\mathrm{decay2}}}\right)\right].\end{aligned} \tag{16}$$

The mean cumulative latency distribution was obtained after normalizing and aligning on the 50% point the cumulative latency distributions obtained for each cell from the 20% rise time point of the successes.

6. To test for the single-site connections, the measured mean amplitudes of successes at the different $[Ca^{2+}]$ are compared with the predicted amplitude (A_s) assuming one or two release sites ***(18,25)***. According to a binomial model of release, this can be calculated for a given N, knowing the quantal size and the failure rate f:

$$A_s = \frac{I}{1-f} = \frac{NQP_R}{1-f} = \frac{NQ\left(1-f^{1/N}\right)}{1-f}. \tag{17}$$

7. For MPFA, the number of traces recorded in each condition could be as low as 30–50. However, the distribution of variances obtained at extreme P_R from few synaptic currents at synapses with few release sites can become skewed, and the use of weighted least squares optimization routine under these conditions might lead to biased results ***(17)***.

Acknowledgments

Supported by The Wellcome Trust, the MRC, and the EC. RAS is in receipt of a Wellcome Trust Senior Fellowship.

References

1. del Castillo, J., and Katz, B. (1954) Quantal components of the end-plate potential. *J. Physiol.* **124**, 560–573.
2. Katz, B. (1969) *The Release of Neural Transmitter Substances*, Liverpool University Press, Liverpool, UK.
3. Vere-Jones, D. (1966) Simple stochastic models for the release of quanta of transmitter from a nerve terminal. *Aust. J. Stat.* **8**, 53–63.
4. Walmsley, B. (1995) Interpretation of "quantal" peaks in distributions of evoked synaptic transmission at central synapses. *Proc. R. Soc. Lond. B Biol. Sci.* **261**, 245–250.
5. Heinemann, S. H., and Conti, F. (1992) Nonstationary noise analysis and application to patch clamp recordings. *Methods Enzymol.* **207**, 131–148.
6. Sigworth, F. J. (1980) The variance of sodium current fluctuations at the node of Ranvier. *J. Physiol.* **307**, 97–129.
7. Reid, C. A., and Clements, J. D. (1999) Postsynaptic expression of long-term potentiation in the rat dentate gyrus demonstrated by variance-mean analysis. *J. Physiol.* **518**, 121–130.
8. Silver, R. A., Momiyama, A., and Cull-Candy, S. G. (1998) Locus of frequency-dependent depression identified with multiple-probability fluctuation analysis at rat climbing fibre-Purkinje cell synapses. *J. Physiol.* **510**, 881–902.
9. Meyer, A. C., Neher, E., and Schneggenburger, R. (2001) Estimation of quantal size and number of functional active zones at the calyx of held synapse by nonstationary EPSC variance analysis. *J. Neurosci.* **21**, 7889–7900.
10. Scheuss, V., and Neher, E. (2001) Estimating synaptic parameters from mean, variance, and covariance in trains of synaptic responses. *Biophys. J.* **81**, 1970–1989.
11. Neher, E., and Sakaba, T. (2001) Estimating transmitter release rates from postsynaptic current fluctuations. *J. Neurosci.* **21**, 9638–9654.
12. Clements, J. D., and Silver, R. A. (2000) Unveiling synaptic plasticity: a new graphical and analytical approach. *Trends Neurosci.* **23**, 105–113.
13. Silver, R. A. (2003) Estimation of nonuniform quantal parameters with multiple-probability fluctuation analysis: theory, application and limitations. *J. Neurosci. Methods* **130**, 127–141.
14. Frerking, M., and Wilson, M. (1996) Effects of variance in mini amplitude on stimulus-evoked release: a comparison of two models. *Biophys. J.* **70**, 2078–2091.
15. Walmsley, B., Edwards, F. R., and Tracey, D. J. (1988) Nonuniform release probabilities underlie quantal synaptic transmission at a mammalian excitatory central synapse. *J. Neurophysiol.* **60**, 889–908.
16. Murthy, V. N., Sejnowski, T. J., and Stevens, C. F. (1997) Heterogeneous release properties of visualized individual hippocampal synapses. *Neuron* **18**, 599–612.
17. Saviane, C., and Silver, R. A. (2006) Errors in the estimation of the variance: implications for multiple-probability fluctuation analysis. *J. Neurosci. Methods* **153**, 250–260.

18. Sargent, P. B., Saviane, C., Nielsen, T. A., DiGregorio, D. A., and Silver, R. A. (2005) Rapid vesicular release, quantal variability and spillover contribute to the precision and reliability of transmission at a glomerular synapse. *J. Neurosci.* **25**, 8173–8187.
19. Saviane, C., and Silver, R. A. (2006) Fast vesicle reloading and a large pool sustain high bandwidth transmission at a central synapse. *Nature* **439**, 983–987.
20. Wadiche, J. I., and Jahr, C. E. (2001) Multivesicular release at climbing fiber-Purkinje cell synapses. *Neuron* **32**, 301–313.
21. Oertner, T. G., Sabatini, B. L., Nimchinsky, E. A., and Svoboda, K. (2002) Facilitation at single synapses probed with optical quantal analysis. *Nat. Neurosci.* **5**, 657–664.
22. Christie, J. M., and Jahr, C. E. (2006) Multivesicular release at Schaffer collateral-CA1 hippocampal synapses. *J. Neurosci.* **26**, 210–216.
23. Harrison, J., and Jahr, C. E. (2003) Receptor occupancy limits synaptic depression at climbing fiber synapses. *J. Neurosci.* **23**, 377–383.
24. Silver, R. A., Lubke, J., Sakmann, B., and Feldmeyer, D. (2003) High-probability uniquantal transmission at excitatory synapses in barrel cortex. *Science* **302**, 1981–1984.
25. Silver, R. A., Cull-Candy, S. G., and Takahashi, T. (1996) Non-NMDA glutamate receptor occupancy and open probability at a rat cerebellar synapse with single and multiple release sites. *J. Physiol.* **494**, 231–250.
26. DiGregorio, D. A., Nusser, Z., and Silver, R. A. (2002) Spillover of glutamate onto synaptic AMPA receptors enhances fast transmission at a cerebellar synapse. *Neuron* **35**, 521–533.
27. Foster, K. A., and Regehr, W. G. (2004) Variance-mean analysis in the presence of a rapid antagonist indicates vesicle depletion underlies depression at the climbing fiber synapse. *Neuron* **43**, 119–131.
28. Clements, J. (2003) Variance-mean analysis: a simple and reliable approach for investigating synaptic transmission and modulation. *J. Neurosci. Methods* **130**, 115–125.
29. Feldmeyer, D., Lubke, J., Silver, R. A., and Sakmann, B. (2002) Synaptic connections between layer 4 spiny neurone-layer 2/3 pyramidal cell pairs in juvenile rat barrel cortex: physiology and anatomy of interlaminar signalling within a cortical column. *J. Physiol.* **538**, 803–822.
30. Nielsen, T. A., DiGregorio, D. A., and Silver, R. A. (2004) Modulation of glutamate mobility reveals the mechanism underlying slow-rising AMPAR EPSCs and the diffusion coefficient in the synaptic cleft. *Neuron* **42**, 757–771.
31. Bekkers, J. M., and Stevens, C. F. (1996) Cable properties of cultured hippocampal neurons determined from sucrose-evoked miniature EPSCs. *J. Neurophysiol.* **75**, 1250–1255.

Index

Lightning Source UK Ltd.
Milton Keynes UK
UKOW040431100413

208963UK00001B/9/P